Selected Titles in This Series

27 **I. M. Sigal and C. Sulem, Editors,** Nonlinear dynamics and renormalization group, 2001

26 **J. Harnad, G. Sabidussi, and P. Winternitz, Editors,** Integrable systems: From classical to quantum, 2000

25 **Decio Levi and Orlando Ragnisco, Editors,** SIDE III—Symmetries and integrability of difference equations, 2000

24 **B. Brent Gordon, James D. Lewis, Stefan Müller-Stach, Shuji Saito, and Noriko Yui, Editors,** The arithmetic and geometry of algebraic cycles, 2000

23 **Pierre Hansen and Odile Marcotte, Editors,** Graph colouring and applications, 1999

22 **Jan Felipe van Diejen and Luc Vinet, Editors,** Algebraic methods and q-special functions, 1999

21 **Michel Fortin, Editor,** Plates and shells, 1999

20 **Katie Coughlin, Editor,** Semi-analytic methods for the Navier-Stokes equations, 1999

19 **Rajiv Gupta and Kenneth S. Williams, Editors,** Number theory, 1999

18 **Serge Dubuc and Gilles Deslauriers, Editors,** Spline functions and the theory of wavelets, 1999

17 **Olga Kharlampovich, Editor,** Summer school in group theory in Banff, 1996, 1998

16 **Alain Vincent, Editor,** Numerical methods in fluid mechanics, 1998

15 **François Lalonde, Editor,** Geometry, topology, and dynamics, 1998

14 **John Harnad and Alex Kasman, Editors,** The bispectral problem, 1998

13 **Michel C. Delfour, Editor,** Boundaries, interfaces, and transitions, 1998

12 **Peter C. Greiner, Victor Ivrii, Luis A. Seco, and Catherine Sulem, Editors,** Partial differential equations and their applications, 1997

11 **Luc Vinet, Editor,** Advances in mathematical sciences: CRM's 25 years, 1997

10 **Donald E. Knuth,** Stable marriage and its relation to other combinatorial problems: An introduction to the mathematical analysis of algorithms, 1997

9 **D. Levi, L. Vinet, and P. Winternitz, Editors,** Symmetries and integrability of difference equations, 1996

8 **J. Feldman, R. Froese, and L. M. Rosen, Editors,** Mathematical quantum theory II: Schrödinger operators, 1995

7 **J. Feldman, R. Froese, and L. M. Rosen, Editors,** Mathematical quantum theory I: Field theory and many-body theory, 1994

6 **Guido Mislin, Editor,** The Hilton Symposium 1993: Topics in topology and group theory, 1994

5 **D. A. Dawson, Editor,** Measure-valued processes, stochastic partial differential equations, and interacting systems, 1994

4 **Hershy Kisilevsky and M. Ram Murty, Editors,** Elliptic curves and related topics, 1994

3 **Rémi Vaillancourt and Andrei L. Smirnov, Editors,** Asymptotic methods in mechanics, 1993

2 **Philip D. Loewen,** Optimal control via nonsmooth analysis, 1993

1 **M. Ram Murty, Editor,** Theta functions: From the classical to the modern, 1993

Volume 27

CRM PROCEEDINGS & LECTURE NOTES

Centre de Recherches Mathématiques
Université de Montréal

Nonlinear Dynamics and Renormalization Group

I. M. Sigal
C. Sulem
Editors

The Centre de Recherches Mathématiques (CRM) of the Université de Montréal was created in 1968 to promote research in pure and applied mathematics and related disciplines. Among its activities are special theme years, summer schools, workshops, postdoctoral programs, and publishing. The CRM is supported by the Université de Montréal, the Province of Québec (FCAR), and the Natural Sciences and Engineering Research Council of Canada. It is affiliated with the Institut des Sciences Mathématiques (ISM) of Montréal, whose constituent members are Concordia University, McGill University, the Université de Montréal, the Université du Québec à Montréal, and the Ecole Polytechnique. The CRM may be reached on the Web at www.crm.umontreal.ca.

American Mathematical Society
Providence, Rhode Island USA

The production of this volume was supported in part by the Fonds pour la Formation de Chercheurs et l'Aide à la Recherche (Fonds FCAR) and the Natural Sciences and Engineering Research Council of Canada (NSERC).

2000 *Mathematics Subject Classification.* Primary 35L05; Secondary 35J05, 35Q55, 35Q99, 70K99, 82C28, 81T15, 37K40.

Library of Congress Cataloging-in-Publication Data
Nonlinear dynamics and renormalization group / I. M. Sigal, C. Sulem.
    p. cm. — (CRM proceedings & lecture notes, ISSN 1065-8580 ; v. 27)
  Includes bibliographical references.
  ISBN 0-8218-2802-9 (alk. paper)
  1. Renormalization (Physics)—Congresses. 2. Differential equations, Nonlinear—Congresses.
I. Sigal, Israel Michael, 1945– II. Sulem, C. (Catherine), 1957– III. Series.

QC20.7.R43 N66 2001
530.15—dc21
                                                                                                      2001022066

Copying and reprinting. Material in this book may be reproduced by any means for educational and scientific purposes without fee or permission with the exception of reproduction by services that collect fees for delivery of documents and provided that the customary acknowledgment of the source is given. This consent does not extend to other kinds of copying for general distribution, for advertising or promotional purposes, or for resale. Requests for permission for commercial use of material should be addressed to the Assistant to the Publisher, American Mathematical Society, P. O. Box 6248, Providence, Rhode Island 02940-6248. Requests can also be made by e-mail to reprint-permission@ams.org.

Excluded from these provisions is material in articles for which the author holds copyright. In such cases, requests for permission to use or reprint should be addressed directly to the author(s). (Copyright ownership is indicated in the notice in the lower right-hand corner of the first page of each article.)

© 2001 by the American Mathematical Society. All rights reserved.
The American Mathematical Society retains all rights
except those granted to the United States Government.
Printed in the United States of America.

∞ The paper used in this book is acid-free and falls within the guidelines
established to ensure permanence and durability.
This volume was submitted to the American Mathematical Society
in camera ready form by the Centre de Recherches Mathématiques.
Visit the AMS home page at URL: http://www.ams.org/

10 9 8 7 6 5 4 3 2 1   06 05 04 03 02 01

Contents

Preface	vii
Analysis of Some Macroscopic Models of High-T_c Superconductivity *Stan Alama and Lia Bronsard*	1
The Effect of Distribution In Space In Ostwald Ripening *Nicholas D. Alikakos and Giorgio Fusco*	17
Mathematical Problems in the Control of Underactuated Systems *David Auckly and Lev Kapitanski*	29
Axially and Helically Symmetric Global Plasma Equilibria *Oleg I. Bogoyavlenskij*	41
On the Complete Ionization of a Periodically Perturbed Quantum System *O. Costin, J. L. Lebowitz, and A. Rokhlenko*	51
The sine-Gordon Model at $\beta = 4\pi$ *J. Dimock*	63
Ground States of Supersymmetric Matrix Models *G. M. Graf*	69
Some Mathematical Problems in the Ginzburg–Landau Theory of Superconductivity *Stephen J. Gustafson*	77
Renormalization in Radiation Reaction: New Developments in Classical Electron Theory *Michael K.-H. Kiessling*	87
Singular Limit of the Modified Nonlinear Schrödinger Equation *Chi-Kun Lin*	97
Dynamics of Quantum Resonances *Marco Merkli*	111
Nelson Diffusions and Blow-Up Phenomena in Solutions of the Nonlinear Schrödinger Equation With Critical Power *Hayato Nawa*	117
Embedded Solitons of the DSII Equation *Dmitri E. Pelinovsky and Catherine Sulem*	135

On the Blow up Phenomenon for the Critical Nonlinear Schrödinger Equation in 1D
    Galina Perelman   147

Vorticity for the Ginzburg–Landau Model of Superconductors in a Magnetic Field
    Sylvia Serfaty   165

Dissipation through Dispersion
    A. Soffer   175

Quantum Tunneling at Positive Temperature
    B. Vasilijevic   185

Preface

This volume contains the lectures delivered at the workshop 'Nonlinear Dynamics and Renormalization Group' that was held at the Centre de recherches mathématiques in Montréal in August 1999. This workshop was part of year program at the CRM devoted to Mathematical Physics. It brought together a prominent group of researchers working on Nonlinear Partial Differential Equations and on Renormalization Group Methods in the hope of developing exchanges of information, ideas and problems between both fields.

We wish to take this opportunity to express our thanks to the lecturers for their stimulating presentations and to all the participants who contributed to the success of the meeting.

We are grateful to the CRM for their financial and administrative support, and for the publication of these proceedings. Our special thanks go to Yvan Saint-Aubin for giving us help and guidance in organizing the event, to the staff of the CRM and in particular to Louis Pelletier for providing us with an incredibly efficient infrastructure.

I. M. Sigal
C. Sulem

Montréal, 2001

Analysis of Some Macroscopic Models of High-T_c Superconductivity

Stan Alama and Lia Bronsard

1. Introduction

The discovery of superconductors with high critical temperature (T_c) by Bednorz and Müller in 1986 has led to a new burst of activity in this field, both among physicists and by mathematicians. While the mechanism of high-T_c superconductivity remains unknown to physicists, they have found that the classical Ginzburg–Landau theories used to describe the type-II superconductors model many high-T_c phenomena quite well. This is a wonderful opportunity for mathematicians to become acquainted with these macroscopic variational models, and explore both the classical results of Ginzburg–Landau theory and the new variations on these models and the questions which arise in modelling the new high-T_c materials.

The two lectures which we presented at this meeting treat two different questions related to modelling high-T_c superconductors. Most of the high-T_c "cuprate" materials are crystals with a distinctly layered structure. Indeed, a pure monocrystalline sample of cuprate high-T_c material (such as $Bi_2Sr_2CaCu_2O_8$ (BSCCO), $Tl_2Ba_2CaCu_2O_8$ (TBCCO), or $YBA_2Cu_3O_7$ (YBCO)) consists of copper oxide superconducting planes stacked with intervening insulating (or weakly superconducting) planes. Layered superconductors had already been studied by physicists well before the discovery of high-T_c materials: they may be manufactured in the laboratory, and many organic superconductors have layered structure. In 1971 the physicists Lawrence and Doniach introduced a Ginzburg–Landau type model for superconducting materials with a planar layered structure. A (two-dimensional) Ginzburg–Landau equation determines the superconducting order parameter within each plane, and the order parameters in adjacent planes are coupled via the Josephson tunnelling effect.

Due to the layered structure one expects high-T_c materials to be highly anisotropic. If we model these materials with an anisotropic Ginzburg–Landau model in which the sample is thought of as a three-dimensional solid with anisotropic material parameters, then we can measure the anisotropy by an "effective mass ratio". It is given by $\Gamma := \xi_{ab}^2/\xi_c^2$, where ξ_c measures coherence length along the

---

2000 *Mathematics Subject Classification.* Primary: 35J50; Secondary: 35Q99.
The research is partially supported by NSERC Research Grants.
This is the final form of the paper.

perpendicular to the planes and ξ_{ab} within the planes and experiments suggest $\Gamma_{\text{YBCO}} \simeq 49$, $\Gamma_{\text{BSCCO}} \simeq 3025$, and $\Gamma_{\text{TBCCO}} > 90000$ (See Iye [**15**].) For certain materials (such as YBCO) and temperatures close to the critical temperature T_c the anisotropic Ginzburg–Landau model seems valid, but for more anisotropic superconductors it does not give a good qualitative or quantitative description of experimental observations.

The structural anisotropy of these materials is most evident in the situation where the sample is placed in a strong magnetic field oriented *parallel* to the superconducting planes. In this regime experiments reveal a "crossover" between "three-dimensional" behavior (governed by the anisotropic Ginzburg–Landau model) and "two-dimensional" behavior at a critical temperature T_{C0}. The "two-dimensional" regime is characterized by a "magnetically transparent" state, in which the lower critical field H_{C1} is very small and the applied magnetic field penetrates completely between the planes, virtually unscreened by the superconductor. These observations are not in agreement with the anisotropic Ginzburg–Landau model, where the magnetic field is largely expelled from the bulk and isolated vortices appear in a triangular "Abrikosov lattice." For highly anisotropic materials (such as BSCCO and TBCCO) the two-dimensional regime occurs within one degree Kelvin of T_c, and we require a model which addresses the discrete nature of the material: the Lawrence–Doniach model. (For discussion of the experimental evidence of the transparent state and the dimensional crossover see [**17**].)

Our analysis of the Lawrence–Doniach model in a parallel magnetic field confirms the experimental predictions of the transparent state and reveals its detailed structure in various parameter regimes. We discuss these results in Section 2; the details will be published in a forthcoming paper [**1**].

The second question we address relates to the nature of high-T_c superconductivity inside the superconducting layers themselves. The quantum-mechanical mechanism behind conventional superconductivity was explained (microscopically) by Bardeen, Cooper, and Schrieffer (1957), and Gorkov showed in 1959 that the Ginzburg–Landau models arise in a limit of the BCS theory near the critical temperature T_c. However, the BCS theory falls short of explaining the origins of high-T_c superconductivity, since it predicts a much lower value for T_c than has been observed in the cuprates. This shortcoming of the BCS theory has led to the development of several competing quantum-mechanical theories of high-T_c superconductivity. One such theory, proposed by S. C. Zhang [**25**], is based on the observation that many of the cuprate materials exhibit two different kinds of order, depending on temperature and "doping" (oxygen content in the crystal structure.) When the materials are "underdoped" they are *antiferromagnets*, and they become superconducting when they are "overdoped". Zhang posits that the lattice interactions responsible for high-T_c superconductivity are the same as those creating antiferromagnetism, and he introduced a lattice sigma-model which unifies SC and AF with a (broken) SO(5) gauge symmetry.

This coupling of the SC and AF order parameters should in some way alter the structure of the superconducting state described by the SO(5) theory. In a recent paper Arovas, Berlinsky, Kallin, and Zhang [**7**] considered the structure of the vortex state in the SO(5) theory. They introduced a phenomenological Ginzburg–Landau model based on the SO(5)-symmetry and examine a single cylindrically symmetric vortex, with a given quantized magnetic flux imposed axially (i.e., *perpendicular* to the uniform cross-section.) Recall that in a conventional superconductor the magnetic field is expelled from the superconducting bulk, and only penetrates in thin tubes (the vortices) where superconductivity is suppressed. Hence, in the conventional theory the magnetic field is constrained to a small *core* of normal (non-SC) phase. Using a simplified model Arovas *et al.* predicted a new kind of vortex structure in the SO(5) model: vortices with antiferromagnetic cores, which should be observed for small values of the chemical potential (which regulates the temperature and doping.) They also predicted that as the chemical potential is gradually decreased, the transition from normal core to AF core vortices occurs in a discontinuous fashion. In other words, AF cores should be produced via a *first order* phase transition. In Section 3 we will describe our analytical and numerical results concerning the transition between the normal core and AF core vortex structures. The full analysis of this problem is contained in the paper by Alama, Bronsard, and Giorgi [4], and numerical and modelling considerations appear in the work of Alama, Berlinsky, Bronsard, and Giorgi [2].

2. Layered Superconductors in a Parallel Magnetic Field

We consider the case of a layered superconductor in a uniform magnetic field imposed *parallel* to the superconducting planes. It is in this configuration that the discrete nature of the layered superconductors is most readily observed. We assume that the planes are parallel to the xy-plane, and the external field lies along the y-direction, $\vec{H} = H\hat{y}$. We will take the planes to be of infinite extent in the y-direction, which implies that the local magnetic field will be everywhere independent of y and point in the y-direction,

$$\vec{h}(x,y,z) = h(x,z)\hat{y}.$$

The vector potential $\vec{A}$ may then be chosen to lie in the xz-plane,

$$\vec{A}(x,y,z) = A_x(x,z)\hat{x} + A_z(x,z)\hat{z}, \quad \vec{h} = \mathrm{curl}\vec{A} = \left(\frac{\partial A_x}{\partial z} - \frac{\partial A_z}{\partial x}\right)\hat{y}.$$

2.1. Finite width samples. Assume that there are a finite number of parallel superconducting planes, numbered $n = 0, \ldots, N$, each separated by a distance p. We will first consider the case where the planes have finite width along the x-axis, $x \in [-L, L]$. (We will also consider solutions which are periodic in x and/or n later.) In each plane we define a (complex-valued) superconducting order parameter $\psi_n(x)$, $n = 0, \ldots, N$. In our units, $|\psi_n| = 1$ represents a purely superconducting state. In the parameter regime corresponding to a transparent state, a straightforward energy estimate shows that $|\psi_n(x)| \sim 1$ uniformly, and hence we may define f_n, ϕ_n via $\psi_n(x) = f_n(x) \exp(i\phi_n(x))$. In this setting, the Lawrence–Doniach model consists of the following Gibbs free energy functional:

$$\Omega_r(f_n, \phi_n, \vec{A}) = \int_{-L}^{L} p \sum_{n=0}^{N} \left[\frac{1}{2}(f_n^2 - 1)^2 + \frac{1}{\kappa^2}(f_n')^2 + \frac{1}{\kappa^2}\left(\phi_n' - A_x(x,z_n)\right)^2 f_n^2\right] dx$$

$$+ \frac{r}{2}\int_{-L}^{L} p \sum_{n=1}^{N} \left(f_n^2 + f_{n-1}^2 - 2f_n f_{n-1} \cos(\Phi_{n,n-1})\right) dx$$

$$+ \frac{1}{\kappa^2} \int_{-L}^{L} \int_{0}^{Np} \left(\frac{\partial A_x}{\partial z} - \frac{\partial A_z}{\partial x} - H \right)^2 dz\, dx,$$

where the *gauge-invariant phase difference*

$$\Phi_{n,n-1}(x) := \phi_n - \phi_{n-1} - \int_{z_{n-1}}^{z_n} A_z(x,z)\, dz,$$

$z_n = np$ and we have chosen units such that the in-plane correlation depth $\lambda_{\text{ab}} = 1$. The parameter $\kappa = \lambda_{\text{ab}}/\xi_{\text{ab}}$ is the Ginzburg–Landau parameter, r is the *interlayer coupling parameter* (or *Josephson coupling parameter*), and the magnetic fields are measured in units of H_c/κ, where H_c is the thermodynamic critical field. (See [**24**].)

The coupling between the superconducting planes given by the second sum in Ω_r produces the Josephson proximity effect, by which superconducting electrons travel from one superconducting region to another by quantum mechanical tunnelling. This may be seen explicitly in the Euler–Lagrange equations, where the currents in the gaps between planes will be determined by the sine of the gauge-invariant phase difference. Indeed the supercurrent within the nth superconducting plane is given by

$$j_x^{(n)} = \left(\phi_n' - A_x(x, z_n) \right) f_n^2,$$

while the current between the $n-1$ and the n superconducting planes is given by

$$j_z^{(n)} = j_c f_n f_{n-1} \sin \Phi_{n,n-1},$$

where $j_c = rp\kappa^2/2$. The interlayer coupling parameter r gives the strength of the Josephson effect, and in terms of the other physical parameters,

$$r = \frac{2\xi_{\text{ab}}^2 \lambda_{\text{ab}}^2}{\lambda_J^2 p^2} = \frac{2}{\Gamma^2 \kappa^2 p^2},$$

(in our units) where λ_J is the *Josephson penetration depth*. The length scale λ_J in the Lawrence–Doniach model is directly related to the effective mass ratio of the anisotropic Ginzburg–Landau model via $\Gamma = \lambda_J^2/\lambda_{\text{ab}}^2$. For the high-$T_c$ cuprates ξ_{ab} and p are of the same order of magnitude, and hence for highly anisotropic superconductors r is a small parameter in the Lawrence–Doniach model. In particular, for BSCCO or TBCCO $r \simeq 10^{-4}$–10^{-3}. We note that Chapman, Du, and Gunzburger [**11**] have proven that solutions of the Lawrence–Doniach model converge to solutions of the anisotropic Ginzburg–Landau model (and in particular the convergence of energy minimizers) under the limit $p \to 0$ with κ, Γ fixed. This limit does not correspond to our "two-dimensional" regime, since it would send $r \to \infty$.

In [**1**] we study the minimizers (and low-energy solutions) of the Lawrence–Doniach system for r near zero. We consider three boundary-value problems: a finite sample in the x, z variables; a sample periodic in x with finitely many planes in z; and a doubly periodic configuration, representing an infinite sample in both x, z. The crucial observation is that when $r = 0$ the planes decouple, and the energy may be minimized explicitly. Even after gauge symmetries have been removed, the minimization problem at $r = 0$ is still degenerate: the absolute minimizers occupy a finite dimensional hyperplane in function space. By a Lyapunov–Schmidt decomposition we may then reduce the problem of finding solutions with $r \simeq 0$ to a finite dimensional variational problem on this hyperplane. In all three settings, the minimum value of energy is $O(r)$, and we indeed recover the "transparent state" observed by physicists. The local magnetic field $h(x,z) = H + O(r)$ inside the sample, and superconductivity is hardly affected in each plane, $|\psi_n(x)| = 1 - O(r)$.

In particular the order parameters are never zero: "vortices" correspond to local maxima of the local magnetic field, and lie between the layers. In the physics literature these are referred to as *Josephson vortices*, as opposed to the Abrikosov vortices typically observed in the Ginzburg–Landau theory.

We provide a sketch of the general method in the context of a sample of *finite width*. The periodic problems are solved in a similar way, but the variational formulation involves suitable interpretation of the periodicity condition, using *'tHooft boundary conditions* [22], and we will discuss the formulation of the periodic problems in a later section.

The first variation of Ω_r within the class

$$\mathcal{E} := \left\{ \begin{array}{c} (f_n, \phi_n, \vec{A}) \colon f_n \in H^1([-L,L]), \phi_n \in H^1([-L,L]),\ n = 0, \ldots, N \\ \vec{A} = (A_x, A_z) \in H^1(D, \mathbb{R}^2) \end{array} \right\},$$

where $D := [-L, L] \times [0, Np]$, produces a system of coupled ordinary and partial differential equations for the unknowns. Variation with respect to f_n (for each $n = 0, \ldots, N$) yields coupled nonlinear magnetic Schrödinger equations for the f_n, with Neumann boundary conditions at $x = \pm L$. Variations with respect to A_x, A_y produce first order PDE's (Maxwell's equation for $\mathrm{curl} \vec{h}$), with a jump condition for the local magnetic field h at each superconducting plane and the boundary condition: $h(x,z) = H$ for $x = \pm L$. Variation with respect to ϕ_n produces a system of current–conservation laws with a boundary condition which expresses the physical fact that current should not flow past the edge of the material.

There is a large degree of degeneracy of Ω_r in $\mathcal{E}$ due to the gauge invariance: if $\chi \in H^2(D)$, and

$$\hat{f}_n = f_n, \quad \widehat{\phi}_n(x) = \phi_n(x) - \chi(x, z_n), \quad \widehat{\vec{A}} = \vec{A} - \nabla\chi,$$

then $\Omega_r(\hat{f}_n, \widehat{\phi}_n, \widehat{\vec{A}}) = \Omega_r(f_n, \phi_n, \vec{A})$. As usual, we eliminate this troublesome degeneracy by fixing a gauge. The most convenient choice is the *Coulomb gauge*, which allows us to control the H^1 norm of the vector potential $\vec{A}$ by its curl. We define a subspace of $\mathcal{E}$ to incorporate this choice of gauge,

$$E := \left\{ (f_n, \phi_n, \vec{A}) \in \mathcal{E} \colon \int_{-L}^{L} \phi_0(x)\, dx = 0, \mathrm{div}\, \vec{A} = 0 \text{ in } D, \text{ and } \vec{A} \cdot \vec{n} = 0 \text{ on } \partial D \right\}.$$

This choice is made with no loss of generality: for every $(f_n, \phi_n, \vec{A}) \in \mathcal{E}$, there exists $\chi \in H^2(D)$ so that $(f_n, \phi_n - \chi(\cdot, z_n), \vec{A} - \nabla\chi) \in E$.

When $r = 0$, the superconducting planes decouple, and we may solve the minimization problem explicitly. In particular, the minimizers with $r = 0$ form an N-dimensional hyperplane $\mathcal{S}$ in E, with

$$f_n^0 \equiv 1, \quad \mathrm{curl}\vec{A}^0 = H\hat{y}, \quad \phi_n(x) = \alpha_n + \int_0^x A_x(s, z_n)\, ds,$$

with free parameters $\alpha_1, \ldots, \alpha_n \in \mathbb{R}$. (Note that $\alpha_0 = 0$ for configurations in E.) In particular, the gauge invariant phase difference is given by $\Phi_{n,n-1} = \delta_n + Hnpx$, where $\delta_n := \alpha_n - \alpha_{n-1}$, $n = 1, \ldots, N$. Note that $\mathcal{T} := T_s\mathcal{S}$ is independent of $s \in \mathcal{S}$.

We now perturb away from the degenerate minima of Ω_0, using a variational Lyapunov–Schmidt procedure. This method has been used by Ambrosetti, Coti-Zelati, and Ekeland [6], Ambrosetti and Badiale [5] in studying heteroclinic solutions of Hamiltonian systems, and similar finite-dimensional reduction techniques have been used by various authors in studying spiked solutions to reaction–diffusion systems, Ginzburg–Landau vortices, and blow-up for critical exponent problems.

We define orthogonal projections P, $P^\perp \colon E \to E$, with $P(E) = \mathcal{T}$ and $P^\perp(E) = W := \mathcal{T}^\perp$ and proceed according to the usual Lyapunov–Schmidt reduction. For $u = (f_n, \phi_n, \vec{A}) \in E$, since $\mathcal{S}$ is a hyperplane in E we may decompose $u = s + w$ with $s \in \mathcal{S}$, $w \in W$. Then we project the equation $\nabla \Omega_r(f_n, \phi_n, \vec{A}) = 0$ into the two linear subspaces $\mathcal{T}$ and W,

(2.1) $$F_1(r, s, w) := P\big[\nabla \Omega_r(s + w)\big] = 0;$$

(2.2) $$F_2(r, s, w) := P^\perp\big[\nabla \Omega_r(s + w)\big] = 0.$$

The second equation can be solved uniquely in a neighborhood of $\mathcal{S}$ for r small, using the Implicit Function Theorem to obtain a unique $w = w(r, s)$ with $F_2\big(r, s, w(r, s)\big) = 0$. We define
$$\mathcal{S}_r := \{s + w(r, s) \colon s \in \mathcal{S}\}.$$

$\mathcal{S}_r$ is a smooth manifold parametrized by the hyperplane $\mathcal{S}$. The important role played by $\mathcal{S}_r$ is that it is a natural constraint for Ω_r (see [5, Lemma 4]), and hence the equation (2.1) may be solved variationally:

LEMMA 2.1. *There exists $0 < \widetilde{r}_0 < r_0$ such that when $|r| < \widetilde{r}_0$ the following hold:*

(a) *If $(f_n, \phi_n, \vec{A}) \in \mathcal{S}_r$ satisfies $\nabla(\Omega_{r|\mathcal{S}_r})(f_n, \phi_n, \vec{A}) = 0$, then $\nabla \Omega_r(f_n, \phi_n, \vec{A}) = 0$ in E.*

(b) *Conversely, if $(f_n, \phi_n, \vec{A}) \in E$ is a critical point of Ω_r with $\Omega_r(f_n, \phi_n, \vec{A}) \leq 2pN(L+1)r$, then $(f_n, \phi_n, \vec{A}) \in \mathcal{S}_r$.*

We now apply the theory of the previous paragraphs to determine the minimizer (and other stationary states) of the Lawrence–Doniach energy for a finite width sample, for $r \ll 1$. Since the solution of (2.2), $w(r, s) = O(r)$ uniformly for $s \in \mathcal{S}$, we can expand

(2.3)
$$\Omega_r\big(s + w(r,s)\big) = r \sum_{n=1}^{N} \int_{-L}^{L} \big[1 - \cos(\delta_n + Hpx)\big]\, dx + O(r^2),$$
$$= r\left(2NL - \frac{\sin(HpL)}{Hp} \sum_{n=1}^{N} \cos \delta_n\right) + O(r^2).$$

By an application of the Implicit Function Theorem together with Lemma 2.1, the critical points of the Lawrence–Doniach model with energy of order r coincide with the critical points of

$$G(s) = 2NL - \frac{\sin(HpL)}{Hp} \sum_{n=1}^{N} \cos \delta_n,$$

where $\delta_n = \alpha_n - \alpha_{n-1}$, and we identify $s = (\alpha_1, \ldots, \alpha_N)$. This determines all low-energy critical points of Ω_r for r small, as long as $\sin(HpL) \neq 0$: choosing $\delta_n \in \{0, \pi\}$ for each $n = 1, \ldots, N$ we obtain a (nondegenerate) critical point of Ω_r. In particular, $G(s)$ is minimized when

(2.4) $$\delta_n = \begin{cases} 0, & \text{when } \frac{\sin(HpL)}{Hp} > 0, \text{ for every } n = 1, \ldots, N; \\ \pi, & \text{when } \frac{\sin(HpL)}{Hp} < 0, \text{ for every } n = 1, \ldots, N. \end{cases}$$

When $\sin(HpL) = 0$, $\Omega_{r|\mathcal{S}_r}$ is degenerate at order r, and we should go to a higher order in the expansion to determine the stationary configurations.

Expanding the solutions in Taylor's series near $r = 0$ produces an approximation of each solution given above. In particular, when all δ_n are chosen equal we see that (with the exception of the top and bottom layers) all gauge-invariant quantities (the modulus of the order parameter, the currents, and the local field) are each independent of z (or n) up to order r. Expansion to order r^2 in the solutions will show these quantities to be independent of z (or n) except for the top and bottom *two* planes and the top and bottom gaps. We call this configuration "vortex planes"—see Fig. 1.

Unlike the Ginzburg–Landau case, the order parameter need not vanish at the "core" where the local field attains its maximal value and around which supercurrents circulate. The "vortices" are then the planes $\{x = \text{const.}\}$ over which $h(x)$ attains its relative maxima. The values of H with $HpL = \pi\mathbb{Z}$ indicate first-order phase transitions (maxima of h "flip" to minima of h) as vortex planes are nucleated in the sample from the boundaries.

Other critical points may be determined asymptotically by some other permutation of $\delta_n \in \{0, \pi\}$. For example, Fig. 2 illustrates a period-2 lattice obtained by choosing $\delta_n = n\pi \pmod{2\pi}$.

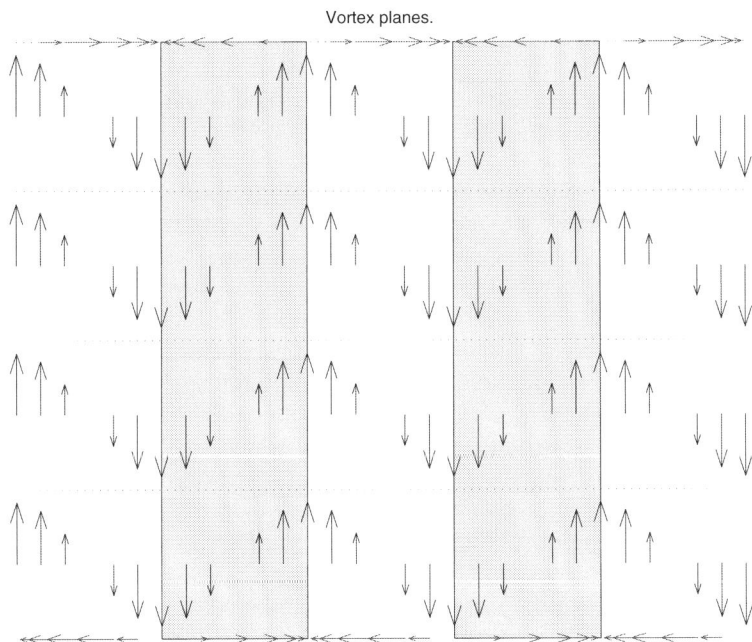

FIGURE 1. Vortex planes, for a sample with a finite number of superconducting planes. Arrows indicate the intensity and direction of the currents. Note that (to highest order in r) there are no intralayer currents in the interior planes, only Josephson currents parallel to the vortex planes.)

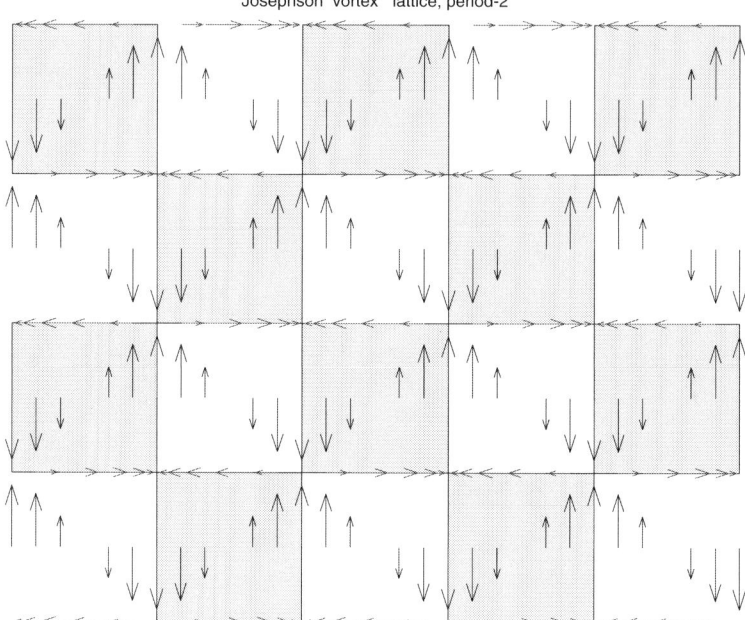

FIGURE 2. Vortex lattice, with period-two in the layering direction, for a sample with a finite number of superconducting planes. Arrows indicate the intensity and direction of the currents. The Josephson vortices lie along the vertical segments along which the current j_z vanishes, and where the magnetic field intensity attains a local maximum (indicated by shading.)

2.2. Periodic vortex lattices. In the previous part we described the global minimizers of the Lawrence–Doniach free energy for any finite width sample with finitely many superconducting layers, in the asymptotic regime $0 < r \ll 1$. However, in an actual physical sample of layered superconductor the value of r is fixed, and it is interesting to ask whether (for given values of the other physical parameters) the above solutions really do represent the minimizers for a real sample. By inspection of the form of the vortex plane solutions described above we see that the expansion may be expected to be valid only when $p\kappa^2 Lr/H$ is small enough. Indeed, at order r^2 there appears a secular term in the expansion for $h(x,z)$, due to the order r correction to the slope of $\Phi_{n,n-1}$. In short, we cannot expect our expansion to be valid uniformly in r for large L, large κ, or small H. (See [**1**] for details.)

To address the question of the behavior of minimizers for large sample widths we consider periodic solutions in an infinitely wide sample. We will consider two periodic problems: a singly-periodic setting in which the sample has finitely many superconducting layers (as in the previous sections) but infinite extent in the x-direction, and a bi-periodic problem in which the sample is infinite both in x and z.

In the first case, we fix a period $2q$, and seek solutions for which the observables $h(x,z)$, $f_n(x)$, $V_n(x) := \phi'_n(x) - A_x(x,z_n)$, and $\sin\Phi_{n,n-1}(x)$ will be $2q$-periodic functions in x. In the second case, we seek solutions for which in additions $h(x, z)$ is periodic in z of period Np and $f_n(x)$, $V_n(x)$, and $\sin \Phi_{n,n-1}(x)$ are periodic in n of period N. One subtlety of the periodic case is that periodic magnetic fields and currents do not generally arise from periodic potentials $\vec{A}$ or phases ϕ_n. The physicist 'tHooft has introduced boundary conditions which correspond to potentials and phases which are periodic "up to a continuous gauge transformation". We will describe these briefly, for the bi-periodic case.

We say $(f_n, \phi_n, \vec{A}) \in t\mathcal{H}$ if $f_n \in H^1_{\text{per}}(\mathbb{R})$, $\phi_n \in H^1_{\text{loc}}(\mathbb{R})$, $\vec{A} \in H^1_{\text{loc}}(\mathbb{R}^2; \mathbb{R}^2)$ and there exist two functions ω_x and ω_z in $H^2_{\text{loc}}(\mathbb{R}^2)$ such that:

$$f_n(x + 2q) = f_n(x) = f_{n+N}(x), \tag{2.5}$$

$$\phi_n(x + 2q) = \phi_n(x) + \omega_x(x, z_n) \bmod 2\pi, \quad \phi_{n+N}(x) = \phi_n(x) + \omega_z(x, z_n), \tag{2.6}$$

$$\vec{A}(x + 2q, z) = \vec{A}(x, z) + \nabla \omega_x(x, z), \quad \vec{A}(x, z + Np) = \vec{A}(x, z) + \nabla \omega_z(x, z). \tag{2.7}$$

Since the gauge invariant quantities are periodic in both x and z we may define the free energy over exactly one period cell, $\Pi = [-q, q] \times [0, Np]$. The Euler–Lagrange equations under 'tHooft conditions are virtually identical to those derived in the first part except that we must take into account the periodicity in n and use periodic boundary conditions both in x and z. One new feature in the periodic problem is that the magnetic flux through each period cell is quantized.

However, the space $t\mathcal{H}$ is rather complicated from the point of view of analysis: by gauge invariance it is very large, and we must choose a norm to control the functions ω_x, ω_z. In addition, ω_x, ω_z must satisfy a system of constraints which ensure that the phases ϕ_n are consistently defined from one period cell to another. Moreover, the "mod 2π" condition in (2.6) decomposes the space into infinitely many connected components, determined by the degree of winding of the phase in each plane. Fortunately we may make a choice of gauge which simplifies the space enormously, and a choice of connected component which gives the smallest energy. The space we obtain is as follows: $(f_n, \phi_n, \vec{A}) \in \mathcal{BP}$ provided there are constants ω, and d in $\mathbb{R}$ such that

$$\vec{A} \in H^1_{\text{loc}}(\mathbb{R}^2; \mathbb{R}^2), \quad f_n \in H^1_{\text{per}}(\mathbb{R}), \quad \phi_n \in H^1_{\text{loc}}(\mathbb{R}), \tag{2.8}$$

$$\phi_n(x + 2q) = \phi_n(x) + \omega + 2\pi n, \tag{2.9}$$

$$\phi_{n+N}(x) = \phi_n(x) + \frac{N\pi}{q} x + d, \tag{2.10}$$

$$f_n(x + 2q) - f_n(x) - f_{n+N}(x), \tag{2.11}$$

$$\vec{A}(x, z) = \langle h \rangle (z, 0) + (\partial_z \xi, -\partial_x \xi), \quad \text{where} \tag{2.12}$$

$$\Delta \xi = h - \langle h \rangle, \quad \xi(x + 2q, z) = \xi(x, z + Np) = \xi(x, z). \tag{2.13}$$

As mentioned above, the flux is prescribed by the choice of the space, and hence the average value of the magnetic field is fixed:

$$\langle h \rangle := \frac{1}{2qNp} \iint_\Pi h(x, z)\, dx\, dz = \frac{\pi}{pq},$$

where we recall $\Pi = [-q, q] \times [0, Np]$.

As in the discussion of the finite width case, when $r = 0$ the superconducting planes decouple. It is at this stage that we will see which period q we should choose,

since
$$\inf_{\mathcal{BP}} \Omega_0^{\text{per}} = 0 \quad \text{if and only if} \quad Hpq = k\pi, \ k = 1, 2, 3, \ldots.$$
Hence, we will fix the period at $2q$, with
$$q = \pi H p.$$
In other words, we take the smallest q for which the minimum energy is $O(r)$ for r small.

As in the previous discussion, we use a Lyapunov–Schmidt decomposition to reduce the problem to a finite dimensional variational problem on the hyperplane of (degenerate) solutions to the $r = 0$ problem. Again we expand the energy in powers of r, and study the critical points of the leading nontrivial term. For the periodic problems it is necessary to go to order r^2 in this expansion:

$$\Omega_r(f_n, \phi_n, \vec{A}) = 2qNr + r^2 \left\{ Nq \frac{\kappa^2}{2H^2p^2} \left[\frac{3}{2} p^2 - 1 - \frac{H^2 p^2}{H^2 p^2 + 2\kappa^2} \right] - Nq \right.$$
$$\left. + q \frac{\kappa^2}{2H^2 p^2} \left(1 - \frac{H^2 p^2}{H^2 p^2 + 2\kappa^2} \right) \sum_{n=1}^{N} \cos(\delta_n - \delta_{n+1}) \right\} + O(r^3).$$

Recalling that (δ_n) parametrize the natural constraint manifold $\mathcal{S}_r$, we see that we would like to minimize the finite dimensional expression $\sum_{n=1}^{N} \cos(\delta_n - \delta_{n+1})$ by choosing the free parameters $\delta_n = \alpha_n - \alpha_{n-1}$ with $\delta_n - \delta_{n+1} = \pm \pi$ for each n. This is possible when the period N is *even*. The resulting solution is a period-$2p$ in z diamond-shaped lattice, described by Bulaevskii and Clem [10]. (See Fig. 2.) The vortex plane solution of Theorodakis–Kuplevakhsky [18, 23] is obtained by taking $\delta_n - \delta_{n+1} = 0$ for all n. (It is obtained by *maximizing* the $O(r^2)$ term with respect to these parameters.) The solution obtained is the same for *any* even N.

When N is *odd* we cannot choose δ_n in this way, and the lattice must contain some sort of dislocation. Since we cannot make $\delta_n - \delta_{n+1} = \pm \pi$ for each n the minimizing configuration must necessarily have larger energy per plane, and hence we conclude that the "true" minimizing state is obtained by taking N even.

When the sample consists of a *finite* number of planes, but is infinite in x, we follow the same procedure and again obtain the period-$2p$ diamond lattice, except for an edge effect which affects the current and order parameter in top and bottom planes at order r in the solution.

2.3. Finite width samples revisited. Recall that for a finite width $L < \infty$ we obtained the contrary conclusion—that vortex planes give the minimizing configuration—but that the validity of the expansion was limited by the observation that the approximations made are only valid when $rp\kappa^2 L/H \ll 1$, and the condition $HpL \neq \pi k$, k an integer. The difference between the two expansions is that when $HpL \neq \pi k$, k an integer, there is a contribution from the term at order r which favors vortex planes.

Here is one possible argument: take a large fixed sample of width $2L$ with all parameters fixed, except H. In the bulk, energy minimization prefers the diamond lattice. The savings in energy is small (order r^2) but scales with area. However, the vortex lattice costs energy at the boundary, at a larger order r, but independent of the width of the sample. If the width is large the two terms are of the same order of magnitude, and competition is possible.

To get a rough idea of what may be happening, take the two *periodic* solutions, for the vortex lattice and for the vortex planes, and simply evaluate their energy on a fixed finite interval $-L \leq y \leq L$. We obtain:

$$\Omega_{L,\text{planes}} - \Omega_{L,\text{lattice}} = 2NLr \left[r \frac{\kappa^4}{H^4 p^4 (1 + (2\kappa^2)/(H^2 p^2))} - \frac{|\sin(HpL)|}{HpL} \right],$$

$$= \frac{2Nr}{\kappa x} \left[\frac{\ell r}{x^3 (1 + 2/x^2)} - |\sin(\ell x)| \right],$$

with $x = Hp/\kappa$ and $\ell = \kappa L$. Whenever $x < \ell r/2$ the vortex lattice has lower energy. This can happen for small H, large L, large κ, large r. However, the graph of the bracketed expression oscillates infinitely often as $x \to \infty$ which suggests an infinite number of first order transitions between vortex planes and diamond lattices. The lattices appear when the length $2L$ is very close to an exact multiple of the period $2q$. The vortex planes win when the length is most incommensurate with the natural period. The period of oscillation between the two is very rapid, which makes the question of which configuration is the minimizer impossible in practice. If we fix all parameters (including x), and let r decrease, for every x which is not an integer multiple of the period we eventually see vortex planes for r sufficiently small. This confirms what we already know from the asymptotic expansion for finite intervals (valid only when r is sufficiently small compared to the various other parameters.)

3. The SO(5) Model

The full SO(5) Ginzburg–Landau free energy is written in terms of the SC order parameter $\psi \in \mathbb{C}$ and the AF order parameter (Néel vector) $\vec{m} = (m_1, m_2, m_3)$. In nondimensional form, the free energy is:

$$\mathcal{F} = \frac{1}{2} \int_\Omega \left\{ \frac{\kappa^2}{2} (1 - |\psi|^2 - |\vec{m}|^2)^2 + g\kappa^2 |\vec{m}|^2 + \left| \left(\frac{1}{i} \nabla - \vec{A} \right) \psi \right|^2 + |\nabla \vec{m}|^2 + |\nabla \times \vec{A}|^2 \right\} dx.$$

(We refer to the paper by Alama, Berlinsky, Bronsard and Giorgi [2] where the free energy is written in dimensional form.) In these variables, the penetration depth $\lambda = 1$, and the Ginzburg–Landau parameter κ is the reciprocal of the correlation length ξ. The parameter g measures the strength of doping (chemical potential) of the material. It is this term which breaks the SO(5) symmetry of the potential term. We take $g > 0$: with this assumption superconductivity is preferred in the bulk of the sample.

To study isolated vortex solutions in the plane $\Omega = \mathbb{R}^2$ we seek critical points of $\mathcal{F}$ of the form

(3.1) $$\psi = f(r) e^{id\theta}, \quad \vec{A} = S(r) \left(\frac{-y}{r^2}, \frac{x}{r^2} \right), \quad \vec{m} = m(r) \vec{m}_0,$$

where $\vec{m}_0$ a fixed unit vector, and $d \in \mathbb{Z} \setminus \{0\}$, the degree of the vortex, which also determines the quantized flux through the sample. As for conventional SC vortices, we expect that only the solutions with $d = \pm 1$ will be energy minimizers (see Gustafson and Sigal [14], Ovchinnikov and Sigal [19].) Under this ansatz the free energy becomes

(3.2) $\mathcal{E}_{\kappa,g}(f,S,m)$
$$= \frac{1}{2}\int\left\{(f')^2 + \left[\frac{S'}{r}\right]^2 + (m')^2 + \kappa^2 g m^2 + \frac{(d-S)^2 f^2}{r^2} + \frac{\kappa^2}{2}(1-f^2-m^2)^2\right\} r\,dr.$$

$\mathcal{E}_{\kappa,g}$ will be a smooth functional an the affine space
$$Y = \{(f,S,m)\colon f = f_0 + u,\ S = S_0 + rv,\ u,v \in X,\ m \in H\},$$
where f_0, S_0 are smooth functions satisfying the appropriate asymptotic limits for f, S,
$$f_0(0) = 0,\quad S_0(0) = 0,\quad f_0(r) \to 1,\quad S_0(r) \to d\quad \text{as } r \to \infty,$$
and
$$X = \left\{u \in H^1_r\colon \int \frac{u^2}{r^2} r\,dr < \infty\right\}.$$

Standard arguments show that the infimum of $\mathcal{E}_{\kappa,g}$ over Y is attained for any value of $g > 0$, $\kappa \neq 0$, $0 \neq d \in \mathbb{Z}$.

Critical points of $\mathcal{E}_{\kappa,g}$ in Y solve the system of equations

$(GL)_{\kappa,g}$
$$\begin{cases} -f'' - \frac{1}{r}f' + \frac{(d-S)^2}{r^2}f = \kappa^2(1 - f^2 - m^2)f, \\ -S'' + \frac{1}{r}S' = (d-S)f^2, \\ -m'' - \frac{1}{r}m' + \kappa^2 g m = \kappa^2(1 - f^2 - m^2)m, \end{cases}$$

with $f(0) = 0$, $S(0) = 0$; $f(r) \to 1$, $S(r) \to d$ as $r \to \infty$; and $m'(0) = 0$, $m(r) \to 0$ as $r \to \infty$. The definition of $f = |\psi|$ leads us to introduce a smaller class of *admissible solutions*, consisting of all critical points of $\mathcal{E}_{\kappa,g}$ in Y with the additional properties:

(3.3) $\qquad f(r) \geq 0,\quad m(r) \geq 0\quad \text{for all } r \geq 0.$

From the form of $\mathcal{E}_{\kappa,g}$ it is clear that the global minimizer satisfies (3.3).

Using the maximum principle we may prove the following facts about admissible solutions:
  (a) $0 < f(r) < 1$, $0 < S(r) < d$ for all $r > 0$.
  (b) $1 - f(r)$, $d - S(r)$ tend to zero exponentially rapidly.
  (c) The magnitude of the local magnetic field, $h(r) = S'(r)/r > 0$ for all $r \geq 0$, and is strictly monotonically decreasing, $h'(r) < 0$ for all $r > 0$.
  (d) Either $m \equiv 0$ or $0 < m(r) < 1$ for all $r \geq 0$.

Note that item (d) provides a clean distinction between "normal core solutions" (when $m \equiv 0$) and "AF core solutions" (when $m(r) > 0$ for all $r \geq 0$.) We prove monotonicity of f and m only for locally minimizing solutions: if (f_*, S_*, m_*) is an admissible solution of $(GL)_{\kappa,g}$ and the Hessian $D^2\,\mathcal{E}_{\kappa,g}(f_*, S_*, m_*) \geq 0$ (as a quadratic form) then $f'_*(r) > 0$, $m'_*(r) < 0$ for all $r > 0$. The proof uses integration by parts, in the spirit of the maximum principle for weak solutions of elliptic PDE. (See [**4**, Theorem 2.9].) We note that all of the above properties were derived for *global* minimizers by Berger and Chen [**8**] in the Ginzburg–Landau setting, but their proofs do not in general carry through for arbitrary solutions (or even local minimizers.)

3.1. Normal cores revisited.
As remarked earlier, the normal core solutions ($m \equiv 0$) are none other than the conventional Ginzburg–Landau symmetric vortices studied by Plohr, Berger and Chen, etc., which appear in standard physics texts. Moreover, they are independent of the parameter g, which only appears in the equation for m. A simple integration by parts in the third equation of (GL)$_{\kappa,g}$ shows that when the parameter $g \geq 1$ any solution must have $m \equiv 0$, and hence the vortex cores are indeed normal for large doping. The question is whether AF core solutions ever appear, are they energy minimizers, and if so what is the nature of the transition. We will see that the AF core solutions bifurcate from the "trivial" curve of normal core solutions.

In order to perform this analysis we must know more about the normal core solutions apart from their existence and other properties listed above. We observe that normal core solutions are critical points of the functional

$$(3.4) \quad E_\kappa(f, S) := \frac{1}{2} \int \left\{ (f')^2 + \left[\frac{S'}{r}\right]^2 + \frac{(d-S)^2 f^2}{r^2} + \frac{\kappa^2}{2}(1-f^2)^2 \right\} r \, dr,$$

on the affine space

$$Y_0 := \{(f, S) \colon f = f_0 + u, \ S = S_0 + rv, \ u, v \in X\}.$$

Note that $E_\kappa(f, S) = \mathcal{E}_{\kappa,g}(f, S, 0)$, for any g. For large values of κ we have the following useful result, proven by Alama, Bronsard, and Giorgi [3]:

THEOREM 3.1. *Assume $\kappa^2 \geq 2d^2$. Then:*

(i) *Any admissible normal core solution is a strict local minimizer of E_κ, that is there exists $\sigma_\kappa > 0$ such that*

$$D^2 E_\kappa(f_*, S_*)[u, v] \geq \sigma_\kappa \left(\|u\|_X^2 + \|v\|_X^2 \right),$$

holds for all $u, v \in X$;

(ii) *There is exactly one admissible normal core solution.*

REMARK 3.2. (a) We use part (i) to prove the uniqueness statement (ii). If there were two admissible normal core solutions then the Mountain-pass Theorem would guarantee the existence of a third solution, which could not be a local minimizer. This contradicts (i), as long as we can conclude that the solution found by the Mountain-pass Theorem is *admissible*. Since the class of admissible solutions defines a closed convex subset of our Hilbert manifold Y_0, the min-max argument may be carried out in that convex set, by generalizing the definition of "critical point" suitably. (See [21, Section II.11].) The constraint $f \geq 0$ behaves like a subsolution constraint, and we can prove that there is indeed a nonminimizing admissible solution and complete the argument. While the Mountain-pass Theorem is a well-known tool for proving existence or multiplicity results, it is a true test of its versatility that it can also be used to show uniqueness!

(b) The restriction on κ is technical, and is certainly not optimal. Indeed, a recent paper by Clemons [13] proves uniqueness for Ginzburg–Landau symmetric vortices for κ near $1/\sqrt{2}$, the value at which the Ginzburg–Landau model is self-dual and the exact solvability of arbitrary vortex configurations is known by a reduction technique (see Jaffe and Taubes [16].)

3.2. Bifurcation of AF core solutions.
We now linearize the equations $(GL)_{\kappa,g}$ around the normal core solution $(f_*, S_*, 0)$, and show that they become unstable at a critical value of g. Indeed, the Hessian of $\mathcal{E}_{\kappa,g}$ splits into two blocks, which we express as a quadratic form,

$$D^2\mathcal{E}_{\kappa,g}(f_*, S_*, 0)[u,v,w] = D^2 E_\kappa(f_*, S_*)[u,v] + \kappa^2 \langle (L+g)w, w\rangle_{L^2_r},$$

where E_κ is the energy defined in (3.4) and L is the radial linear Schrödinger operator,

$$Lw := -\frac{1}{\kappa^2}\frac{1}{r}\frac{d}{dr}(rw'(r)) - (1-f_*^2)w.$$

By Theorem 3.1 (i) the first term in the sum is strictly positive definite. Since $0 < f_* < 1$ the potential term in L is strictly negative and decaying to zero as $r \to \infty$, and it is a well-known fact that in two dimensions such operators have a *negative ground state* eigenvalue $\lambda_\kappa = -g_\kappa^* < 0$, with corresponding positive eigenfunction w_κ^*. In conclusion, when $g > g_\kappa^*$ the second variation of $\mathcal{E}_{\kappa,g}$ around the normal core solution is positive definite, and the normal core is a strict local minimizer. However, when $0 < g < g_\kappa^*$ the normal core solution loses stability, and hence the global minimizer must have an AF core. Antiferromagnetic core solutions bifurcate from the normal core solutions at the simple eigenvalue $-g_\kappa^*$ of L according to the theorem of Crandall and Rabinowitz. By Rabinowitz's Global Bifurcation Theorem (together with some *a priori* estimates) one can show the existence of a global branch of AF core solutions which exist for all $g < g_\kappa^*$, before becoming unbounded as $g \to 0^+$. (See [4] for details.)

Physically, an important question concerns the nature of the transition to AF core solutions: is it discontinuous (first-order) or continuous (second-order)? Analytically this amounts to determining the direction of bifurcation or (equivalently) the stability of the solution curve. If the bifurcating solutions are unstable, the curve bends to $g > g_\kappa^*$ before turning back to smaller g, creating a hysteresis loop and a first-order transition. If the AF core solutions nucleated at g_κ^* are stable, the branch bends forward to smaller g and the transition is second-order. Since the energy $\mathcal{E}_{\kappa,g}$ is smooth, we may use the Implicit Function Theorem to calculate any number of derivatives of the parametrized curve of AF core solutions $(f(t), S(t), m(t); g(t))$ emanating from the normal core solutions at $g(0) = g_\kappa^*$. We may readily show that $g'(0) = 0$ and we obtain an explicit formula for $g''(0)$, but this latter quantity is expressed as the difference of two terms which can be derived in principle from (f_*, S_*), but are not known in closed form. Unfortunately this means that we cannot determine the nature of the transition analytically, and must resort to numerical approximations. These computations, done in [2] indicate that the transition is *second order*, and contradict the assumption of a discontinuous transition implicit in [7].

3.3. An extreme type II model.
In addition, we study the following "extreme type II" model,

$$(GL)_{\infty,g} \quad \begin{cases} -f'' - \frac{1}{r}f' + \frac{d^2}{r^2}f = (1-f^2-m^2)f, \\ -m'' - \frac{1}{r}m' + gm = (1-f^2-m^2)m. \end{cases}$$

The system $(GL)_{\infty,g}$ is obtained in the limit $\kappa \to \infty$ after rescaling solutions to $(GL)_{\kappa,g}$ by the correlation length $\xi = 1/\kappa$. For high T_c superconductors κ is very large, and hence the vortex cores are very narrow compared to the penetration depth, which measures the length scale for magnetic fields. By rescaling we capture the structure of the vortex cores and decouple the magnetic field, which lives on a much larger length scale. Indeed, the calculations which led Arovas et al. [7] to predict AF vortex cores are mostly based on $(GL)_{\infty,g}$ and its associated free energy functional.

The analysis of the extreme type II model proceeds along the lines of the finite κ case: there is a corresponding free energy

$$(3.5) \quad \mathcal{E}_{\infty,g}(f,m) = \frac{1}{2}\int\left\{(f')^2+(m')^2+gm^2+\frac{d^2}{r^2}[f^2-f_0^2]+\frac{1}{2}(1-f^2-m^2)^2\right\}r\,dr,$$

which defines a smooth functional on the affine space

$$Z = \{(f,m)\colon f = f_0 + u,\ u \in X,\ m \in H\}.$$

Minimization yields a solution for all $g > 0$, and the "admissible" solutions with $f(r) \geq 0$, $m(r) \geq 0$ satisfy all properties enumerated in Subsection 3.1 except (b). The normal core solutions are again obtained by setting $m = 0$ in the energy, and are the same for every value of g. They are unique (a fact proven using shooting arguments by Chen, Elliot, and Qi [12]) and are nondegenerate minimizers of an appropriate energy functional. Linearization of the equations $(GL)_{\infty,g}$ around the normal cores solution again leads to bifurcation at a simple eigenvalue $g = g_\infty^*$, the ground state of a radial Schrödinger operator in two dimensions. However, we obtain more thanks to the following:

THEOREM 3.3. *Let (f,m) be any admissible solution with $m(r) > 0$ for $r > 0$. Then (f,m) is a nondegenerate local minimizer of $\mathcal{E}_{\infty,g}$; that is $D^2\mathcal{E}_{\infty,g}(f,m) > 0$ in the sense of quadratic forms.*

Since all AF core solutions are stable, we may eliminate the possibility of bending or secondary bifurcation and we conclude that the transition is *second-order*. Moreover, we obtain an exact multiplicity result for admissible solutions to $(GL)_{\infty,g}$:

THEOREM 3.4.   (i) *If $g \geq g_\infty^*$ the only admissible solution is the (unique) normal core solution;*
  (ii) *If $0 < g < g_\infty^*$ there exist exactly two admissible solutions: the normal core solution and a unique admissible AF core solution. Furthermore, the AF core solution is the global minimizer of $\mathcal{E}_{\infty,g}$.*

References

1. S. Alama, J. Berlinsky, and L. Bronsard, *Vortex lattices for the Lawrence–Doniach model of layered superconductors in a parallel field*, preprint in preparation.
2. S. Alama, A. J. Berlinsky, L. Bronsard, and T. Giorgi, *Vortices with antiferromagnetic cores in the SO(5) model of high temperature superconductivity*, Phys. Rev. B **60** (1999), no. 9, 6901–6906.
3. S. Alama, L. Bronsard, and T. Giorgi, *Uniqueness of symmetric vortex solutions in the Ginzburg-Landau model of superconductivity*, J. Funct. Anal. **167** (1999), 399–424.
4. ———, *Vortex structures for an SO(5) model of high-T_C superconductivity and antiferromagnetism*, Proc. Roy. Soc. Edinburgh Sect. A, (to appear); Preprint `math.AP/9903127`.
5. A. Ambrosetti and M. Badiale, *Homoclinics: Poincaré–Melnikov type results via a variational approach*, Ann. Inst. H. Poincaré Anal. Non Linéaire, **15** (1998), no. 2, 233–252.
6. A. Ambrosetti, V. Coti-Zelati, and I. Ekeland, *Symmetry breaking in Hamiltonian systems*, J. Differential Equations **67** (1987), 165–184.
7. D. P. Arovas, A. J. Berlinsky, C. Kallin, and S.-C. Zhang, *Superconducting vortex with antiferromagnetic core*, Phys. Rev. Lett. **79** (1997), 2871–2874.

8. M. S. Berger and Y. Y. Chen, *Symmetric vortices for the Ginzburg–Landau equations of superconductivity and the nonlinear desingularization phenomenon*, J. Funct. Anal. **82** (1989), 259–295.
9. L. Bulaevskii, *Magnetic properties of layered superconductors with weak interaction between the layers*, Soviet Phys. JETP **37** (1973), 1133–1136.
10. L. Bulaevskii and J. Clem, *Vortex lattice of highly anisotropic layered superconductors in strong, parallel magnetic fields*, Phys. Rev. B **44** (1991), 10234–10238.
11. S. Chapman, Q. Du, and M. Gunzburger, *On the Lawrence–Doniach and anisotropic Ginzburg–Landau models for layered superconductors*, SIAM J. Appl. Math. **55** (1995), 156–174.
12. X. Chen, C. M. Elliot, and T. Qi, *Shooting method for vortex solutions of a complex-valued Ginzburg–Landau equation*, Proc. Roy. Soc. Edinburgh Sect. A **124** (1994), 1075–1088.
13. C. Clemons, *An existence and uniqueness result for symmetric vortices for the Ginzburg–Landau equations of superconductivity*, J. Differential Equations **157** (1999), 150–162.
14. S. Gustafson and I. M. Sigal, *The stability of magnetic vortices*, Preprint.
15. Y. Iye, *How anisotropic are the cuprate high T_c superconductors?* Comments Cond. Mat. Phys. **16** (1992), 89–111.
16. A. Jaffe and C. Taubes, *Monopoles and vortices*, Birkhäuser, Boston, 1980.
17. P. Kes, J. Aarts, V. Vinokur, and C. van der Beek, *Dissipation in highly anisotropic superconductors*, Phys. Rev. Lett. **64** (1990), 1063–1066.
18. S. Kuplevakhsky, *Microscopic theory of weakly couple superconducting multilayers in an external magnetic field*, Preprint `cond-mat/9812277`.
19. Yu. N. Ovchinnikov and I. M. Sigal, *Ginzburg–Landau equation I. Static vortices*, Partial Differential Equations and Their Applications (Toronto, 1995), CRM Proc. Lecture Notes, vol. 12, Amer. Math. Soc., Providence, RI, 1997, pp. 199–220.
20. P. Rabinowitz, *Some global results for nonlinear eigenvalue problems*, J. Funct. Anal. **7** (1971), pp. 487–513.
21. M. Struwe, *Variational methods*, Springer, Berlin, 1990.
22. G. t'Hooft, *A property of electric and magnetic flux in non-Abelian gauge theories*, Nuclear Phys. B **153** (1979), no. 1-2, 141–160.
23. S. Theorodakis, *Theory of vortices in weakly-Josephson-coupled layered superconductors*, Phys. Rev. B **42** (1990), 10172–10177.
24. M. Tinkham, *Introduction to superconductivity*, 2nd ed., Mc Graw-Hill, New York, 1996.
25. S.-C. Zhang, *A unified theory based on SO(5) symmetry of superconductivity and antiferromagnetism*, Science, vol. 275 1997, pp. 1089–1096.

DEPT. OF MATHEMATICS AND STATISTICS, MCMASTER UNIV., HAMILTON, ONTARIO, CANADA L8S 4K1.

E-mail address, Stan Alama: `alama@mcmaster.ca`

E-mail address, Lia Bronsard: `bronsard@mcmaster.ca`

The Effect of Distribution In Space In Ostwald Ripening

Nicholas D. Alikakos and Giorgio Fusco

We consider a two-phase system in $3d$ during a stage of evolution known as **Ostwald Ripening** [19], or alternatively called **Coarsening** or **Aging**. In this stage the system is far from equilibrium. The driving force is the Gibbs–Thomson condition which states that the chemical potential on the interface is proportional to its mean curvature. Therefore matter flows from high to low curvature, resulting in the reduction of the interfacial area. In this paper we consider initially dilute mixtures, and the **particles** stand for the minority phase. The system evolves under quasistatic dynamics:

$$(1) \quad \begin{cases} \Delta u = 0 \text{ off } \Gamma(t), \\ u = H \text{ on } \Gamma(t), \\ \frac{\partial u}{\partial \nu} = 0 \text{ on } \partial \Omega, \\ V = -\left[\frac{\partial u}{\partial n}\right] \text{ on } \Gamma(t), \end{cases}$$

where Ω is a bounded, smooth domain in $\mathbb{R}^3$ (the container of the mixture), $\Gamma = \cup_{i=1}^n \Gamma_i$, the union of the boundaries of the finite system of particles, $u = u(x,t)$ is the chemical potential, H is the mean curvature of Γ, and V is the normal velocity, and finally $[\partial u/\partial n]$ is the jump of the normal derivative across $\Gamma(t)$ ($[\partial u/\partial n] = \partial u^+/\partial n - \partial u^-/\partial n$, u^+, u^- the restrictions of u on $\Omega^+(t)$, $\Omega^-(t)$ the exterior and interior of $\Gamma(t)$ in Ω where n is the interior unit normal to $\Omega^-(t)$).

Equation (1) is the sharp interface limit of the celebrated Cahn–Hilliard equation [3, 13, 14, 34]. It is also known as the Mullins–Sekerka problem in the context of solidification [30].

Equation (1) is a volume preserving, perimeter shortening law,

$$(2) \quad \begin{cases} \frac{d}{dt}\operatorname{Per}(\Gamma(t)) = 2\int_\Gamma HV = -2\int_\Omega |\nabla u|^2 \leq 0, \\ \frac{d}{dt}\operatorname{Vol}(\Omega^-(t)) = \int_\Gamma V = 0. \end{cases}$$

2000 *Mathematics Subject Classification.* Primary: 35R35, 35B35; Secondary: 82C24.

The work of the first author was partially supported by ΠΕΝΕΔ 99/527, and a grant from the UA.

We would like to thank Professor J. Lebowitz and Professor E. Presutti for many very stimulating discussions. Also we would like to thank Professor O. Penrose for making available to us his notes on Ostwald ripening.

This is the final form of the paper.

©2001 American Mathematical Society

These calculations presuppose well posedness of (1). We refer the reader to [**16, 17, 21**] for the local theory, and to [**15, 40**] for global weak solutions.

System (1) can be reduced to a problem that lives entirely on the interface via potential theory [**6, 50**]:

$$(*) \qquad \int_\Gamma g(x,y)V(y)dy - \frac{1}{|\Gamma|}\int_\Gamma \int_\Gamma g(x,y)V(y)\,dy\,dx = H(x) - \overline{H}, \quad x \in \Gamma,$$

where $g(x,y)$ is the Green's function:

$$\Delta_y g(x,y) = \delta_x(y) - \frac{1}{|\Omega|}, \quad \frac{\partial g}{\partial n_y} = 0 \text{ on } \partial\Omega,$$

$\overline{H} := |\Gamma|^{-1}\int_\Gamma H(y)dy$, $|\Gamma| :=$ surface area of Γ.

In [**5**] we considered an initial configuration of n, almost spherical, particles. We showed that to each particle we can associate a center $\xi \in \mathbb{R}^3$, a radius $\rho \in \mathbb{R}^+$, and a shape function $r(\cdot) \in C^{3+a}(S^2)$ such that

$$(3) \qquad \Gamma_i(0) = \big\{x | x = \xi^i + \rho_i\big[1 + r_i(u)\big]u, u \in S^2\big\},$$

with r_i satisfying the orthogonality conditions

$$(4) \qquad \begin{cases} \int_{S^2} r_i(u)\,du = 0, & i = 1,\ldots,n, \\ \int_{S^2} r_i(u)\langle u, e_j\rangle\,du = 0, & j = 1,2,3, \end{cases}$$

where $\langle\,,\,\rangle$ is the Euclidean inner product, $e_j \in \mathbb{R}^3$, $j = 1, 2, 3$ is the standard unit vector and S^2 is the unit sphere. Under the scaling hypothesis

$$(5) \qquad \begin{cases} \frac{\text{size of particle}}{\text{distance between particles}} = O(\varepsilon), \\ \frac{\text{deviation from sphericity}}{\text{size of particle}} = O(\varepsilon), \end{cases}$$

and for $0 < \varepsilon \ll 1$, we have shown that the coordinate system above is preserved globally in time,

$$(6) \qquad \Gamma_i(t) = \big\{x | x = \xi^i(t) + \rho_i(t)\big[1 + r_i(u,t)\big]u,\ u \in S^2\big\}.$$

Moreover we proved that for $0 < \varepsilon \ll 1$, (1) is equivalent to a system of equations for ρ_i, ξ^i, and r_i, with the ρ-equations, to **principal order**, given by

$$(7) \qquad \dot{\rho}_i = \left(\frac{1}{\bar{\rho}} - \frac{1}{\rho_i}\right)\frac{1}{\rho_i}, \quad i = 1,\ldots,n,$$

$$\bar{\rho} = \frac{1}{n}\sum_{i=1}^n \rho_i.$$

The equations (7) state that at each given t, the radius $\rho_i(t)$ of the ith particle $\Gamma_i(t)$, decreases or increases according to whether at that time its value is below or above the average value $\bar{\rho}(t)$. It can be checked that

$$(8) \qquad \begin{cases} \frac{d}{dt}\sum_{i=1}^n \rho_i^3 = 0, \\ \frac{d}{dt}\sum_{i=1}^n \rho_i^2 \leq 0, \end{cases}$$

reflecting the volume preserving, perimeter shortening properties of (1) (cf. (2)). Equations (7) are order preserving(cf. [**5**]): $\rho_1(0) < \rho_2(0) < \cdots < \rho_n(0) \Rightarrow \rho_1(t) < \rho_2(t) < \cdots < \rho_n(t)$, and it can be shown that there are times $\hat{\tau}_1 < \hat{\tau}_2 < \cdots < \hat{\tau}_{n-1}$

at which the 1st, 2nd, ..., $n-1$ particle get extinct. At time $\widehat{\tau}_i$, $\bar{\rho}$ has a positive jump:

$$\bar{\rho}(\widehat{\tau}_i^+) - \bar{\rho}(\widehat{\tau}_i^-) = \frac{1}{(n-i)(n-i+1)} \sum_{j=i+1}^{n} \rho_j(\widehat{\tau}_i).$$

As a result of these jumps, $\bar{\rho}$ eventually increases in spite of the fact that from (7) and Jensen's inequality it follows that

$$\dot{\bar{\rho}}(t) \leq 0, \quad t \in (\widehat{\tau}_i, \widehat{\tau}_{i+1}).$$

In [5] we have shown that also for the original, complete, ρ, ξ, r equations, there are times $\tau_1 < \tau_2 < \cdots < \tau_{n-1}$, close to $\widehat{\tau}_i$'s, such that

(9) $$\rho_i(t) \to 0, \text{ as } t \to \tau_i.$$

Moreover we showed that $r_i(u,t) \to 0$, as $t \to \tau_i$, $i = 1, \ldots, n-1$, $r_n(u,t) \to 0$, as $t \to \infty$. Our global solution was obtained by removing the extinct particle and continuing. It is a piecewise smooth solution and also a global weak solution in the sense of [16, 40]. We refer the reader to the expository paper [6] for an account of these results.

Equations (7) provide a reduction of (1) and they are closely related to the celebrated Lifschitz–Slyosov–Wagner theory of coarsening [25, 47] which appeared in 1961 and was the first to provide a successful quantitative analysis of the phenomenon of Ostwald ripening (known since 1901) under the hypothesis of infinite dilution. Under various simplifying assumptions the LSW derives effective equations for the growth of a spherical particle coupled with an external field representing the effect of the rest of the particles. The point of departure in [25, 47] is (1). Their results concern the long time behavior of the radius distribution $n(R,t)$, where $\int_{R_1}^{R_2} n(R,t)\,dR$ gives the fraction of particles in $[R_1, R_2]$. Specifically LSW provides the equation

(10) $$\frac{\partial n}{\partial t}(R,t) + \frac{\partial}{\partial R}\left(\frac{dR}{dt} n(R,t)\right) = 0,$$

with (cf. Eq. (7) above)

(11) $$\frac{dR}{dt} = \left(\frac{1}{\overline{R}(t)} - \frac{1}{R(t)}\right)\frac{1}{R(t)},$$

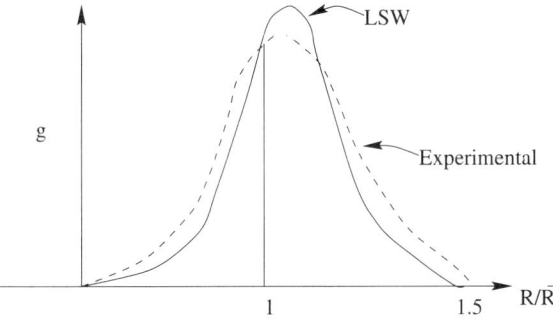

FIGURE 1. Self-similar distribution: Experimental and according to LSW theory.

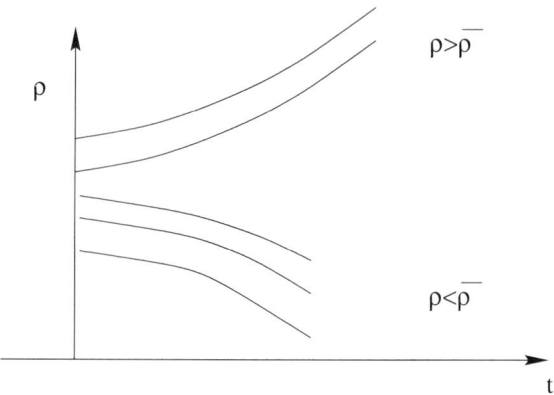

FIGURE 2. Particles time-lines according to (7).

where
$$\overline{R} = \frac{\int R n(R,t) dR}{\int n(R,t) \, dR}.$$

System (10), (11) is analysed in [**25, 47**] where it is argued that it possesses a self-similar solution,

(12) $$n(R,t) \cong \frac{1}{t^{4/3}} g\left(\frac{R}{\overline{R}(t)}\right),$$

that is capturing the typical behavior of the system for large times (cf. Fig. 1).

The theory predicts also the following temporal laws for the average radius and the total number of particles:

(13) $$\begin{cases} \overline{R}(t) = \left(\overline{R}^3(0) + \frac{4}{9}t\right)^{1/3}, \\ N(t) = c\left(\overline{R}^3(0) + \frac{4}{9}t\right)^{-1}. \end{cases}$$

The 1/3 is a scaling fact of the equations (see for example [**6**]) and is pretty robust. The self-similar distribution however does not quite agree with the experimental curve, which is broader [**42–44**]. The uniqueness of the self-similar solution has been questioned recently [**33**]. From the very beginning it was realized that the infinite dilution limit is not realistic and corrections for small but nonzero volume fraction have been worked out [**27, 28, 35**]. Also it has been realized [**46**] that the geometric configurations affect the way a particle evolves: identical particles in different locations evolve differently, depending on their neighbors. The argument against the classical LSW can be based on equations (11) whose time lines do not cross: On the other hand in [**46**] it is reported that the experimental time lines can in general cross (Fig. 3)

In [**8**] we present a rigorous refinement of equation (7) (see system (28), (29) below) that takes into account the geometry of the configuration. The analysis in [**8**] shows that, if ρ_{io} is sufficiently small and Γ_{io} is sufficiently close to a sphere, then the influence of the distortion r_i on the evolution of ρ_i, ξ_i can be disregarded

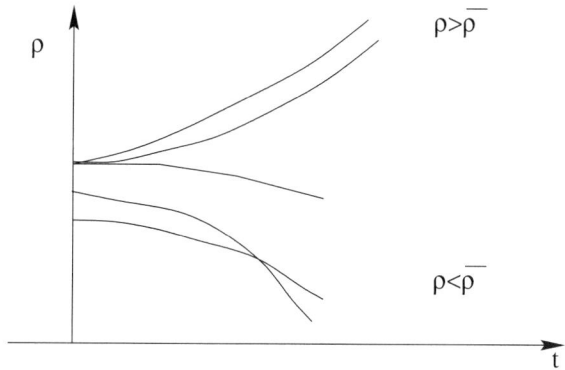

FIGURE 3. Crossing of particles time-lines.

and $\dot{\rho}_i$, $\dot{\xi}_i$ can be computed to leading order under the assumption
$$r_i \equiv 0, \quad i = 1, \ldots, n.$$
In the present paper we present formally this computation and thus derive equations for $\dot{\rho}_i$, $\dot{\xi}_i$ that quantify the effect of the space distribution of the $\Gamma'_i s$ on the evolution of ρ_i. These equations, which are rigorously justified in [**8**], agree well with various effects observed in experiments like the crossing of time lines and shielding. Our main interest here is to determine the self and mutual influence of particles. For keeping the computation to a minimum, we disregard the effect of the boundary. This allows us to replace the Green's function in (*) with its singular part. For expressions of $\dot{\rho}_i$, $\dot{\xi}_i$ which also include the effect of the boundary we refer to [**8**]. The very recent preprint [**32**] is also on the effect of the Geometry on the coarsening rates.

If we set $r_i = 0$ and rescale by $\rho_i \to \varepsilon \rho_i$, then the interface of the ith particle Γ_i is represented, as in (3), by

(14) $\quad x = X^i(u) = \xi_i + \varepsilon \rho_i u, \quad u \in S^2$, the unit sphere, $i = 1, \ldots, n.$

Equalities Below are Ment up to Higher Order Terms

The normal velocity V_i of the ith particle, positive for a shrinking sphere, is

(15) $$V_i = -\varepsilon \dot{\rho}_i - \sum_{j=1}^{3} \dot{\xi}_{ij} \langle u, e_j \rangle,$$

where
$$\dot{\xi}_{ij} = \langle \dot{\xi}_i, e_j \rangle.$$
Substituting (14) in (*) for $x \in \Gamma_i$, we obtain

(16) $$\varepsilon \rho_i \int_{S^2} \frac{V_i(v) dv}{4\pi |u-v|} + \varepsilon^2 \rho_k^2 \sum_{k \neq i} \int_{S^2} \frac{V_k(v) \, dv}{4\pi |\xi_k - \xi_i + \varepsilon(\rho_k v - \rho_i u)|} = \frac{1}{\varepsilon \rho_i} - E,$$

where
$$E = \overline{H} - \frac{1}{|\Gamma|} \int_\Gamma \int_\Gamma g(x,y) V(y) \, dy \, dx.$$

We calculate

$$
\text{(17)} \quad \int_{S^2} \frac{V_k(v)\,dv}{4\pi|\xi_k - \xi_i + \varepsilon(\rho_k v - \rho_i u)|}
$$
$$
= \frac{1}{3}\sum_j \frac{\varepsilon\rho_k}{d_{ik}^2}\dot{\xi}_{kj}\langle \bar{\xi}_{ki}, e_j\rangle - \frac{\varepsilon\dot{\rho}_k}{d_{ik}}\left(1 + \frac{\varepsilon\rho_i}{d_{ik}}\langle u, \bar{\xi}_{ki}\rangle\right),
$$

$$
\text{(18)} \quad d_{ij} = |\xi_i - \xi_j|, \quad \bar{\xi}_{ij} = \frac{\xi_i - \xi_j}{|\xi_i - \xi_j|},
$$

where we have utilized (15), together with the computations

$$
\text{(19)} \quad |\xi_i - \xi_j + \varepsilon(\rho_i v - \rho_j u)|^{-1} = \frac{1}{d_{ij}}\left[1 - \varepsilon\left(\frac{\rho_i}{d_{ij}}\langle u, \bar{\xi}_{ij}\rangle - \frac{\rho_j}{d_{ij}}\langle v, \bar{\xi}_{ij}\rangle\right)\right],
$$
$$
\int_{S^2}\langle v, e_j\rangle\,dv = 0, \quad \int_{S^2}\langle v, e_i\rangle\langle v, e_j\rangle\,dv = \frac{4\pi}{3}\delta_{ij}.
$$

Therefore (16) takes the form:

$$
\text{(20)} \quad \varepsilon\rho_i \int_{S^2} \frac{V_i(v)\,dv}{4\pi|u - v|} + \varepsilon^2\sum_{k\neq i}\rho_k^2\left[\frac{1}{3}\frac{\varepsilon\rho_k}{d_{ik}^2}\dot{\xi}_{kj}\langle \bar{\xi}_{ki}, e_j\rangle - \frac{\varepsilon\dot{\rho}_k}{d_{ik}}\left(1 + \frac{\varepsilon\rho_i}{d_{ki}}\langle u, \bar{\xi}_{ki}\rangle\right)\right]
$$
$$
= \frac{1}{\varepsilon\rho_i} - E.
$$

Next we operate with T_0, the operator of Dirichlet–Neumann type that is closely related to the definition of V in (1), which we now define. Let B be the unit ball, and for a given smooth $\mathcal{X}$ define ϕ_i, ϕ_e, via

$$
\text{(21)} \quad \begin{cases} -\Delta\phi_i = 0, & \text{in } B, \\ \phi_i = \mathcal{X}, & \text{on } S^2, \\ -\Delta\phi_e = 0, & \text{in } \mathbb{R}^3 - B, \\ \phi_e = \mathcal{X}, & \text{on } S^2, \end{cases}
$$

Then set

$$
\text{(22)} \quad T_0\mathcal{X} := \frac{\partial\phi_i}{\partial n} + \frac{\partial\phi_e}{\partial n}.
$$

From well known properties of spherical harmonics we have that

$$
\text{(23)} \quad T_0(1) = 1, \quad T_0\big(\langle u, e_i\rangle\big) = 3\langle u, e_i\rangle,
$$

and also by classical potential theory that

$$
\text{(24)} \quad T_0 \int_{S^2} \frac{V_i(v)\,dv}{4\pi|u - v|} = V_i(u).
$$

Thus from (20) we obtain

$$
\text{(25)} \quad \varepsilon\rho_i V_i + \sum_{k\neq i}\frac{1}{3}\frac{\varepsilon^3\rho_k^3}{d_{ik}^2}\dot{\xi}_{kj}\langle \bar{\xi}_{ki}, e_j\rangle - \sum_{k\neq i}\frac{\varepsilon^3\dot{\rho}_k\rho_k^2}{d_{ik}} - \sum_{k\neq i}3\frac{\varepsilon^4\dot{\rho}_k\rho_k^2\rho_i}{d_{ik}^2}\langle u, \bar{\xi}_{ki}\rangle
$$
$$
= \frac{1}{\varepsilon\rho_i} - E.
$$

By (15) we can replace V_i with

$$\text{(26)} \qquad -\varepsilon \dot{\rho}_i - \sum_{j=1}^{3} \dot{\xi}_{ij} \langle u, e_j \rangle.$$

The Projections

We will L^2 project (25), after we substitute (26) for V_i, on the (eigen) functions 1, $\langle u, e_i \rangle$ of T_0 (cf. (23)):

PROJECTING ON 1.

$$\text{(i)} \qquad -\varepsilon^2 \rho_i \dot{\rho}_i + \varepsilon^3 \sum_{k \neq i} \frac{1}{3} \frac{\rho_k^3}{d_{ik}^2} \dot{\xi}_{kj} \langle \bar{\xi}_{ki}, e_j \rangle - \varepsilon^3 \sum_{k \neq i} \frac{\dot{\rho}_k \rho_k^2}{d_{ik}} = \frac{1}{\varepsilon \rho_i} - E.$$

The 2nd term on the left can be ignored because, as we will see from the $\dot{\xi}$-equations, it is higher order. The 3rd term is smaller than the 1st. Therefore we can estimate it dy assuming $\varepsilon^3 \dot{\rho}_k \rho_k^2 = \rho_k/\bar{\rho} - 1$ which follows from (7) after rescaling by $\rho_i \to \varepsilon \rho_i$. Thus we obtain

$$\text{(27)} \qquad -\varepsilon^2 \rho_i \dot{\rho}_i - \sum_{k \neq i} \frac{1}{d_{ik}} \left(\frac{\rho_k}{\bar{\rho}} - 1 \right) = \frac{1}{\varepsilon \rho_i} - E.$$

To determine E we can utilize conservation of volume that to leading order reads: $\sum_{i=1}^{n} \dot{\rho}_i \rho_i^2 = 0$

After some calculations we obtain

$$\text{(28)} \qquad \dot{\rho}_i = \frac{1}{\varepsilon^3 \rho_i} \left\{ \left(\frac{1}{\bar{\rho}} - \frac{1}{\rho_i} \right) + \varepsilon \left[\frac{1}{n\bar{\rho}} \sum_{\substack{k,h \\ k \neq h}} \frac{\rho_h}{|\xi_h - \xi_k|} \left(\frac{\rho_k}{\bar{\rho}} - 1 \right) \right. \right.$$
$$\left. \left. - \sum_{j \neq i} \frac{1}{|\xi_j - \xi_i|} \left(\frac{\rho_j}{\bar{\rho}} - 1 \right) \right] \right\}.$$

PROJECTING ON $\langle u, e_i \rangle$.

$$-\varepsilon \rho_i \dot{\xi}_{ij} - \varepsilon^4 \sum_{k \neq i} 3 \frac{\dot{\rho}_k \rho_k^2 \rho_i}{d_{ik}^2} \langle \bar{\xi}_{ki}, e_j \rangle = 0, \quad j = 1, 2, 3,$$

and therefore again utilizing that $\varepsilon^3 \dot{\rho}_k \rho_k^2 = (1/\bar{\rho} - 1/\rho_k)\rho_k$ (to principal order) we obtain

$$\text{(29)} \qquad \dot{\xi}_i = -3 \sum_{k \neq i} \left(\frac{1}{\bar{\rho}} - \frac{1}{\rho_k} \right) \rho_k \frac{\xi_k - \xi_i}{|\xi_k - \xi_i|^3}.$$

This concludes the derivation.

NOTES 1. (1) We remark that the sums above are single sums, and also are taken over the first of the two indices that appear. For example $\sum_{k \neq j}$ means $\sum_{k \neq j, k=1,\ldots,n}$.
(2) System (28) conserves volume. It is also perimeter shortening for two particles and this under no extra conditions. For more than two particles however a condition on the largeness of relative distance to size in needed.

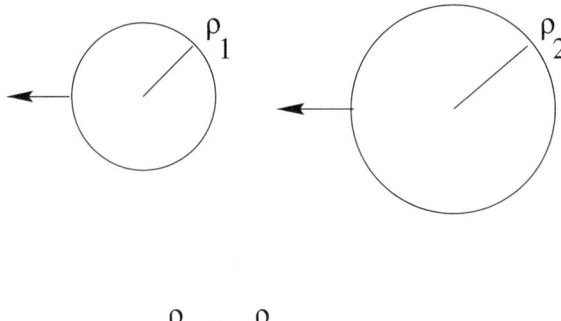

FIGURE 4. The 2-particles cases.

(3) Special cases of system (28), (29) have already appeared in the literature. For 2 particles in 2d in [**17**] one can find the analog of (7) together with the analog of (29). For 2 particles in 3d the system (28), (29) is derived in [**46**]. In this paper one can find a detailed examination of the translational motion (particle migration). The method of images plays a role in these works. Equations (28) are due to Weins and Cahn [**48**] and have been derived in various places [**41, 45**], under various hypotheses including that the Gibbs–Thomson is satisfied in an average sense, and that the centers stay fixed in space. We refer to [**42–44**] where more references can be found.

We now illustrate the use of system (28) by considering special scenaria. We rescale time by setting $t = \varepsilon^3 \tau$.

The 2-particle case. System (28), (29) gives

$$\text{(30)} \quad \frac{d\xi_1}{d\tau} = \frac{d\xi_2}{d\tau} = 3\varepsilon^3 \frac{\xi_1 - \xi_2}{|\xi_1 - \xi_2|^3} \frac{\rho_2 - \rho_1}{\rho_1 + \rho_2},$$

$$\text{(31)} \quad \begin{cases} \frac{d\rho_1}{d\tau} = \frac{1}{\rho_1^2} \frac{\rho_1 - \rho_2}{\rho_1 + \rho_2} \left(1 + \frac{2\varepsilon}{|\xi_1 - \xi_2|} \frac{\rho_1 \rho_2}{\rho_1 + \rho_2}\right) \\ \frac{d\rho_2}{d\tau} = \frac{1}{\rho_2^2} \frac{\rho_2 - \rho_1}{\rho_1 + \rho_2} \left(1 + \frac{2\varepsilon}{|\xi_1 - \xi_2|} \frac{\rho_1 \rho_2}{\rho_1 + \rho_2}\right) \end{cases}$$

Both particles move with the same velocity in the direction of the smaller particle. Notice also that the inclusion of the ε–term increases the coarsening rate.

The effect of neighbors on coarsening. A) Consider the arrangement of Fig. 5. If the third ball were not present then $dA/d\tau = 0$ for balls 1, 2.

We calculate (for $a \to 0^+$):

$$\bar{\rho} \cong \frac{2}{3} A.$$

$$\frac{\rho_k}{\bar{\rho}} \cong \frac{3}{2}, \quad k = 1, 2; \qquad \frac{\rho_3}{\bar{\rho}} \cong 0.$$

Hence the equation for 1 is:

$$\text{(32)} \quad \frac{dA}{d\tau} = \frac{1}{2A^2}\left(1 - \varepsilon \frac{A}{2d}\right).$$

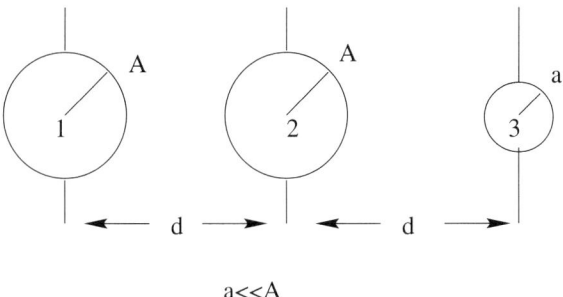

FIGURE 5. The middle ball grows faster.

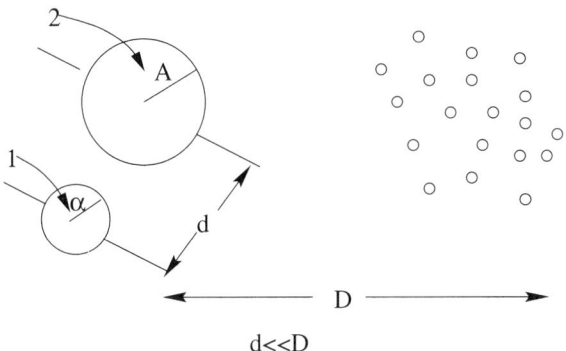

FIGURE 6. Particles 2 screens particle 1.

On the other hand the equation for 2 is:

$$\frac{dA}{d\tau} = \frac{1}{2A^2}\left(1 + \varepsilon\frac{A}{2d}\right). \tag{33}$$

The conclusion is that the ball in the middle grows faster.

B) Next we show that the coarsening rate of a particle varies with the size of its neighbors and it increases when the size of the neighboring particle exceeds the average. In Fig. 6 we consider $n-2$ equal particles at a distance $\geq D$-from each other and from two particles 1 and 2 with radii α, A within distance $d \ll D$ from each other. The particle 2 screens the neighboring particles for particle 1. We assume n large so that the radius ρ of the $n-2$ particles is approximately equal to $\bar{\rho}$. We calculate:

$$\frac{1}{n\bar{\rho}}\sum_{\substack{h,k \\ h \neq k}} \frac{\rho_h}{|\xi_h - \xi_k|}\left(\frac{\rho_k}{\bar{\rho}} - 1\right) = 0\left(\frac{1}{n} + \frac{1}{D}\right).$$

Thus

$$\frac{d\alpha}{d\tau} \simeq \frac{1}{\alpha}\left\{\left(\frac{1}{\bar{\rho}} - \frac{1}{\alpha}\right) - \frac{\varepsilon}{d}\left(\frac{A}{\bar{\rho}} - 1\right)\right\}, \tag{34}$$

that quantifies the effect of $A \gtrless \bar{\rho}$ (cf. Fig. 7).

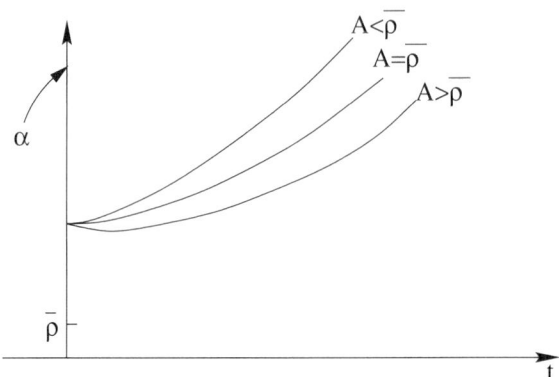

FIGURE 7. The shielding effect.

Remark on taking the limit on the number of Particles. In this paper we have mainly considered a fixed number of particles. In [**8**] we study certain limits. In the following we present here a basic estimate that plays a role in the analysis. We consider the scalings
$$\Omega = \Omega_\varepsilon = O\left(\frac{1}{\varepsilon}\right).$$
The size of the particle is $O(1)$.

The distance between particles is $O(1/\varepsilon^\beta)$, $0 < \beta \leq 1$.

By taking $\beta < 1$ we can pack more and more particles as $\varepsilon \to 0$. $\beta = 1$ corresponds to the finite case treated in this paper.

$$\text{(35)} \qquad \text{\# of particles} \sim \frac{(1/\varepsilon)^3}{(1/\varepsilon)^{3\beta}} = \varepsilon^{3(\beta-1)}.$$

Notice that the relative capacity of the minority phase is given by

$$\text{(36)} \qquad \frac{(\text{\# particles})\,(\text{Radius})}{\text{diam}\,\Omega_\varepsilon} \sim \frac{\varepsilon^{3(\beta-1)}}{1/\varepsilon} = \varepsilon^{3\beta-2}.$$

The basic quantity that has to be estimated is
$$\sum_{i\neq j,\ i=1,\ldots,n} \frac{1}{|\xi_i - \xi_j|}.$$

It is not difficult to show the estimate

$$\text{(37)} \qquad \sum_{i\neq j,\ i=1,\ldots,n} \frac{1}{|\xi_i - \xi_j|} = O(\varepsilon^{3\beta-2}),$$

and thus $\beta > 2/3$ guarantees that this sum goes to zero as $\varepsilon \to 0$. This coincides with the condition that the relative capacity tends to zero as $\varepsilon \to 0$ (cf. [**31**]), $\beta = 2/3$ is critical and leads to a qualitatively different continuum limit [**8**] (cf. [**32**]).

References

1. S. Agmon, A. Douglis and L. Nirenberg, *Estimates near the boundary for solutions of elliptic partial differential equations satisfying general boundary conditions.* I., Comm. Pure Appl. Math. **12** (1959), 623- 727.

2. V. Alexiades and A. D. Solomon, *Mathematical modeling of melting and freezing processes*, Hemisphere Publ., 1993.
3. N. D. Alikakos, P. W. Bates, and X. Chen., *The convergence of solutions of the Cahn–Hilliard equation to the solution of the Hele–Shaw model*, Arch. Rational Mech. Anal. **128** (1994), 165–205.
4. N. D. Alikakos, P. W. Bates, X. Chen, and G. Fusco, *Mullins–Sekerka motion of small droplet on a fixed boundary*, J. Geom. Anal. **10** (2000), no. 4, 575–596.
5. N. D. Alikakos and G. Fusco, *Ostwald ripening for dilute systems under quasistationary dynamics*, Preprint, 1998.
6. _____ , *The equations of Ostwald ripening for dilute systems*, J. Statist. Phys. **95** (1999), no. 5-6, 851–866.
7. _____ , (in progress).
8. N. D. Alikakos and G. Fusco, and G. Karali, *Continuum limits of particle systems evolving under quasistationary Stephan dynamics* (in preparation).
9. H. Amann, *Linear and quasilinear parabolic. I. Abstract linear theory*, Monogr. Mathem., vol. 89. Birkhäuser Verlag, Basel, 1995.
10. S. B. Angement, *Nonlinear analytic semigroups*, Proc. Roy. Soc. Edinburgh Sect. A **115** (1990), 91–107.
11. P. W. Bates and G. Fusco, *Equilibria with many nuclei for the Cahn–Hilliard equation*, J. Differential Equations **160** (2000), 283–356.
12. G. Bellettini and G. Fusco, *Some aspect of the dynamics of $V = H - \bar{H}$*, J. Differential Equations **157** (1999), 206–246.
13. J. W. Cahn, *On the spinodal decomposition*, Acta Metall. **9** (161), 795–801.
14. J. W. Cahn and J. E. Hilliard, *Free energy of a nonuniform system I. Interfacial free energy*, J. Phys. Chem. **28** (1958), 258–267.
15. X. Chen, *Global asymptotic limit of the solutions of the Cahn–Hilliard equations*, J. Differential Geom. **44** (1996), no. 2, 262–311.
16. _____ , *The Hele–Shaw problem and area-preserving curve-shortening motions*, Arch. Rational Mech. Anal. **123** (1993), 117–151.
17. X. Chen, X. Hong, and F. Yi, *Existence, uniqueness, and regularity of solutions of Mullins–Sekerka problem*, Preprint, 1997.
18. P. Constantine and M. Pugh, *Global solutions for small data to the Hele–Shaw problem*, Nonlinearity **6** (1993), 393–415.
19. C. K. L. Davies, P. Nash, and R. N. Stevens, *The effect of volume fraction of precipitate on Ostwald ripening*, Acta Metall. **28** (1980), 179–189.
20. J. Escher and G. Simonett, *A center manifold analysis for the Mullins-Sekerka model*, J. Differential Equations **143** (1998), 267–292.
21. _____ , *Classical solutions for Hele-Shaw models with surface tension*, Adv. Differential Equations **2** (1997), 619–642.
22. P. Fife, *Dynamical aspects of the Cahn–Hilliard equations*, Barrett lectures, University of Tennessee, Spring 1991.
23. G. Huisken, *The volume preserving mean curvature flow*, J. Reine Angew. Math. **382** (1987), 35-48, (1987).
24. G. Huisken and S.T. Yau, *Definition of center of mass for isolated physical systems and unique foliations by stable spheres with constant mean curvature*, Invent. Math. **24** (1996), 281-311, (1996).
25. I. M. Lifschitz and V. V. Slyosov, *The kinetics of precipitation from supersaturated solid solutions*, J. Phys. Chem. Solids **19** (1961), 35–50.
26. A. Lunardi, *Analytic semigroups and optimal regularity for parabolic problems*, Progr. Nonlinear Differential Equations Appl., vol. 16. Birkhäuser Verlag, Basel, 1995.
27. M. Marder, *Correlations and Ostwald ripening*, Phys. Rev. A **36** (1987), 858–874.
28. J. A. Marqusee and J. Ross, *Theory of Ostald ripening: Competitive growth and its dependence on volume fraction*, J. Chem. Phys. **80** (1984), 536–543.
29. C. Miranda, *Sulle proprietà di regolarità di certe trasformazioni integrali*, Atti Accad. Naz. Lincei Mem. Cl. Sci. Fis. Mat. Natur. Sez. I (8) **7** (1965), 303–336.
30. W. W. Mullins and R. F. Sekerka, *Morphological stability of a particle growing by diffusion and heat flow*, J. Appl. Physics **34** (1963), 323–329.

31. B. Niethammer, *Derivation of the LSW-theory for Ostwald ripening by homogenization methods*, Arch. Ration. Mech. Anal. **147** (1999), no. 2, 119–178.
32. B. Niethammer and F. Otto, *Ostwald ripening: The screening length revisited*, Preprint, 2000.
33. B. Niethammer and R. Pego, *Non self-similar behavior in the LSW theory of Ostwald ripening*, J. Statist. Phys. **95** (1999), no. 5-6, 867–902.
34. R. L. Pego, *Front migration in the nonlinear Cahn–Hilliard*, Proc. Roy. Soc. London, Ser. A **422** (1989), 261–278.
35. O. Penrose, Private Communication.
36. O. Penrose, J. L. Lebowitz, J. Marro, M. Kalos, and J. Tobochnik, *Kinetics of a first-order phase transition: Computer simulations and theory*, J. Statist. Phys. **34** (1984), no. 3-4, 399–426.
37. O. Penrose, J. L. Lebowitz, J. Murro, M. H. Kalos, and A. Sur, *Growth of clusters in a first-order phase transition*, J. Statist. Phys. **19** (1978), 243–267.
38. G. da Prato and P. Grisvard, *Equations d'évolution abstraites non linéaires de type parabolique*, Ann. Mat. Pura Appl. (4) **120** (1979), 329–396.
39. J. Rubinstein and P. Sternberg, *Nonlocal reaction-diffusion equations and nucleation*, IMA J. Appl. Math. **48** (1992), no. 3, 249–264.
40. H. M. Soner, *Convergence of the phase-field equations to the Mullins–Sekerka problem with kinetic undercooling*, Arch. Rational Mech. Anal. **131** (1995), no. 2, 139–197.
41. J. J Velázquez, *On the effect of stochastic fluctuations in the dynamics of the Lifshitz–Slyozov–Wagner model*, J. Statist. Phys. **99** (2000), no. 1-2, 57–113.
42. P. W. Voorhees, *The theory of Ostwald ripening*, J. Statist. Physics **38** (1985), 231–252.
43. _____, *Ostwald ripening of two-phases mixtures*, Ann. Rev. Mater. Sc. **22** , (1992), ???–???.
44. P. W. Voorhees and M. E. Glicksman, *Solution to the multi-particle diffusion problem with applications to Ostwald ripening. I. Theory*, Acta Metall. **32** (1984), 2001–2011.
45. _____, *Solution to the multi-particle diffusion problem with applications to Ostwald ripening. II. Computer simulations*, Acta Metall. **32** (1984), 2001–2011.
46. P. W. Voorhees and Schaffer, *In situ observation of particle motion and diffusion interactions during coarsening*, Acta Metall. **33** (1987), 327–339.
47. C. Wagner, *Theory der Alterung von Niedersclagen durch umlosen*, Z. Electrochem. **65** (1961), 581–594.
48. J. J. Weins and J. W. Cahn, *The effect of size and distribution on second phase particles and voids on sintering*, Sinterins and Related Phenomena (G. C. Kuczynski, eds.) Plenum, New York, 1973.
49. R. Ye, *Foliation by constant mean curvature spheres*, Pacific J. Math. **147** (1991), no. 2, 381–396.
50. J. Zhu, X. Chen and T.X. Hou, *An efficient boundary integral method for the Mullins-Sekerka problem*, J. Comput. Phys. **127** (1996), no. 2, 246–267.

DEPARMENT OF MATHEMATICS, UNIVERSITY OF ATHENS, PANEPISTIMIOPOLIS, ATHENS, GR-15784, GREECE.
E-mail address: nalikako@cc.uoa.gr

DIPARTIMENTO DI MATEMATICA, UNIVERSITA DI L' AQUILA, ITALY.
E-mail address: fusco@axp.mat.uniroma2.it

Mathematical Problems in the Control of Underactuated Systems

David Auckly and Lev Kapitanski

There are many interesting mathematical problems in control theory. In this paper we will discuss problems and techniques related to underactuated systems. An underactuated system is one with fewer control inputs than degrees of freedom. Balancing a ruler on the tip of a finger is a good example of an underactuated system. This system has five degrees of freedom (three for the fingertip and two angles for the ruler). However, only the three degrees of freedom for the fingertip are directly controlled. In fact, any system requiring balance is an underactuated system. A bicycle is an obvious example. An airplane is a less obvious example (six degrees of freedom, underactuated by two).

We will give a mathematical formulation of several problems arising from applications, review some standard and new techniques, and pose some interesting and challenging open questions. We wish to thank Fedor Andreev, Atul Kelkar, and Warren White for useful conversations.

Stabilization of Underactuated Systems

To describe a mechanical system we start with a manifold, Q, representing all possible configurations of the system. The configuration space Q is equipped with a Riemannian metric, g, so that the kinetic energy is $1/2g(\dot{x}, \dot{x})$. It also comes with a function $V: Q \to \mathbb{R}$, and two fiber preserving maps $c, f: TQ \to TQ$. The function V represents potential energy, c represents dissipation, and f represents applied external forces. The equations of motion are given by

$$(1) \qquad \nabla_{\dot{\gamma}}\dot{\gamma} + c(\dot{\gamma}) + \operatorname{grad}_\gamma V = f(\dot{\gamma}).$$

The external forces, f, are used to control the system. The system obtained by setting the control input to zero is called *the open loop system*. Equation (1) including the control input is called *the closed loop system*. The system is *underactuated* if f is restricted to be 0 in some directions. In other words, there is a g-orthogonal projection P onto the subspace of unactuated directions, and $P(f)$ must vanish.

2000 *Mathematics Subject Classification*. Primary: 93A99.

This work was partially supported by a grant from the National Science Foundation CMS 9813182.

This is the final form of the paper.

©2001 American Mathematical Society

The basic problem is to find a function f in some class so that solutions to equation (1) have some desired properties. Physically one may not always be able to measure the full state $(x, \dot{x}) \in TQ$ of the mechanical system. In this situation the function f must only depend on the observable variables. For now, we will consider the case when all variables may be observed. This is referred to as full state feedback control.

The stabilization problem is to find a control input, so that some point $(x_0, 0)$ will be an asymptotically stable equilibrium. Other notions of stability may also be considered, however, asymptotic stability is the most useful in applications. We will next review several approaches to the stabilization problem.

The most commonly employed technique used to address this problem is linearization. Choosing f so that the eigenvalues of the linearized equation lie in the left half plane, will ensure that the desired point is a locally asymptotically stable equilibrium. This reduces the problem to an algebraic question that can be easily solved and implemented.

First, note that it is not possible to stabilize every system, for example,

$$\begin{cases} \ddot{x}^1 - x^1 = 0, \\ \ddot{x}^2 + x^2 = u. \end{cases}$$

For a linear $n \times n$ system of the form $\dot{y} = Ay + Bu$ necessary and sufficient conditions for the existence of a linear stabilizing control law $u = Cy$ are well known [14, 18]: the rank of the matrix $[sI - A, B]$ must be n for all $\operatorname{Re} s \geq 0$. If a linear stabilizing control law exists, there is a finite dimensional family of linear stabilizing control laws.

Once it is known that a stabilizing control law exists, one must choose a specific control law. The problem of finding a matrix C given the eigenvalues of $A + BC$ is called pole placement. Engineers use various rules of thumb to decide where to place the poles. These rules of thumb are based upon the behavior of solutions to a constant coefficient second order ODE. For higher order systems one purposefully places two dominant poles, z_1 and z_2, with the remaining poles near the real axis and far to the left. This enables one to approximate solutions of the higher order system by solutions of a second order system.

Example 1. The Inverted Pendulum Cart

With appropriate scaling the metric g is given by $g = d\theta^2 + 2b\cos(\theta)\, dx\, d\theta + dx^2$, where b is a physical parameter, $0 < b < 1$. The potential energy is given by $V = \cos(\theta)$. Since no torques can be applied directly to the pendulum, $P = \big(b\cos(\theta)\, dx + d\theta\big) \otimes \partial/\partial\theta$ is the orthogonal projection onto the direction $\partial/\partial\theta$. Assuming that there is no dissipation, $c = 0$. The $\partial/\partial\theta$- and $\partial/\partial x$-components of the equations of motion read:

$$\ddot{\theta} + b\cos(\theta)\ddot{x} - \sin(\theta) = 0, \quad b\cos(\theta)\ddot{\theta} + \ddot{x} - b\sin(\theta)\dot{\theta}^2 = u,$$

where $u = g(\partial/\partial x, f)$ represents the external force applied to the base of the cart, and it is the control input.

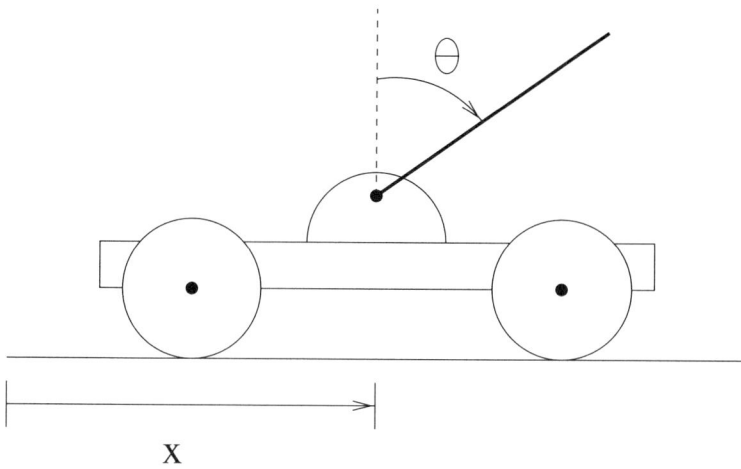

FIGURE 1.

The linearization around $\theta = 0$, $x = 0$, $\dot\theta = 0$, $\dot x = 0$ reads

$$\frac{d}{dt}\begin{pmatrix}\theta\\x\\\dot\theta\\\dot x\end{pmatrix} = \begin{pmatrix} 0 & 0 & 1 & 0 \\ 0 & 0 & 0 & 1 \\ \frac{1-a_1b}{1-b^2} & \frac{-a_2b}{1-b^2} & \frac{-a_3b}{1-b^2} & \frac{-a_4b}{1-b^2} \\ \frac{a_1+b}{1-b^2} & \frac{a_2}{1-b^2} & \frac{a_3}{1-b^2} & \frac{a_4}{1-b^2} \end{pmatrix}\begin{pmatrix}\theta\\x\\\dot\theta\\\dot x\end{pmatrix},$$

where $u = a_1\theta + a_2 x + a_3\dot\theta + a_4\dot x$ is the linearized control input. The characteristic polynomial of the above matrix is

$$\lambda^4 + \frac{a_3b - a_4}{1-b^2}\lambda^3 + \frac{1+a_2-a_1b}{1-b^2}\lambda^2 + \frac{a_4(1+b^2)}{(1-b^2)^2}\lambda + \frac{a_2(1+b^2)}{(1-b^2)^2}.$$

Once the desired eigenvalues are specified, it is an easy matter to solve for the a_k and get the control law.

Another standard technique employed in control design is linear quadratic optimal control, [18]. For linearized systems $\dot y = Ay + Bu$, one looks for control laws which will minimize the functional

$$J(y,u) = \int_0^\infty \langle y(t), Qy(t)\rangle + \langle u(t), Nu(t)\rangle\, dt,$$

subject to the condition $\dot y = Ay + Bu$. Here Q and N are positive definite quadratic matrices which are usually specified by engineering rules of thumb.

If a stabilizing control law exists, the resulting control input can be expressed as a linear function of the state y.

Linearization works very well for many practical applications. Unfortunately, the limitations of linearization are seldomly discussed. Continuing in this tradition we will now address nonlinear methods without stating why.

Many nonlinear methods employ the notion of a Lyapunov function. For a system of ODEs, $\dot x = f(x)$, a Lyapunov function is a function $F(x)$ which is bounded from below and decreases along the trajectories, [4]. It can be defined globally or locally. If x_0 is a stationary solution of $\dot x = f(x)$, and a unique local minimizer of the Lyapunov function F, then x_0 is Lyapunov stable. If F is nonconstant along nonstationary trajectories then x_0 will be locally asymptotically stable. If, in addition, F is defined globally, then x_0 is a global asymptotically stable equilibrium.

For the stabilization problem, $\dot{x} = f(x, u)$, one wishes to find a control input, u, as a function of the state, x, and a Lyapunov function, $F(x)$, so that the closed loop system admits $F(x)$ as a Lyapunov function. In the linear case, $\dot{y} = Ay + Bu$, with $u = Cy$, one looks for a quadratic Lyapunov function $F(y) = \langle y, Ky \rangle$. The function F will be a Lyapunov function if and only if

$$(2) \qquad D = (A + BC)^* K + K(A + BC),$$

is negative definite. Given any negative definite D one may solve equation (2) for K if and only if all the eigenvalues of $A + BC$ are in the left half-plane. The solution is unique:

$$K = -\int_0^\infty e^{t(A+BC)^*} D e^{t(A+BC)} \, dt.$$

Note, that in the optimal control approach discussed previously, the value function

$$F(x) = \inf_u J(x, u),$$

is a Lyapunov function. We also remark that, if there is a locally asymptotically stabilizing control law, then there exists a local Lyapunov function, [4].

Several recent papers propose to find control inputs so that the closed-loop system (1) would have a natural candidate for a Lyapunov function, [1, 2, 3, 5, 6, 7, 8, 9, 13, 17]. In [3] we introduce the following approach to the stabilization problem for underactuated systems.

We consider the control problem (1)
subject to the constraint $P(f) = 0$. We wish to find f and a Lyapunov function simultaneously. In fact, we will describe an infinite dimensional family of control inputs.

Our approach to this question is to find functions $\hat{g}$, $\hat{c}$, $\widehat{V}$ and f so that solutions to equation (1) are automatically solutions to

$$(3) \qquad \widehat{\nabla}_{\dot{\gamma}} \dot{\gamma} + \hat{c}(\dot{\gamma}) + \widehat{\text{grad}}_\gamma \widehat{V} = 0.$$

This is the matching philosophy. The motivation for this philosophy is that

$$\widehat{H}(X) = \frac{1}{2} \hat{g}(X, X) + \widehat{V},$$

is a natural candidate for a Lyapunov function because $d/dt \widehat{H}(X) = -\hat{g}(\hat{c}(X), X)$. A state, $X_0 \in TQ$ will be an asymptotically stable equilibrium if $\widehat{H}(X) \geq 0$, and $-\hat{g}(\hat{c}(X), X) \geq 0$ with equality only at X_0.

Equations (1) and (3) clearly hold if and only if:

$$(4) \qquad f(X) \equiv \nabla_X X - \widehat{\nabla}_X X + \text{grad}_\gamma V - \widehat{\text{grad}}_\gamma \widehat{V} + c(X) - \hat{c}(X),$$

for every vector field X. The condition $P(f) = 0$ then becomes a system of nonlinear partial differential equations for $\hat{g}$, $\widehat{V}$, and $\hat{c}$. Notice that constant multiples of g, c, and V satisfy $P(f) = 0$ even when P has full rank. Thus, one would expect many solutions when P does not have full rank. Separating $P(f) = 0$ into terms which are quadratic in the velocity, independent of the velocity or odd functions of the velocity gives:

(5a) $$P(\nabla_X X - \widehat{\nabla}_X X) = 0,$$

(5b) $$P(\text{grad}_\gamma V - \widehat{\text{grad}_\gamma V}) = 0,$$

(5c) $$P\big(c(X) - \hat{c}(X)\big) = 0.$$

We will look for solutions to these matching equations with $\hat{g}$ nondegenerate so that $g(X,Y) = \hat{g}(\lambda X, Y)$ with $\lambda \in \Gamma(T^*Q \otimes TQ)$. It is clear that λ has to be g self-adjoint, i.e., $g(\lambda X, Y) = g(X, \lambda Y)$. We will derive a linear system of partial differential equations for λ which must be satisfied if $\hat{g}$ is to solve equation (5a).

In our previous paper, we described a method to find every solution to the matching equations by solving three linear systems of partial differential equations in a row. We will review this method now.

One first solves the equations

(6) $$\nabla_g \lambda \big|_{\text{Im}} P^{\otimes 2} = 0,$$

for $\lambda|_{\text{Im } P}$. Then one solves

(7) $$L_{\lambda PX} \hat{g} = L_{PX} g,$$

(this is a slight rewrite of our previous paper [**3, Eq. (1.12)**]),

(8) $$L_{\lambda PX} \widehat{V} = L_{PX} V,$$

(this is Eq. (1.13) of our previous paper [**3**]), then after solving equation (5c), the control input will be given by (4).

In the previous paper we explicitly showed that any solution to the matching equations solves equations (6), (7), (8), and (5c) [**3, Propositions 1.1, 1.2, and 1.3**]. Implicit in [**3, Proposition 1.4**] is the fact that any solution of equations (6), (7), (8) and (5.3) is in turn a solution to the matching equation. We will make this argument explicit now. Indeed, taking into account the fact that P is g-selfadjoint, our $\hat{g}$-equation (7) implies the matching equation (5.1). Here is a short proof. For any Z and X we have

$$\begin{aligned} &g\big(P(\widehat{\nabla}_Z Z - \nabla_Z Z), X\big) \\ &= g(\widehat{\nabla}_Z Z - \nabla_Z Z, PX) \\ &= \hat{g}(\widehat{\nabla}_Z Z, \lambda PX) \quad g(\nabla_Z Z, PX) \\ &= Z\hat{g}(Z, \lambda PX) - \hat{g}(Z, \widehat{\nabla}_Z \lambda PX) - Zg(Z, PX) + g(Z, \nabla_Z PX) \\ &= -\hat{g}(Z, \widehat{\nabla}_{\lambda PX} Z) + \hat{g}\big(Z, [\lambda PX, Z]\big) + g(Z, \nabla_{PX} Z) - g\big(Z, [PX, Z]\big) \\ &= -\frac{1}{2}\big(L_{\lambda PX} \hat{g}(Z, Z)\big) + \frac{1}{2}\big(L_{PX} g(Z, Z)\big) \\ &= 0. \end{aligned}$$

Since this is true for all X, $P(\widehat{\nabla}_Z Z - \nabla_Z Z) = 0$.

Equations (6) and (7) imply additional compatibility conditions. Even though we do not know all the compatibility conditions in general, we do know all the compatibility conditions for systems with two degrees of freedom. Let us summarize our method in the case of two degrees of freedom one of which is unactuated. Since the unactuated subspace is one dimensional, it can be locally expressed as the span of a unit length vectorfield, PX. Choose coordinates x^1, x^2 so that $PX = \partial/(\partial x^1)$. In these coordinates $g_{11} = 1$. We will always write $\lambda PX = \sigma \partial/(\partial x^1) + \mu \partial/(\partial x^2)$, where σ and μ are yet to be found. The λ-equation may be rewritten as

$$(9) \quad \frac{\partial}{\partial x^1}(g_{11}\sigma + g_{12}\mu) - 2[11,2]\mu = 0, \quad \frac{\partial}{\partial x^2}(g_{11}\sigma + g_{12}\mu) - 2[12,2]\mu = 0.$$

Here,

$$(10) \quad [ij,k] = g(\nabla_{\partial_i}\partial_j, \partial_k) = \frac{1}{2}(g_{ik,j} + g_{jk,i} - g_{ij,k}).$$

For these equations to be consistent the following compatibility condition must hold:

$$(11) \quad \frac{\partial}{\partial x^2}\left([11,2]\mu\right) = \frac{\partial}{\partial x^1}\left([12,2]\mu\right).$$

Notice that this is a first order partial differential equation equation. Theoretically, it can be solved for μ via the method of characteristics. Generically, a solution will include an arbitrary function of a single variable. Once μ is known, σ is given by

$$\sigma(x^1,x^2) = g_{12}(x^1,x^2)\mu(x^1,x^2) + 2\int\left([11,2]\mu(x^1,x^2)\,dx^1 + [12,2]\mu(x^1,x^2)\,dx^2\right).$$

The next step is to solve equations (6) for $\hat{g}$. It turns out that it is easiest to solve first for $\hat{g}_{11}$ and then find the remaining components from the algebraic system

$$g = \hat{g}\lambda.$$

First note that the $\{11\}$ component of the right side of (7) is

$$(L_{PX}g)\left(\frac{\partial}{\partial x^1}, \frac{\partial}{\partial x^1}\right) = \frac{\partial}{\partial x^1}g_{11} = 0.$$

Next,

$$(L_{\lambda PX}\hat{g})\left(\frac{\partial}{\partial x^1}, \frac{\partial}{\partial x^1}\right) = \lambda PX(\hat{g}_{11}) - 2\hat{g}\left(\left[\lambda PX, \frac{\partial}{\partial x^1}\right], \frac{\partial}{\partial x^1}\right).$$

Since

$$\left[\lambda PX, \frac{\partial}{\partial x^1}\right] = -\frac{\partial \sigma}{\partial x^1}\frac{\partial}{\partial x^1} - \frac{\partial \mu}{\partial x^1}\frac{\partial}{\partial x^2} = \left(-\frac{\partial \sigma}{\partial x^1} + \frac{\sigma}{\mu}\frac{\partial \mu}{\partial x^1}\right)\frac{\partial}{\partial x^1} - \frac{1}{\mu}\frac{\partial \mu}{\partial x^1}\lambda PX,$$

and $\hat{g}\lambda = g$, we obtain,

$$\lambda PX\hat{g}_{11} - 2\left(-\frac{\partial \sigma}{\partial x^1} + \frac{\sigma}{\mu}\frac{\partial \mu}{\partial x^1}\right)\hat{g}_{11} + 2\frac{1}{\mu}\frac{\partial \mu}{\partial x^1}g_{11} = 0.$$

Thus, $\hat{g}_{11}$ satisfies the following first order partial differential equation

$$\sigma\frac{\partial \hat{g}_{11}}{\partial x^1} + \mu\frac{\partial \hat{g}_{11}}{\partial x^2} + 2\left(\frac{\partial \sigma}{\partial x^1} - \frac{\sigma}{\mu}\frac{\partial \mu}{\partial x^1}\right)\hat{g}_{11} + 2\frac{(\partial \mu/\partial x^1)}{\mu} = 0.$$

The general solution to this first order PDE has an arbitrary function of a single variable in it. It is, once again, possible to solve this equation by the method of characteristics. Denote by $y(x^1, x^2)$ any solution of the homogeneous equation

$$(12) \quad \sigma\frac{\partial y}{\partial x^1} + \mu\frac{\partial y}{\partial x^2} = 0,$$

such that $\partial y/\partial x^2 \neq 0$. Let $\bar{\sigma}$ and $\bar{\mu}$ be σ and μ considered as functions of x^1 and y, i.e.,
$$\bar{\sigma}(x^1, y(x^1, x^2)) = \sigma(x^1, x^2), \quad \bar{\mu}(x^1, y(x^1, x^2)) = \mu(x^1, x^2).$$
Then the solution to equation (7) is given explicitly by

$$(13) \qquad \hat{g}_{11}(x^1, x^2) = \frac{\mu^2}{\sigma^2}\left[-2\int_0^{x^1} \frac{\bar{\sigma}}{\bar{\mu}^3}\frac{\partial \bar{\mu}}{\partial x^1}\, dx^1\bigg|_{y=y(x^1,x^2)} + h(y(x^1, x^2))\right],$$

where $h(y)$ is an arbitrary function of a single variable. After $\hat{g}_{11}$ is found, we have

$$(14) \qquad \hat{g}_{12} = \frac{1}{\mu}(g_{11} - \sigma\hat{g}_{11}), \quad \hat{g}_{22} = \frac{1}{\mu}(g_{12} - \sigma\hat{g}_{12}).$$

The equation for $\widehat{V}$ reads
$$\sigma\frac{\partial \widehat{V}}{\partial x^1} + \mu\frac{\partial \widehat{V}}{\partial x^2} = \frac{\partial V}{\partial x^1}.$$
Again, this equation can be solved by the method of characteristics. Once a solution y of (12) is known, $\widehat{V}$ is given by:

$$(15) \qquad \widehat{V}(x^1, x^2) = -\int_0^{x^1} \frac{\overline{V}_{x^1}}{\bar{\sigma}}\, dx^1 \bigg|_{y=y(x^1,x^2)} + w(y(x^1, x^2)),$$

where $\overline{V}_{x^1}(x^1, y(x^1, x^2)) = ((\partial V(x^1, x^2)/\partial x^1))$. Also,
$$\hat{c}_1 = c_1 + g_{12}(\hat{c}_2 - c_2).$$

It is convenient to write the nonzero control input as a sum:

$$(16) \quad u = g\left(f, \frac{\partial}{\partial x^2}\right)$$
$$= g\left(\nabla_X X - \widehat{\nabla}_X X + \text{grad}_\gamma V - \widehat{\text{grad}}_\gamma \widehat{V} + c(X) - \hat{c}(X), \frac{\partial}{\partial x^2}\right) = u_g + u_V + u_c,$$

where
$$*u_g = g\left(\nabla_X X - \widehat{\nabla}_X X, \frac{\partial}{\partial x^2}\right), \quad u_V = g\left(\text{grad}_\gamma V - \widehat{\text{grad}}_\gamma \widehat{V}, \frac{\partial}{\partial x^2}\right),$$
$$u_c = g\left(c(X) - \hat{c}(X), \frac{\partial}{\partial x^2}\right).$$

In coordinates,
$$(17) \qquad u_g = \left([ij, 2] - g_{k2}\widehat{\Gamma}^k_{ij}\right)\dot{x}^i\dot{x}^j,$$

where $\widehat{\Gamma}^k_{ij}$ are the Christoffel symbols corresponding to $\hat{g}$,
$$\widehat{\Gamma}^k_{ij} = \hat{g}^{kp}\widehat{[ij, p]},$$

$\hat{g}^{kp}$ is the inverse matrix to $\hat{g}_{lm}$, and $\widehat{[ij, p]}$ is defined as in (9) with all g's replaced by $\hat{g}$'s. The next term is:

$$(18) \qquad u_V = V_{x^2} - \hat{g}^{ij}g_{j2}\widehat{V}_{x^i}.$$

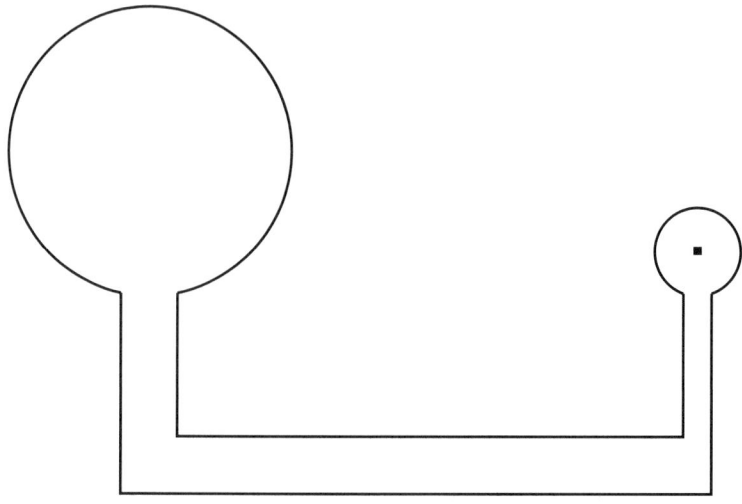

FIGURE 2.

Finally,

(19) $$u_c = \det g(c_2 - \hat{c}_2).$$

From the explicit formulae (16)–(19) one sees that the first order germs of our control inputs contain every possible linear control law:

THEOREM. *If a system with 2 degrees of freedom underactuated by 1 is linearly stabilizable, then, by choosing appropriate solutions of our matching equations, the linearization of the controlled system will have prescribed eigenvalues in the left halfplane. Even if the system is not stabilizable, any linear control law may be obtained as the first order germ of some control law in our family.*

Open Problems

One of the reasons that the limitations of linearization are seldomly discussed is that there is no effective criteria to compare arbitrary control laws. To be more specific we will restrict our discussion to the stabilization problem.

A good stabilizing control law will produce a large basin of attraction, send solutions to the equilibrium in a short period of time, and will have low cost. Of course, the precise meaning of "large", "short", and "low" depends on the concrete engineering problem. It is not clear how to quantify these concepts. For linear systems the size of the basin of attraction is irrelevant since the whole space is the basin of attraction of a stable equilibrium. For nonlinear systems this question is subtle. One could just use the volume or diameter as a measure of the size of the basin of attraction. These are not, however, usually appropriate measures of size, see Figure 2. In addition, they are difficult to compute.

Alternatively, one could measure the radius of the largest inscribed ball centered at the equilibrium. It is not, however, clear which metric should be used in state space. It is difficult to analytically or numerically estimate this radius. Even if one could compute this radius, it might not be the most relevant measure of performance. Real systems have an operating range: roller blades are not designed

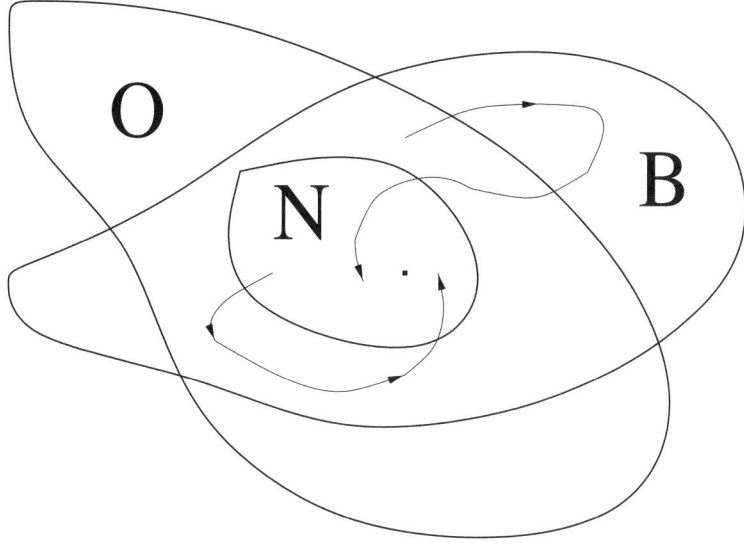

Figure 3.

to handle Mach 2. So, let $\mathcal{B}$ be the basin of attraction and $\mathcal{O}$ be the operating range. It is more reasonable to measure the size of the subset $\mathcal{N} \subset \mathcal{B}$ consisting of all states whose forward trajectories remain in the operating range, see Figure 3. We will call $\mathcal{N}$ *the normal operating range*. It is just as difficult to form a reasonable measure of the size of $\mathcal{N}$.

PROBLEM. Find an effective analytical or numerical method to estimate the injectivity radius of the normal operating range or define a better measure of "size" together with an effective analytical or numerical estimate.

In engineering applications the time it takes to drive the states to the equilibrium is very important. This is usually characterized by the rise time, peak time, settling time, etc., [**11, p. 222**]. These notions are only well defined for solutions of $m\ddot{z} + c\dot{z} + kz = f$ with m, c, k, and f constants. Considering the initial conditions with $\dot{z} = 0$ only, one rescales the initial value problem to $\ddot{z} + 2\zeta\dot{z} + z = 1$,$z(0) = 0$, $\dot{z}(0) = 0$. The rise time, T_r, is the first time $z(t)$ reaches 1. The time constant is $1/\zeta$. The settling time is the time it takes $z(t)$ to get and stay within 2% of the final value 1. These notions are used for linear constant coefficient ODEs which have two dominant poles.

One can define analogues of these notions for nonlinear systems. An engineer may choose a *target operating range* $\mathcal{D}$, which is a small neighborhood of the equilibrium.

Define the $\mathcal{ND}$-settling time

$$T_{\mathcal{ND}} = \sup_{x_0 \in \mathcal{N}} \{ t \geq 0 \mid x(t; x_0) \notin \mathcal{D} \}.$$

This is the time after which any trajectory starting in the normal operating range $\mathcal{N}$ will settle into the target range $\mathcal{D}$.

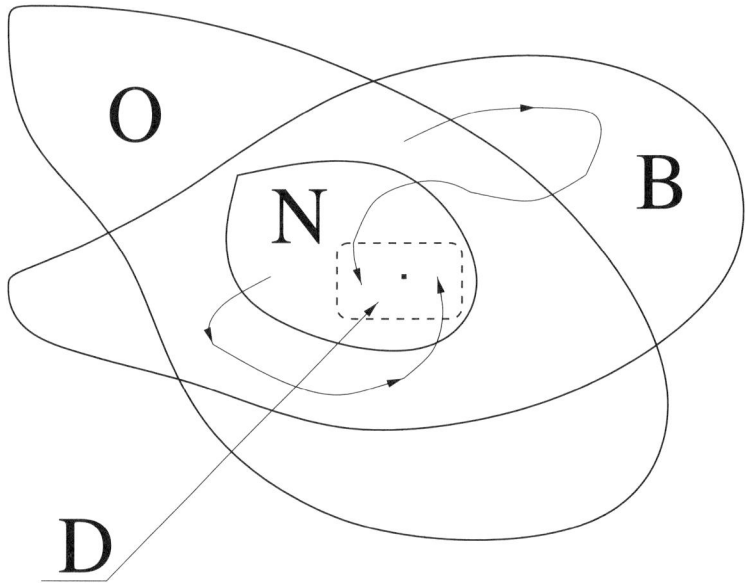

Figure 4.

Problem. Find an effective analytical or numerical method to estimate the $\mathcal{ND}$-settling time or define a better measure of "short time" together with an effective analytical or numerical estimate.

Another important characteristic of a control law is the cost. One often wishes to minimize some function of the trajectory and/or the control input (for example, minimize the energy used to complete a task). This imposes a nontrivial restriction on a control problem. Real life systems have additional physical restrictions. For example, there is a maximal voltage which may be applied to a DC motor before it saturates.

Reiterating, a good stabilizing control law should maximize the basin of attraction and minimize the time and cost while operating within the physical restrictions of the given system.

The problem of maximizing the size of the basin of attraction taken to the extreme leads to the question of finding the topologically best control laws. For example, the state space for the inverted pendulum cart is $S^1 \times \mathbb{R} \times \mathbb{R}^2$. For topological reasons it is impossible to find a control law so that the resulting flow has a globally asymptotically stable equilibrium in this case. The best one could hope for is a flow with a regular compact global attractor, [**12, 16**]. Recall that such an attractor is the union of the unstable manifolds of finitely many hyperbolic fixed points. Also recall that the shape (in the sense of K. Borsuk) of the (global) attractor must be the same as the shape of the state space, [**15**]. We wish to minimize the number of fixed points. For the inverted pendulum cart *the topologically best flow* is depicted in Figure 5.

This flow must have one index 0 fixed point and one index 1 fixed point. The basin of attraction of the index 0 fixed point is the complement of a properly embedded $\mathbb{R}^3$.

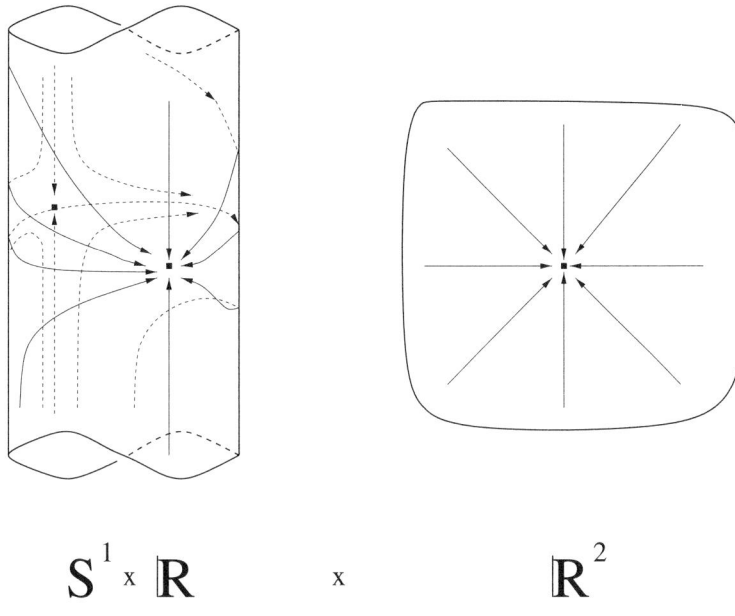

FIGURE 5.

PROBLEM. When is it possible to find a control input producing a topologically best flow? Is there a method to find such control laws?

At the moment we do not even know whether a topologically best control law exists for the inverted pendulum cart. Furthermore, we do not know whether a control law with a global attractor exists!

A related problem is to describe the behavior of the system far away from the equilibrium. It is possible to compactify some dynamical systems with algebraic nonlinearities, [**10**]. The advantage of a compactification is that overall dynamics may be well understood by linearization about each of the critical submanifolds. The problem is that for nonalgebraic nonlinearities it is not known when a compactification exists. It is even interesting and instructive to try to compactify the system $\ddot{\theta} + c\dot{\theta} + \sin(\theta) = 0$.

We wrote this paper to encourage other people to think about nonlinear control theory. It has a great many unanswered fundamental questions. The field is ready for new ideas.

References

1. D. Auckly, L. Kapitanski, A. Kelkar, and W. White, *Matching and pole placement for underactuated systems*, Preprint (1999).
2. _____, *Matching and digital control implementation for underactuated systems*, Preprint (1999).
3. D. Auckly, L. Kapitanski, and W. White, *Control of nonlinear underactuated systems*, Comm. Pure Appl. Math. **53** (2000), no. 3, 354–369.
4. N. P. Bhatia and G. P. Szegö, *Dynamical systems: Atability theory and applications*, Lecture Notes in Mathematics, vol. 35, Springer-Verlag, Berlin-New York, 1967.
5. A. Bloch, N. Leonard, and J. Marsden, *Stabilization of mechanical systems using controlled Lagrangians*, Proc. of the 36th IEEE Conference on Decision and Control, vol. 36, San Diego, California, 1997, pp. 2356–2361.

6. _____, *Matching and stabilization by the method of controlled Lagrangians*, Proc. of the 37th IEEE Conference on Decision and Control, vol. 37, Tampa, Florida, 1998, pp. 1446–1451.
7. _____, *Stabilization of the pendulum on a rotor arm by the method of controlled Lagrangians*, Proc. IEEE Internat. Conf. on Robotics and Automation, Detroit, Michigan, 1999, pp. 500–505.
8. _____, *Controlled Lagrangians and a stabilization of mechanical systems I: The first matching theorem*, Preprint (1999).
9. _____, *Potential shaping and the method of controlled Lagrangians*, Preprint (1999).
10. O. I. Bogoyavlensky, *Methods in the qualitative theory of dynamical systems in astrophysics and gas dynamics*, Springer Series in Soviet Mathematics, Springer-Verlag, Berlin-New York, 1985.
11. R. C. Dorf and R. H. Bishop, *Modern control systems*, Addison-Wesley, 1995.
12. J. K. Hale, *Asymptotic behavior of dissipative systems*, Mathematical Surveys and Monographs, vol. 25, Amer. Math. Soc., Providence, RI, 1987.
13. J. Hamberg, *General matching conditions in the theory of controlled Lagrangians*, Proc. of the 38th Conference on Decision and Control, vol. 38, Phoenix, Arizona, 1999, pp. 2519–2523.
14. T. Kailath, *Linear systems*, Prentice-Hall, Englewood Cliffs, N.J., 1980.
15. L. Kapitanski and I. Rodnianski, *Shape and Morse theory of attractors*, Comm. Pure Appl. Math. **53** (2000), no. 2, 218–242.
16. O. A. Ladyzhenskaya, *Attractors for semigroups and evolution equations*, (Lezioni Lincee), Cambridge University Press, Cambridge, 1991.
17. A. J. van der Schaft, *Stabilization of Hamiltonian systems*, Nonlinear Anal. **10** (1986), no. 10, 1021–1035.
18. E. D. Sontag, *Mathematical control theory*, Texts in Applied Mathematics, vol. 6, Springer-Verlag, New York, 1990.

DEPARTMENT OF MATHEMATICS, KANSAS STATE UNIVERSITY, MANHATTAN, KS 66506–2602, USA
E-mail address: `dav@math.ksu.edu`

DEPARTMENT OF MATHEMATICS, KANSAS STATE UNIVERSITY, MANHATTAN, KS 66506–2602, USA
E-mail address: `levkapit@math.ksu.edu`

Axially and Helically Symmetric Global Plasma Equilibria

Oleg I. Bogoyavlenskij

ABSTRACT. Exact global axially symmetric and helically symmetric plasma equilibria are derived. All constructed plasma equilibria have no singularities and are localized in the sense that they have finite magnetic energy in any layer $c_1 < z < c_2$.

1. Introduction

The problem of finding global plasma equilibria is well-known since the plasma equilibrium equations

(1.1) $$\operatorname{curl} \mathbf{B} \times \mathbf{B} = \mu \operatorname{grad} p, \quad \operatorname{div} \mathbf{B} = 0,$$

were first applied to the controlled thermonuclear fusion problem [7,8,11,13] and to the astrophysical problems [4,5]. However, all found during the last four decades exact solutions to equations (1.1) which are not translationally invariant either are not localized and unboundedly grow at infinity or have singularities [2,3,6,9,12]. Such solutions have a very restricted applicability in astrophysics. The physically relevant global plasma equilibria have to satisfy the following conditions in the cylindrical coordinates r, ϕ, z:

(a) The magnetic field $\mathbf{B}$, the electric current $\mathbf{J} = \operatorname{curl} \mathbf{B}/\mu$ and the plasma pressure p are smooth and bounded functions everywhere in the Euclidean space $\mathbb{R}^3$.
(b) At $r \to \infty$, the asymptotics $\mathbf{B} \to 0$, $\mathbf{J} \to 0$, $p \to p_1 > 0$ hold.
(c) All magnetic field lines and current lines are bounded in the radial variable r.

In this paper, we derive the exact global plasma equilibria that model astrophysical jets in the co-moving frame of reference. The asymptotic value of pressure p_1 in condition (b) is the average pressure in the astrophysical outflow. As usual, the gravitational force $-\rho \operatorname{grad} \Psi$ is included into the pressure gradient in (1.1), in the approximation of constant density ρ. The constructed plasma equilibria are

---

2000 *Mathematics Subject Classification.* Primary: 2906; Secondary: 1202.

The work of the author was partially supported by the Natural Sciences and Engineering Research Council of Canada.

This is the final form of the paper.

localized in the sense that the total magnetic energy in any layer $c_1 < z < c_2$ is finite.

In Section 2, we present the axially symmetric global plasma equilibria. The most important feature of the obtained exact solutions is that they are quasi-periodic in variable z. That means that they never repeat themselves in the z-direction but appear arbitrarily close to any initial data when variable z evolves. The same is true for the corresponding magnetic field lines. Therefore the pattern of winding of the magnetic lines about each other *does vary* in the z-direction in the newly constructed exact global plasma equilibria.

For the exact plasma equilibria with axial symmetry, the magnetic surfaces are either cylinders or nested tori with circular magnetic axes. The distribution of these toroidal magnetic surfaces and their magnetic axes is quasi-periodic in variable z.

In Section 3, we derive the global plasma equilibria with helical symmetry. We construct a family of z- and ϕ-invariant global plasma equilibria, each equilibrium is contained in a 3-dimensional linear space of perturbations which are neither z- nor ϕ-invariant but possess a helical symmetry. These solutions are periodic in variable z with period $2\pi\gamma$ and decrease at $r \to \infty$ as rapidly as $c_N \exp(-\beta r^2) r^{2N}$. Hence the localization property follows.

For the helically symmetric solutions, the magnetic surfaces are nested cylinders that rotate as helices around the axis z. The innermost cylinders are their magnetic axes which all together form a finite system of helices.

The search for the global solutions to equations (1.1) is important also for fluid dynamics and magnetohydrodynamics because the MHD equilibrium equations

$$(1.2) \qquad \mathbf{V} \times \operatorname{curl} \mathbf{V} - \mathbf{B} \times \operatorname{curl} \mathbf{B}/\rho\mu = \operatorname{grad}(p/\rho + V^2/2),$$
$$\operatorname{div} \mathbf{V} = 0, \quad \operatorname{div} \mathbf{B} = 0, \quad [\mathbf{V}, \mathbf{B}] = 0,$$

for the aligned vector fields $\mathbf{V} = \lambda \mathbf{B}$ are equivalent to equations (1.1). For $\lambda = 0$, we obtain equilibrium equations for ideal hydrodynamics.

The derived axially and helically symmetric solutions form finite-dimensional linear spaces. Any linear combination of the corresponding magnetic fields is again an exact solution. However, the law of addition of the plasma pressure has more complex quadratic form. These properties resemble those of the N-soliton solutions of integrable equations of plasma physics: the KdV, KP and NLS equations.

2. Global Axially Symmetric Plasma Equilibria

To derive the exact plasma equilibria, we consider equations (1.1) for an axially symmetric magnetic field $\mathbf{B}$ [7, 10, 11]:

$$(2.1) \qquad \mathbf{B} = \frac{\psi_z}{r}\hat{\mathbf{e}}_r - \frac{\psi_r}{r}\hat{\mathbf{e}}_z + \frac{I}{r}\hat{\mathbf{e}}_\phi,$$

where $\psi(r,z)$ is the flux function, $\psi_x = \partial\psi/\partial x$ and vectors $\hat{\mathbf{e}}_r$, $\hat{\mathbf{e}}_z$, $\hat{\mathbf{e}}_\phi$ are the coordinate unit orts. Then the current density $\mathbf{J}$ has the form

$$(2.2) \qquad \mathbf{J} = \frac{1}{\mu}\operatorname{curl}\mathbf{B} = -\frac{I_z}{\mu r}\hat{\mathbf{e}}_r + \frac{I_r}{\mu r}\hat{\mathbf{e}}_z + \frac{\Phi}{\mu r}\hat{\mathbf{e}}_\phi,$$

where $\Phi = \psi_{rr} - r^{-1}\psi_r + \psi_{zz}$. The plasma equilibrium equations (1.1) are equivalent to the equalities $I = I(\psi)$, $p = p(\psi)$ and the Grad–Shafranov equation [7, 11]

$$(2.3) \qquad \frac{\partial^2 \psi}{\partial r^2} - \frac{1}{r}\frac{\partial \psi}{\partial r} + \frac{\partial^2 \psi}{\partial z^2} + I(\psi)I'(\psi) + \mu r^2 p'(\psi) = 0,$$

where $I(\psi)$ and $p(\psi)$ are arbitrary functions. The equation (2.3) implies $\Phi = -I(\psi)I'(\psi) - \mu r^2 p'(\psi)$.

THEOREM 2.1. *There exist an infinite family of z-invariant global plasma equilibria parametrized by an integer $N \geq 0$ and a real $\beta > 0$. For $N \geq 2$, each equilibrium is contained in a $(2N-1)$-dimensional linear space of global plasma equilibria which are not translationally invariant. These exact solutions are either periodic or quasi-periodic in variable z.*

PROOF. (1) We suppose that function $I(\psi)$ is linear $I(\psi) = \alpha\psi$ and function $p(\psi)$ is quadratic $p(\psi) = p_1 - 2\beta^2\psi^2/\mu$, where $p_1 > (2\beta^2/\mu)\max(\psi^2(x,y,z))$ and α and $\beta > 0$ are arbitrary constants. It is evident that constant p_1 does not enter the plasma equilibrium equations (1.1). Under these assumptions, equation (2.3) takes the form

$$(2.4) \qquad \psi_{rr} - \frac{1}{r}\psi_r + \psi_{zz} + \alpha^2\psi - 4\beta^2 r^2\psi = 0.$$

The current (2.2) becomes

$$(2.5) \qquad \mathbf{J} = -\frac{\alpha}{\mu}\mathbf{B} + \frac{4\beta^2 r\psi}{\mu}\hat{\mathbf{e}}_\phi.$$

The formulae (2.1) and (2.5) imply the asymptotics $\mathbf{J} = -\alpha\mathbf{B}/\mu$ at $r \to 0$ and $\alpha \neq 0$.

A direct verification proves that the Gaussian distribution

$$(2.6) \qquad \psi(r,z) = \exp(-\beta r^2),$$

is an exact solution to equation (2.4) for $\alpha = 0$. The corresponding parameters $\mathbf{B}$, $\mathbf{J}$ and p have the form

$$(2.7) \qquad \mathbf{B} = 2\beta e^{-\beta r^2}\hat{\mathbf{e}}_z, \quad \mathbf{J} = \frac{4\beta^2 r}{\mu}e^{-\beta r^2}\hat{\mathbf{e}}_\phi, \quad p = p_1 - \frac{2\beta^2}{\mu}e^{-2\beta r^2}.$$

Hence the Gaussian distribution (2.6) defines a global solution to equations (1.1) that satisfies the three conditions above.

Substituting $\psi(r,z) = \exp(-\beta r^2)u(x,z)$ into equation (2.4) and using new variable $x = 2\beta r^2$, we derive

$$(2.8) \qquad xu_{xx} - xu_x + \frac{1}{8\beta}(\alpha^2 u + u_{zz}) = 0.$$

Separating variables by $u(x,z) = P(x)T(z)$, we get

$$(2.9) \qquad xP'' - xP' + \frac{\alpha^2 + \lambda}{8\beta}P = 0, \quad T'' = \lambda T.$$

For $\lambda = -\omega^2$, we have $T(z) = a\cos(\omega z) + b\sin(\omega z)$. For the first equation (2.9), we are interested only in the polynomial solutions $P(x)$. Such solutions exist if and only if $(\alpha^2 - \omega^2)/8\beta = n$, where $n \geq 0$ is an integer. Hence we obtain a finite spectrum of admissible values of $\omega = \omega_0, \ldots, \omega_N$:

$$(2.10) \qquad \omega_n = \sqrt{\alpha^2 - 8\beta n}, \quad n = 0, 1, \ldots, N, \quad N = \left[\frac{\alpha^2}{8\beta}\right].$$

The first equation (2.9) results in the form $xP'' - xP' + nP = 0$. Differentiating, we obtain $xL'' + (1-x)L' + (n-1)L = 0$, where $L(x) = P'(x)$. This equation defines the Laguerre polynomials $L_{n-1}(x)$ [1, 14]. Hence the polynomials $P(x)$ are primitive functions of the Laguerre polynomials. We denote them $L_n^*(x)$:

$$(2.11) \quad L_n^*(x) = \int_0^x L_{n-1}(t)\mathrm{d}t = -\frac{x}{n}e^x \frac{\mathrm{d}^n}{\mathrm{d}x^n}(e^{-x}x^{n-1}), \quad L_0^*(x) = -1,$$

$$L_n^*(x) = (n-1)!x + (n-1)!\sum_{k=1}^{n-1} \frac{(-1)^k \binom{n-1}{k}}{(k+1)!} x^{k+1},$$

where $\binom{n-1}{k}$ are the binomial coefficients.

Thus we obtain the exact solutions to equation (2.4):

$$(2.12) \quad \psi_n(z,r) = \exp(-\beta r^2)\big(a_n \cos(\omega_n z) + b_n \sin(\omega_n z)\big) L_n^*(2\beta r^2).$$

For any fixed constants α and β, the formulae (2.10) and (2.12) define N exact solutions to the linear equation (2.4). Hence we get the $2N$-dimensional linear space of exact solutions

$$\psi(r,z) = \exp(-\beta r^2) \sum_{n=1}^{N} \big(a_n \cos(\omega_n z) + b_n \sin(\omega_n z)\big) L_n^*(2\beta r^2).$$

These solutions satisfy the conditions (a) and (b) above. However, the condition (c) is not met because the magnetic surface $\psi(r,z) = 0$ and magnetic lines on it are unbounded in variable r. For the flux function (2.12), the surface $\psi_n(r,z) = 0$ contains an infinite number of planes $z = z_k = z_0 + 2\pi k/\omega_n$.

(2) To derive the global solutions satisfying the three conditions (a), (b), (c), we let $\alpha^2 = 8\beta N$. Then $\omega_N = 0$ and exact solution (2.12) takes the form

$$(2.13) \quad \psi_N(r) = \exp(-\beta r^2) L_N^*(2\beta r^2).$$

The corresponding magnetic field (2.1), the current (2.5) and the pressure p are

$$(2.14) \quad \mathbf{B} = 2e^{-\beta r^2}\left(\beta\big(L_N^*(x) - 2L_N^{*\prime}(x)\big)\hat{\mathbf{e}}_z + \frac{\sqrt{2\beta N}}{r} L_N^*(x)\hat{\mathbf{e}}_\phi\right),$$

$$\mathbf{J} = -\frac{\sqrt{8\beta N}}{\mu}\mathbf{B} + \frac{4\beta^2 r}{\mu}\psi_N(r)\hat{\mathbf{e}}_\phi, \quad p = p_1 - \frac{2\beta^2}{\mu}e^{-2\beta r^2} L_N^{*2}(x), \quad x = 2\beta r^2.$$

This is a z- and ϕ-invariant global plasma equilibrium. The flux function (2.13) turns into the Gaussian distribution (2.6) at $N = 0$.

Taking a linear combination of the exact solutions (2.12) for $n = 1, \ldots, N-1$ and (2.13), we obtain the $(2N-1)$-dimensional linear space of exact solutions to equation (2.4):

$$(2.15) \quad \psi(r,z) = \exp(-\beta r^2)$$

$$\times \left(a_N L_N^*(2\beta r^2) + \sum_{n=1}^{N-1} \big(a_n \cos(\omega_n z) + b_n \sin(\omega_n z)\big) L_n^*(2\beta r^2)\right),$$

where $\omega_n = \sqrt{8\beta(N-n)}$, the coefficients a_n, b_n are arbitrary and $a_N \neq 0$. The pressure $p(\psi) = p_1 - 2\beta^2 \psi^2/\mu$ is a quadratic function of the coefficients a_n, b_n, a_N.

Formulae (2.11) imply that the exact solutions (2.15) have the asymptotic forms $\psi_0(z,r) = r^2 G(z)$ at $r \to 0$, where

$$G(z) = 2\beta(N-1)!a_N + 2\beta \sum_{n=1}^{N-1} (n-1)!\big(a_n \cos(\omega_n z) + b_n \sin(\omega_n z)\big).$$

Hence the magnetic field (2.1) and the current density (2.5) have the smooth asymptotics at $r \to 0$

$$\mathbf{B} = rG'(z)\hat{\mathbf{e}}_r - 2G(z)\hat{\mathbf{e}}_z + \sqrt{8\beta N}\, rG(z)\hat{\mathbf{e}}_\phi, \quad \mathbf{J} = -\frac{\sqrt{8\beta N}}{\mu}\mathbf{B}.$$

The poloidal projections of the magnetic field lines coincide with the level curves $\psi(r, z) = \mathrm{const}$ [7, 10]. Formula (2.15) implies that these curves approach the straight lines $r = \mathrm{const}$ when $r \gg 1$ because its leading term is $-a_N(-2\beta r^2)^N \times \exp(-\beta r^2)$. Hence all magnetic field lines and all current lines are bounded in the radial variable r. Hence the exact solutions (2.15) at $a_N \neq 0$ satisfy the physical conditions (a), (b), (c). All the derived global plasma equilibria (2.15) (but the exact solutions (2.13)) are not translationally invariant.

For $N = 2$, solutions (2.15) are periodic functions of z with period $T = \pi/\sqrt{2\beta}$. For $N \geq 3$, the generic solutions (2.15) are quasi-periodic functions of z. Indeed, the frequencies $\omega_n = \sqrt{8\beta(N-n)}$ for $n = N-1, N-2, \ldots, 1$ are proportional to the numbers

(2.16) $$1, \sqrt{2}, \sqrt{3}, \ldots, \sqrt{N-1}.$$

Hence solution (2.15) is a quasi-periodic function of z if for example $a_{N-1} \neq 0, a_{N-2} \neq 0$.

$\square$

The exact solutions (2.13), (2.14) possess cylindrical symmetry: they are ϕ- and z-invariant. The exact solutions (2.15) for arbitrarily small $a_n \neq 0$, $b_n \neq 0$ define the global plasma equilibria that are small perturbations of solutions (2.13), (2.14) but are not z-invariant.

3. Helically Symmetric Global Plasma Equilibria

The helically symmetric plasma equilibria first were introduced by Johnson *et al* in [8] and have found important applications in the stellarators theory [13]. In the cylindrical coordinates r, z, ϕ, these solutions depend on the two variables $u = z - \gamma\phi$ and r and hence are invariant with respect to the helical transformations $z \to z + \gamma\phi_0$, $\phi \to \phi + \phi_0$. The magnetic field $\mathbf{B}$, the electric current $\mathbf{J}$ and the plasma pressure p are $2\pi\gamma$-periodic functions of variable u and hence z.

The helically symmetric magnetic field $\mathbf{B}$ has the form

(3.1) $$\mathbf{B} = \frac{\psi_u}{r}\hat{\mathbf{e}}_r - \frac{\psi_r}{r}\hat{\mathbf{e}}_u + \frac{f}{r}\hat{\mathbf{e}}_\phi,$$

where $\psi(u, r)$ and $f(u, r)$ are some smooth functions, $\psi_x = \partial\psi/\partial x$ and $\hat{\mathbf{e}}_r, \hat{\mathbf{e}}_u, \hat{\mathbf{e}}_\phi$ are the unit tangent vectors corresponding to the coordinates r, u, ϕ. The current $\mathbf{J}$ is

(3.2) $$\mathbf{J} = \frac{1}{\mu}\operatorname{curl}\mathbf{B} = \frac{1}{\mu}\left(\frac{I_r}{r}\hat{\mathbf{e}}_u - \frac{I_u}{r}\hat{\mathbf{e}}_r + \Phi\hat{\mathbf{e}}_\phi\right),$$

where

(3.3) $$I = \frac{r^2 + \gamma^2}{r^2}f - \frac{\gamma}{r}\psi_r,$$

(3.4) $$\Phi = \frac{1}{r}\frac{\partial^2\psi}{\partial u^2} + \frac{\partial}{\partial r}\left(\frac{1}{r}\frac{\partial\psi}{\partial r}\right) - \gamma\frac{\partial}{\partial r}\left(\frac{f}{r^2}\right).$$

Johnson *et al* had shown in [**8**] that the plasma equilibrium equations (1.1) for the helically symmetric solutions are equivalent to the equalities $I = I(\psi)$, $p = p(\psi)$ with arbitrary functions $I(\psi)$ and $p(\psi)$, and the equation

$$(3.5) \qquad \frac{1}{r^2}\frac{\partial^2 \psi}{\partial u^2} + \frac{1}{r}\frac{\partial}{\partial r}\left(\frac{r}{r^2+\gamma^2}\frac{\partial \psi}{\partial r}\right) + \frac{II'(\psi)}{r^2+\gamma^2} + \frac{2\gamma I(\psi)}{(r^2+\gamma^2)^2} = -\mu p'(\psi),$$

where prime denotes differentiation with respect to ψ. The equation (3.5) is equivalent to the Johnson–Frieman–Kulsrud–Oberman (JFKO) equation [**8**] where another variable $u = \ell\phi - hz$ was employed. We use the variable $u = z - \gamma\phi$ because only parameter $\gamma = \ell/h$ is essential. For $\gamma = 0$, the JFKO equation (3.5) turns into the Grad–Shafranov equation (2.3) that describes the axially symmetric plasma equilibria.

Formula (3.3) for $I = I(\psi)$ yields

$$(3.6) \qquad f = \frac{\gamma r \psi_r + r^2 I(\psi)}{r^2+\gamma^2}, \quad \psi_r - \frac{\gamma f}{r} = \frac{r^2 \psi_r - \gamma r I(\psi)}{r^2+\gamma^2}.$$

The formulae (3.1) and (3.6) imply the expressions for the magnetic field **B** in the cylindrical coordinates r, z, ϕ:

$$(3.7) \qquad \mathbf{B} = \frac{\psi_u}{r}\hat{\mathbf{e}}_r + B_1 \hat{\mathbf{e}}_z + B_2 \hat{\mathbf{e}}_\phi, \quad B_1 = \frac{\gamma I(\psi) - r\psi_r}{r^2+\gamma^2}, \quad B_2 = \frac{rI(\psi) + \gamma\psi_r}{r^2+\gamma^2},$$

and in the cartesian coordinates x, y, z:

$$\mathbf{B} = \left(\frac{x\psi_u}{r^2} - \frac{y(\gamma\psi_r + rI(\psi))}{r(r^2+\gamma^2)}\right)\hat{\mathbf{e}}_x + \left(\frac{y\psi_u}{r^2} + \frac{x(\gamma\psi_r + rI(\psi))}{r(r^2+\gamma^2)}\right)\hat{\mathbf{e}}_y + B_1\hat{\mathbf{e}}_z.$$

The JFKO equation (3.5) and the first formula (3.6) imply that function Φ (3.4) has the form $\Phi/r = -\mu p'(\psi) - fI'(\psi)/r^2$. Hence we obtain for the current **J** (3.2) in the cylindrical coordinates r, z, ϕ:

$$(3.8) \qquad \mathbf{J} = -\frac{I'(\psi)}{\mu}\mathbf{B} - p'(\psi)(r\hat{\mathbf{e}}_\phi + \gamma\hat{\mathbf{e}}_z).$$

THEOREM 3.1. *There exists an infinite family of global plasma equilibria depending on an arbitrary integer $N \geq 0$ and two reals $\beta > 0$ and γ. The equilibria are z- and ϕ-invariant in the cylindrical coordinates r, z, ϕ. For any two integers $\ell \geq 1$ and $n \geq 0$ satisfying the condition $2n + \ell < N$ and*

$$(3.9) \qquad \beta = \frac{\ell^2}{4\gamma^2(2N - 2n - \ell)},$$

the equilibria are contained in a 3-dimensional linear family of global plasma equilibria which are neither z- nor ϕ-invariant but possess the helical symmetry with parameter γ. These exact solutions are periodic in variable z with period $2\pi\gamma/\ell$ and in variable ϕ with period $2\pi/\ell$.

PROOF. (1) We suppose that $I(\psi) = \alpha\psi$ and $p(\psi) = p_1 - 2\beta^2\psi^2/\mu$, where $p_1 > (2\beta^2/\mu)\max(\psi^2(x,y,z))$. Then the plasma pressure p is > 0 everywhere. It is evident that constant p_1 does not enter the plasma equilibrium equations (1.1). Under these assumptions, the JFKO equation (3.5) becomes linear:

$$(3.10) \qquad \frac{1}{r^2}\frac{\partial^2\psi}{\partial u^2} + \frac{1}{r}\frac{\partial}{\partial r}\left(\frac{r}{r^2+\gamma^2}\frac{\partial\psi}{\partial r}\right) = 4\beta^2\psi - \frac{\alpha^2\psi}{r^2+\gamma^2} - \frac{2\alpha\gamma\psi}{(r^2+\gamma^2)^2}.$$

Separating variables by the substitution

(3.11) $$\psi(r,u) = A(r)\big(a\cos(\omega u) + b\sin(\omega u)\big),$$

we obtain the equation

(3.12) $$\frac{1}{r}\frac{d}{dr}\left(\frac{r}{r^2+\gamma^2}\frac{dA(r)}{dr}\right) = \left(4\beta^2 + \frac{\omega^2}{r^2} - \frac{\alpha^2}{r^2+\gamma^2} - \frac{2\alpha\gamma}{(r^2+\gamma^2)^2}\right)A(r).$$

This equation does not belong to any known class of integrable ones [1]. To find exact solutions, we apply the substitution $A(r) = r^\lambda e^{-\beta r^2} B(x)$, $x = 2\beta r^2$, $\lambda \geq 0$, that transforms equation (3.12) into

(3.13) $$(x^2 + c_1 x)B'' + \big(-x^2 + (\lambda - c_1)x + (\lambda+1)c_1\big)B'$$
$$+ \left(\frac{\alpha^2 - \eta}{8\beta}x + \frac{\alpha^2\gamma^2 - \eta\gamma^2}{4} + \frac{\alpha\gamma - c_1 - \lambda}{2} + \frac{x+c_1}{4x}(\lambda^2 - \gamma^2\omega^2)\right)B = 0,$$

where $B' = dB(x)/dx$, $c_1 = 2\beta\gamma^2$, and $\eta = 4\beta^2\gamma^2 + 4\beta\lambda + \omega^2$. We are interested only in the polynomial solutions $B(x)$ with a nonzero free term. Inspecting the highest and the lowest order terms in (3.13), we obtain the necessary conditions

(3.14) $$\frac{\alpha^2 - \eta}{8\beta} = n, \quad \lambda = |\gamma\omega|,$$

where the integer $n \geq 0$ is the order of the polynomial $B(x)$. The form of the solution (3.11) $\psi(r,u) = A(r)\big(a\cos(\omega z - \gamma\omega\phi) + b(\sin(\omega z - \gamma\omega\phi))\big)$ implies that $\gamma\omega$ must be an integer: $|\gamma\omega| = \ell \geq 0$. Hence we get $\lambda = \ell$, $\omega = \pm\ell/\gamma$, and $\eta = (2\beta\gamma + \ell/\gamma)^2$. The first necessary condition (3.14) becomes the equation

(3.15) $$\alpha^2\gamma^2 = (\ell + c_1)^2 + 4nc_1,$$

for the two unknown integers ℓ and n. We present equation (3.13) in the form

(3.16) $$(x^2 + c_1 x)B'' + \big(-x^2 + (\ell - c_1)x + (\ell+1)c_1\big)B' + n(x + c_1 - k_{\ell n}c_1)B = 0,$$

where $k_{\ell n} = (\ell + c_1 - \alpha\gamma)/(2nc_1)$. Note that equation (3.16) is different from all classical differential equations which define the Chebyshev, Hermite, Laguerre, Legendre or Jacobi polynomials [1, 14].

LEMMA 3.2. *Differential equation* (3.16) *has a polynomial solution*

(3.17) $$B_{\ell n}(x) = \frac{d^\ell}{dx^\ell} L_{\ell+n}(x) - k_{\ell n} x \frac{d^{\ell+1}}{dx^{\ell+1}} L_{\ell+n}(x),$$

where $L_m(x)$ are the Laguerre polynomials, provided that equation (3.15) is satisfied.

PROOF. To find the polynomial solutions $B_{\ell n}(x)$, we make the substitution

(3.18) $$B(x) = P(x) - k_{\ell n} x P'(x).$$

A direct verification proves the identity

$$(x^2 + c_1 x)B'' + \big(-x^2 + (\ell - c_1)x + (\ell+1)c_1\big)B' + n(x + c_1 - k_{\ell n}c_1)B$$
$$= (x + c_1 - k_{\ell n}c_1)Q - k_{\ell n}x(x+c_1)Q' + \big(nc_1 k_{\ell n}^2 - (\ell+c_1)k_{\ell n} - 1\big)xP',$$

where $Q = xP'' + (1 + \ell - x)P' + nP$. The necessary condition (3.15) implies the equation

$$nc_1 k_{\ell n}^2 - (\ell + c_1)k_{\ell n} - 1 = 0,$$

where $k_{\ell n} = (\ell + c_1 - \alpha\gamma)/(2nc_1)$. Equation $Q = 0$ or

$$xP'' + (1 + \ell - x)P' + nP = 0,$$

has polynomial solutions $P_{\ell n}(x) = d^\ell L_{\ell+n}(x)/dx^\ell$ of degree n where $L_m(x)$ are the Laguerre polynomials

$$(3.19) \qquad L_m(x) = \frac{1}{m!} e^x \frac{d^m}{dx^m}(e^{-x} x^m) = \sum_{k=0}^{m} \frac{(-1)^k m!}{(k!)^2 (m-k)!} x^k.$$

Hence using formula (3.18), we obtain that if the two integers ℓ and n satisfy equation (3.15) then equation (3.16) follows from $Q = 0$ and has the polynomial solution (3.17). □

Finally, we obtain that equation (3.10) has the exact solution

$$(3.20) \qquad \psi_{\ell n} = \left(a_{\ell n} \cos\left(\frac{\ell}{\gamma} z - \ell\phi \right) + b_{\ell n} \sin\left(\frac{\ell}{\gamma} z - \ell\phi \right) \right) r^\ell e^{-\beta r^2} B_{\ell n}(2\beta r^2),$$

where $a_{\ell n}$ and $b_{\ell n}$ are arbitrary constants. The linearity of equation (3.10) implies that if the algebraic equation (3.15) has several integral solutions ℓ, n then any linear combination of functions (3.20) is an exact solution to equation (3.10).

(2) For $\ell = n = 0$, the necessary condition (3.15) is $\alpha = 2\beta\gamma$, and equation (3.13) has solution $B = $ const. Hence the JFKO equation (3.5) has an exact solution of the Gaussian function form

$$(3.21) \qquad \psi_0(r) = \exp(-\beta r^2).$$

The corresponding magnetic field (3.7), current (3.8) and pressure p are

$$(3.22) \qquad \mathbf{B} = 2\beta e^{-\beta r^2} \hat{\mathbf{e}}_z, \quad \mathbf{J} = \frac{4\beta^2}{\mu} r e^{-\beta r^2} \hat{\mathbf{e}}_\phi, \quad p = p_1 - \frac{2\beta^2}{\mu} e^{-2\beta r^2}.$$

This is a global plasma equilibria possessing cylindrical symmetry.

For $\ell = 0$ and $n = N \neq 0$, the necessary condition (3.15) is

$$(3.23) \qquad \alpha^2 \gamma^2 = c_1^2 + 4Nc_1,$$

and the exact solution (3.20) takes the form

$$(3.24) \qquad \psi_N(r) = a_N e^{-\beta r^2} B_{0N}(2\beta r^2),$$

where polynomials $B_{0N}(x)$ are

$$(3.25) \qquad B_{0N}(x) = L_N(x) - k_N x L'_N(x).$$

Equation (3.23) implies ($c_1 = 2\beta\gamma^2$)

$$(3.26) \quad \alpha\gamma = c_1 \sqrt{1 + 2N/(\beta\gamma^2)}, \quad k_N = \frac{c_1 - \alpha\gamma}{2Nc_1} = \frac{1 - \sqrt{1 + 2N/(\beta\gamma^2)}}{2N} < 0.$$

Hence polynomials $B_{0N}(x)$ do depend on parameter $\beta\gamma^2$ for all $N \geq 1$.

For the flux function (3.24), the magnetic field (3.7)

$$(3.27) \qquad \mathbf{B} = \frac{\alpha\gamma\psi_N - r\psi'_N}{r^2 + \gamma^2} \hat{\mathbf{e}}_z + \frac{\alpha r \psi_N + \gamma \psi'_N}{r^2 + \gamma^2} \hat{\mathbf{e}}_\phi$$

has no singularities and decreases at $r \to \infty$ as rapidly as $c_N \exp(-\beta r^2) r^{2N}$. Hence the total magnetic energy in any layer $c_1 \leq z \leq c_2$ is finite. The magnetic surfaces $\psi_N(r) = $ const are cylinders $r = $ const.

Thus we have demonstrated that the JFKO equation (3.5) has global plasma equilibrium solutions (3.22) and (3.24), (3.27) where the two parameters $\beta > 0$ and $\gamma \neq 0$ are arbitrary. These equilibria possess cylindrical symmetry for they are z- and ϕ-invariant.

(3) For $\ell \geq 1$, the flux functions $\psi_{\ell n}(r,u)$ (3.20) define the magnetic fields **B** (3.7) and currents **J** (3.8) that are smooth in the cartesian coordinates and decrease at $r \to \infty$ as rapidly as $c_N \exp(-\beta r^2) r^{2n}$. Hence the above conditions (a) and (b) are satisfied. However condition (c) is not met because the magnetic surface $\psi_{\ell n}(r,u) = 0$ is a helicoid $z = \gamma\phi + c$ that is unbounded in variable r.

To find the global equilibria, we consider a linear combination of the exact solutions $\psi_N(r)$ (3.24) and $\psi_{\ell n}(r,u)$ (3.20) which also satisfies equation (3.10) if the two necessary conditions (3.15) and (3.23) hold simultaneously. These two conditions yield the inequality $2N > 2n + \ell$ and the formulae

$$(3.28) \quad c_1 = 2\beta\gamma^2 = \frac{\ell^2}{2(2N - 2n - \ell)}, \quad \alpha\gamma = \frac{\ell\sqrt{(4N-\ell)^2 - 16nN}}{2(2N - 2n - \ell)},$$

$$(3.29) \quad k_{\ell n} = \frac{\ell + c_1 - \alpha\gamma}{2nc_1} = \frac{1}{2\ell n}\left(4(N-n) - \ell - \sqrt{(4N-\ell)^2 - 16nN}\right).$$

The first formula (3.28) implies condition (3.9). Thus we obtain that the JFKO equation (3.5) for any value of parameter γ has the exact solutions

$$(3.30) \quad \psi = e^{-\beta r^2}\left(a_N B_{0N}(x) + \left(a_{\ell n}\cos(\ell u/\gamma) + b_{\ell n}\sin(\ell u/\gamma)\right)r^\ell B_{\ell n}(x)\right),$$

where $N, \ell \geq 1$, $n \geq 0$ are arbitrary integers satisfying the inequality $2N > 2n + \ell$, and $x = 2\beta r^2$. The functions $I(\psi)$ and $p(\psi)$ are $I(\psi) = \alpha\psi$, $p(\psi) = p_1 - 2\beta^2\psi^2/\mu$, where parameters $\alpha(\gamma)$, $\beta(\gamma)$ are given by formulae (3.28). The inequality $2N > 2n + \ell$ implies that function $\psi(r,u)$ (3.30) has the leading term $(-2\beta)^N a_N \exp(-\beta r^2) r^{2N}$ at $r \to \infty$. Hence equilibria (3.30) satisfy the above conditions (a) and (b), and all magnetic surfaces $\psi(r,u) = \text{const}$ asymptotically for $r \gg 1$ are cylinders $r = \text{const}$. Therefore all magnetic field lines and current lines are bounded in variable r. Thus the exact solutions (3.30) define the global plasma equilibria. These equilibria are periodic in variable z with period $2\pi\gamma/\ell$ and in variable ϕ with period $2\pi/\ell$. □

The exact solutions (3.24), (3.27) possess cylindrical symmetry: they are ϕ- and z-invariant. The exact solutions (3.30) for arbitrarily small $a_{\ell n} \neq 0$, $b_{\ell n} \neq 0$ define the global plasma equilibria which are small perturbations of solutions (3.24), (3.27) but are not z-invariant.

References

[1] M. Abramowitz and I. A. Stegun, *Handbook of mathematical functions*, National Bureau of Standards, Washington, D.C., 1964.
[2] D. Biskamp, *Nonlinear magnetohydrodynamics*, Cambridge University Press, Cambridge, 1993.
[3] P. K. Browning, *Magnetohydrodynamics in solar coronal and laboratory plasmas: A comparative study*, Phys. Rep. **169** (1988), 329–384.
[4] S. Chandrasekhar, *Axisymmetric magnetic fields and fluid motions*, Astrophys. J. **124** (1956), 232–243.
[5] S. Chandrasekhar and K. H. Prendergast, *The equilibrium of magnetic stars*, Proc. Nat. Acad. Sc. U.S.A. **42** (1956), 5–9.
[6] J. F. Freidberg, *Ideal magnetohydrodynamics*, Plenum Press, New York, 1987.
[7] H. Grad and H. Rubin, *Hydromagnetic equilibria and force-free fields*, Proc. of the Second United Nations Internat. Conf. on the Peaceful Uses of Atomic Energy, vol. 31, United Nations, Geneva, 1958, pp. 190–197.

[8] J. L. Johnson, C. R. Oberman, R. M. Kulsrud, and E. A. Frieman, *Some stable hydromagnetic equilibria*, Phys. Fluids **1** (1958), 281–296.
[9] B. B. Kadomtsev, *Equilibrium of a plasma with helical symmetry*, JETP **10** (1960), 962–963.
[10] M. D. Kruskal and R. M. Kulsrud, *Equilibrium of a magnetically confined plasma in a toroid*, Phys. Fluids **1** (1958), 265–274.
[11] V. D. Shafranov, *On magnetohydrodynamical equilibrium configurations*, JETP **6** (1958), 545–554.
[12] L. S. Solov'ev, *The theory of hydromagnetic stability of toroidal plasma configurations*, JETP **26** (1968), 400–407 .
[13] L. Spitzer, *The stellarator concept*, Phys. Fluids **1** (1958), 253–264.
[14] G. Szego, *Orthogonal polynomials*, Amer. Math. Soc., Providence, R.I., 1967.

DEPARTMENT OF MATHEMATICS, QUEEN'S UNIV., KINGSTON, ONTARIO, K7L 3N6, CANADA.
E-mail address: bogoyavl@oib.mast.queensu.ca

On the Complete Ionization of a Periodically Perturbed Quantum System

O. Costin, J. L. Lebowitz, and A. Rokhlenko

Dedicated to the memory of Jürgen Moser

ABSTRACT. We analyze the time evolution of a one-dimensional quantum system with zero range potential under time periodic parametric perturbation of arbitrary strength and frequency. We show that the projection of the wave function on the bound state vanishes, i.e. the system gets fully ionized, as time grows indefinitely.

1. Introduction and Results

The ionization of atoms subjected to external time dependent perturbations is an issue of central importance in quantum mechanics which has attracted substantial theoretical and experimental interest [3, 4]. There exists by now a variety of theoretical methods, and a vast amount of literature, devoted to the subject. Beyond the celebrated Fermi's golden rule, approaches include higher order perturbation theory, semi-classical phase-space analysis, Floquet theory, complex dilation, some exact results for small fields and bounds for large fields and numerical integration of the time dependent Schrödinger equation [1, 2, 7, 8, 10–12, 15, 18, 20, 22]. Nevertheless there is apparently no complete analysis of the ionization of any periodically perturbed model with no restrictions on the amplitudes and frequencies of the perturbing field. This is not so surprising considering the very complex behavior we find in even the most elementary of such systems.

In the present paper we show rigorously the full ionization, in all ranges of amplitudes and frequencies, of one of the simplest models with spatial structure which, with a different perturbing potential, is however frequently used as a model system [6, 9, 16, 23]. The unperturbed Hamiltonian we consider is

$$(1.1) \qquad \mathcal{H}_0 = -\frac{h^2}{2m}\frac{d^2}{dx^2} - g\delta(x), \quad g > 0, \quad -\infty < x < \infty.$$

2000 *Mathematics Subject Classification*. Primary: 81Q05; Secondary: 34L40.

The authors would like to thank A. Soffer and M. Weinstein for interesting discussions and suggestions. Work of O.C. was supported by NSF Grant 9704968, that of J.L.L. and A.R. by AFOSR Grant F49620-98-1-0207 and NSF Grant DMR-9813268.

This is the final form of the paper.

©2001 American Mathematical Society

$\mathcal{H}_0$ has a single bound state $u_b(x) = \sqrt{p_0}e^{-p_0|x|}$, $p_0 = m/(\hbar^2)g$ with energy $-\hbar\omega_0 = -\hbar^2 p_0^2/2m$ and a continuous uniform spectrum on the positive real line, with generalized eigenfunctions

$$u(k,x) = \frac{1}{\sqrt{2\pi}}\left(e^{ikx} - \frac{p_0}{p_0 + i|k|}e^{i|kx|}\right), \quad -\infty < k < \infty,$$

and energies $\hbar^2 k^2/2m$.

Beginning at $t = 0$, we apply a parametric perturbing potential, i.e. for $t \geq 0$ we have

(1.2) $$\mathcal{H}(t) = \mathcal{H}_0 - g\eta(t)\delta(x),$$

and solve the time dependent Schrödinger equation for $\psi(x,t)$,

(1.3) $$\psi(x,t) = \theta(t)u_b(x)e^{i\omega_0 t} + \int_{-\infty}^{\infty} \Theta(k,t)u(k,x)e^{-i(\hbar k^2)/(2m)t}\,dk, \quad (t \geq 0),$$

with initial values $\theta(0) = 1$, $\Theta(k,0) = 0$. This gives the survival probability $|\theta(t)|^2$, as well as the fraction of ejected electrons $|\Theta(k,t)|^2 dk$ with (quasi-) momentum in the interval dk.

This problem can be reduced to the solution of an integral equation [**19**]. Setting

(1.4) $$\theta(t) = 1 + 2i\int_0^t Y(s)\,ds,$$

(1.5) $$\Theta(k,t) = 2|k|/\left[\sqrt{2\pi}(1-i|k|)\right]\int_0^t Y(s)e^{i(1+k^2)s}\,ds,$$

$Y(t)$ satisfies the integral equation

(1.6) $$Y(t) = \eta(t)\left\{1 + \int_0^t [2i + M(t-t')]Y(t')\,dt'\right\} = \eta(t)\bigl(1 + (2i + M) * Y\bigr),$$

where $\hbar$, $2m$ and $g/2$ have been set equal to 1 (implying $p_0 = 1$, $\omega_0 = 1$),

$$M(s) = \frac{2i}{\pi}\int_0^\infty \frac{u^2 e^{-is(1+u^2)}}{1+u^2}\,du = \frac{1+i}{2\sqrt{2\pi}}\int_s^\infty \frac{e^{-iu}}{u^{3/2}}\,du,$$

and

$$f * g = \int_0^t f(s)g(t-s)\,ds.$$

THEOREM 1.1. *When $\eta(t) = r\sin\omega t$ the survival probability $|\theta(t)|^2$ tends to zero as $t \to \infty$, for all $\omega > 0$ and $r \neq 0$.*

NOTE 1.2. For definiteness we assume in the following that $r > 0$.

The method of proof relies on the properties of the Laplace transform of Y, $y(p) = \mathcal{L}Y(p) = \int_0^\infty e^{-pt'}Y(t')\,dt'$ (note that $y(p) = i/2(1 - p\mathcal{L}\theta)$). In particular we need to show that $y(p)$ is bounded in the closed right half of the complex p-plane. Before the proof we describe briefly some additional results on this model system, cf. [**5**].

FIGURE 1. Singularities of y and relevant inverse Laplace contours.

1.1. Further results not proven in the present paper. (1) Theorem 1.1 generalizes to the case when $\eta(t)$ is a trigonometric polynomial:

$$\eta(t) = \sum_{j=1}^{K} \left[A_j \sin(j\omega t) + B_j \cos(j\omega t) \right], \tag{1.7}$$

where we assume $|A_K| + |B_K| \neq 0$.

(2) The detailed behavior of the system as a function of t, ω, and r is obtained from the singularities of $y(p)$ in the complex p-plane. We summarize them *for small r*; below $\delta > 0$ is small.

At $p = \{in\omega - i \colon n \in \mathbb{Z}\}$, y has square root branch points and y is analytic in the right half plane and also in an open neighborhood $\mathcal{N}$ of the imaginary axis with cuts through the branch points. As $|\Im(p)| \to \infty$ in $\mathcal{N}$ we have $|y(p)| = O(r\omega|p|^{-2})$. If $|\omega - 1/n| > \mathrm{const}_n O(r^{2-\delta})$, $n \in \mathbb{Z}^+$, then for small r the function y has a unique pole p_m in each of the strips $-m\omega > \Im(p)+1 \pm O(r^{2-\delta}) > -m\omega - \omega$, $m \in \mathbb{Z}$. $\Re(p_m)$ is strictly independent of m and gives the exponential decay of θ. After suitable contour deformation of the inverse Laplace transform, θ can be (uniquely) written in the form

$$\theta(t) = e^{-\gamma(r;\omega)t} F_\omega(t) + \sum_{m=-\infty}^{\infty} e^{(mi\omega - i)t} h_m(t), \tag{1.8}$$

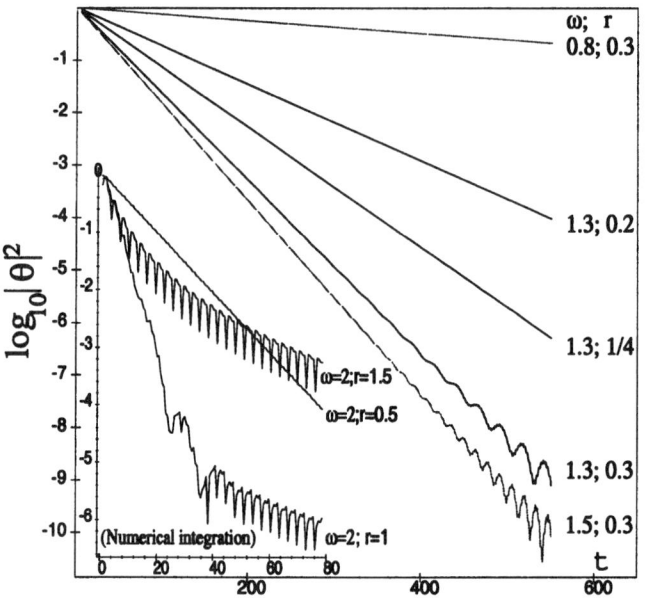

FIGURE 2. Plot of $\log_{10} |\theta(t)|^2$ vs. time in units of ω_0^{-1} for several values of ω and r. The main graph was calculated from (1.8) and the inset used numerical integration of (1.6).

where F_ω is periodic of period $2\pi\omega^{-1}$ and

$$h_m(t) \sim \sum_{j=0}^{\infty} c_{m,j} t^{-3/2-j} \text{ as } t \to \infty, \quad \arg(t) \in \left(-\frac{\pi}{2} - \varepsilon, \frac{\pi}{2} + \varepsilon\right).$$

Not too close to resonances, i.e. when $|\omega - n^{-1}| > O(r^{2-\delta})$, for all integer n, $|F_\omega(t)| = 1 \pm O(r^2)$ and its Fourier coefficients decay faster than $r^{|2m|} |m|^{-|m|/2}$. Also, the sum in (1.8) does not exceed $O(r^2 t^{-3/2})$ for large t, and the h_m decrease with m faster than $r^{|m|}$.

(3) By (1.8), for times of order $1/\Gamma$ where $\Gamma = 2\Re(\gamma)$, the survival probability for ω not close to a resonance decays as $\exp(-\Gamma t)$. This is illustrated in Figure 2 where it is seen that for $r \lesssim 1/2$ the exponential decay holds up to times at which the survival probability is extremely small, after which $|\theta(t)|^2 = O(t^{-3})$ with many oscillations as described by (1.8). Note the slow decay for $\omega = .8$, when ionization requires the absorption of two photons.

(4) When r is larger the polynomial-oscillatory behavior starts sooner. Since the amplitude of the late asymptotic terms is $O(r^2)$ for small r, increased r yields higher late time survival probability. This phenomenon, sometimes referred to as atomic stabilization [8–11], can be associated with the perturbation-induced probability of back-transitions to the well.

(5) Using the continued fraction representation (2.11) Γ can be calculated convergently for any ω and r.

The limiting behavior for small r of the exponent Γ is described as follows. Let n be the integer part of $\omega^{-1}+1$ and assume $\omega^{-1} \notin \mathbb{N}$. Then we have, for $T>0$ ($t = r^{-2n}T$),

$$(1.9) \qquad \widehat{\Gamma} = -T^{-1} \lim_{r \to 0} \ln|\theta(r^{-2n}T)|^2 = \frac{2^{-2n+2}\sqrt{n\omega - 1}}{n\omega \prod_{m<n}(1 - \sqrt{1 - m\omega})^2}.$$

(6) The behavior of Γ is different at the resonances $\omega^{-1} \in \mathbb{N}$. For instance, whereas if ω is not close to 1, the scaling of Γ implied by (1.9) is r^2 when $\omega > 1$ and r^4 when $1/2 < \omega < 1$, by taking $\omega - 1 = r^2/\sqrt{2}$ we find

$$-T^{-1} \lim_{\substack{r \to 0 \\ \omega = 1 + r^2/\sqrt{2}}} \ln|\theta(r^{-3}T)|^2 = \frac{2^{1/4}}{8} - \frac{2^{3/4}}{16}.$$

2. Proofs of Theorem 1.1

LEMMA 2.1. (i) $\mathcal{L}Y$ exists and is analytic in the right half plane $\mathbb{H} = \{p: \Re(p) > 0\}$. Furthermore, $y(p) \to 0$ as $\Im(p) \to \pm\infty$ in $\mathbb{H}$.

(ii) The function $y(p)$ satisfies (and is determined by) the functional equation

$$(2.1) \qquad y = r(T^- - T^+)(h_1 + h_2 y),$$

with

$$(T^\pm f)(p) = f(p \pm i\omega), \quad h_1(p) = -\frac{i}{2p} \quad \text{and} \quad h_2(p) = \frac{1}{2p}(1 + \sqrt{1-ip}),$$

and by the boundary condition $y(p) \to 0$ as $\Im(p) \to \pm\infty$ in $\mathbb{H}$.

The branch of the square root is such that for $p \in \mathbb{H}$, the real part of $\sqrt{1-ip}$ is nonnegative and the imaginary part nonpositive.

PROOF. (i) The time evolution of ψ is unitary and thus $|\langle \psi \mid u_b \rangle| = |\theta(t)| \leq 1$. The stated analyticity is an immediate consequence of the elementary properties of the Laplace transform [1]. The asymptotic behavior follows then from the Riemann–Lebesgue lemma.

(ii) We have in $\mathbb{H}$,

$$(2.2) \qquad \mathcal{L}M = \lim_{a \downarrow 0} \frac{2i}{\pi} \int_0^\infty dx\, e^{-px} \int_0^\infty \frac{u^2 e^{-i(x-ia)(1+u^2)}}{1+u^2} du$$

$$(2.3) \qquad = \frac{i}{\pi} \int_{-\infty}^\infty \frac{u^2}{(1+u^2)(p+i(1+u^2))} du.$$

For $\Re(p) > 0$ we push the integration contour through the upper half plane. At the two poles in the upper half plane $u^2 + 1$ equals 0 and ip respectively, so that

$$(2.4) \qquad \frac{i}{\pi} \int_{-\infty}^\infty \frac{u^2}{(1+u^2)(p+i(1+u^2))} du$$

$$= \frac{i}{\pi}\left(\frac{(-1)}{(2i)(p)} \oint \frac{ds}{s} + \frac{u_0^2}{(ip)(2iu_0)} \oint \frac{ds}{s}\right) = -\frac{i}{p} + \frac{u_0}{p},$$

---

[1] See also the appendix for a proof of analyticity for $\Re(p) > p_0$ (all that is required in the subsequent analysis), relying only on the properties of the convolution equation.

where u_0 is the root of $p + i(1 + u^2) = 0$ in the *upper* half plane. Thus

$$\mathcal{L}M = -\frac{i}{p} + \frac{i\sqrt{1-ip}}{p}, \tag{2.5}$$

with the branch satisfying $\sqrt{1-ip} \to 1$ as $p \to 0$ in $\mathbb{H}$. As p varies in $\mathbb{H}$, $1 - ip$ belongs to the lower half plane $-i\mathbb{H}$ and then $\sqrt{1-ip}$ varies in the fourth quadrant.

For $\Re(p) > 0, \omega > 0$ we have

$$\mathcal{L}(e^{\pm i\omega}M) = -\frac{i}{p \mp i\omega} + \frac{i\sqrt{1 - ip \mp \omega}}{p \mp i\omega}$$

(with $\sqrt{1 - ip - \omega} = -i\sqrt{\omega - 1 + ip}$ if $\omega > 1$)

and relation (2.1) follows. □

After the substitution $y(p) = 2(\sqrt{1-ip} - 1)e^{-(\pi p)/(2\omega)}v(p)$ we get

$$v(p - i\omega) + v(p + i\omega) = \frac{2}{r}(\sqrt{1-ip} - 1)v(p) + \frac{i\omega}{\omega^2 + p^2}. \tag{2.6}$$

REMARK 2.2. It is clear that the functional equation (2.6) only links the points on one dimensional lattice $\{p + i\mathbb{Z}\omega\}$. It is convenient to take p_0 such that $p = p_0 + in\omega$ with $\Re(p_0) = \Re(p)$ and

$$\Im(p_0) \in [0, \omega), \tag{2.7}$$

and write $v(p) = v(p_0 + in\omega) = v_n$ which transforms (2.6) to a recurrence relation:

$$v_{n+1} + v_{n-1} = \frac{2}{r}(\sqrt{1 - ip_0 + n\omega} - 1)v_n + \frac{i\omega}{\omega^2 + (p_0 + in\omega)^2}, \tag{2.8}$$

where v_n depends parametrically on p_0. It will be seen that the asymptotic conditions as well as analyticity in p_0 determine the solution of (2.8) uniquely.

REMARK 2.3. The approach is based on a discrete analog of the Wronskian technique. The regularity of the bounded solution of (2.8) will be a consequence of the absence of a bounded solution of the homogeneous equation

$$v_{n+1} + v_{n-1} = \frac{2}{r}(\sqrt{1 - ip_0 + n\omega} - 1)v_n = D_n v_n, \tag{2.9}$$

a problem which we analyze first.

PROPOSITION 2.4. *For p_0 satisfying (2.7) and $\Re(p_0) \geq 0$ (actually for any $p_0 \in \overline{\mathbb{H}} = \mathbb{H} \cup i\mathbb{R}$) there is no nonzero solution of (2.9) such that $v \in l_2(\mathbb{Z})$.*

PROOF. To get a contradiction, assume $v \not\equiv 0$ is an $l_2(\mathbb{Z})$ solution of (2.9). Multiplying (2.9) by $\overline{v_n}$, and summing with respect to n from $-\infty$ to $+\infty$ we get

$$\sum_{n=-\infty}^{\infty} v_{n+1}\overline{v}_n + \sum_{n=-\infty}^{\infty} v_{n-1}\overline{v}_n \tag{2.10}$$

$$= 2\sum_{n=-\infty}^{\infty} \Re(v_n \overline{v}_{n+1}) = \sum_{n=-\infty}^{\infty} \frac{2}{r}(\sqrt{1 - ip_0 + n\omega} - 1)|v|_n^2.$$

For $p_0 \in \overline{\mathbb{H}}$ the imaginary part of $\sqrt{1 - ip_0 + n\omega}$ is nonpositive, by Lemma 2.1, and is strictly negative for $n < 0$ large enough. Thus if for some such n, v_n is nonzero then the last sum in (2.10) has a strictly negative imaginary part, which is impossible since the left side is real. If on the other hand v_n is zero when n is large negative, then solving (2.9) for v_{n+1} in terms of the v_n, v_{n-1} it would follow inductively that $v \equiv 0$, contradicting the assumption. $\square$

LEMMA 2.5. (i) *There is, up to multiplicative constants, a unique pair of solutions v^+ and v^- of (2.9) such that $v_n^\pm \to 0$ as $n \to \pm\infty$ (respectively). These solutions are related to convergent continued fractions representations:*

$$(2.11) \qquad v_{n\mp 1}^\pm / v_n^\pm =: \frac{1}{\rho_n^\pm} = D_n - \frac{1}{D_{n\pm 1} - 1/(D_{n\pm 2})\cdots}.$$

(ii) *We have the following estimates*

$$(2.12) \qquad \frac{1}{\rho_n^\pm} = \frac{1}{\tilde{\rho}_n^\pm} + O(n^{-3/2}) \quad (n \to \pm\infty)$$

where

$$(2.13) \qquad \frac{1}{\tilde{\rho}_n^+} = \frac{2}{r}\sqrt{n\omega} - \frac{2}{r} - \frac{r^2 - 2 + 2ip_0}{2r\sqrt{n\omega}} - \frac{r}{2\omega n} \quad (n > 0),$$

$$\frac{1}{\tilde{\rho}_n^-} = -\frac{2i}{r}\sqrt{|n|\omega} - \frac{2}{r} + \frac{(2-r^2)i + 2p_0}{2r\sqrt{|n|\omega}} + \frac{r}{2\omega|n|} \quad (n < 0).$$

Let $\tilde{v}_n^\pm$ be solutions of the one step recurrences $\tilde{v}_n^\pm = \tilde{v}_{n\mp 1}^\pm \tilde{\rho}_n^\pm$. Then

$$(2.14) \quad \ln \tilde{v}_n^+ = -\frac{1}{2} n \ln n + n \ln\left(\frac{r}{2}\sqrt{\frac{e}{\omega}}\right)$$
$$+ 2\sqrt{\frac{n}{\omega}} + \left(\frac{2ip_0 + r^2 + \omega}{4\omega}\right) \ln n + o(1) \quad (n \to \infty),$$

and

$$(2.15) \quad \ln(\tilde{v}_n^-) = -\frac{1}{2}|n| \ln |n| + |n| \ln\left(\frac{r}{2}\sqrt{\frac{e}{\omega}}\right) + i\pi|n| - 2i\sqrt{|n|/\omega}$$
$$+ \left(\frac{2ip_0 + r^2 + \omega}{4\omega}\right) \ln |n| + o(1) \quad (n \to -\infty),$$

and, for some constants $K^\pm$,

$$(2.16) \qquad \ln(v_n^\pm) = \ln(\tilde{v}_n^\pm) + K^\pm + o(1),$$

($v_n^\pm$ *decay roughly as $1/\sqrt{|n|!}$ for $n \to \pm\infty$, respectively).*

(iii) *Two special solutions of (2.9), v^+ and v^-, are well defined by:*

$$(2.17) \quad v_n^+ = \tilde{v}_n^+ \prod_{j \geq n+1} \frac{\tilde{\rho}_j^+}{\rho_j^+} \text{ for } n > N, \quad \text{and} \quad v_n^- = \tilde{v}_n^- \prod_{j \leq n-1} \frac{\tilde{\rho}_j^-}{\rho_j^-} \text{ for } n < -N,$$

if N is sufficiently large (this amounts to making a convenient choice of the free multiplicative constant in (i)). These functions do not depend on N. v^+ and v^- are linearly independent for $p_0 \in \overline{\mathbb{H}}$: their discrete Wronskian, defined by $W(v^+, v^-)_n = v_n^+ v_{n+1}^- - v_n^- v_{n+1}^+$, satisfies

$$(2.18) \qquad W(v^+, v^-) = \text{const} \neq 0.$$

As functions of parameters, $v^\pm$ and $W(v^+, v^-)$ are analytic in $p_0 \in \mathbb{H}$. If $\omega \notin \{0, n^{-1} : n \in \mathbb{N}\}$ then $v^\pm$ and $W(v^+, v^-)$ are analytic in some neighborhood of $p_0 = 0$ as well. For any $\omega > 0$, $v_n^\pm$ are Lipschitz continuous of exponent at least $1/2$ in p_0, for $p_0 \in \mathbb{R}$.

PROOF. (i) We look at v^+, the case of v^- being similar. Dropping the $+$ superscript we have from (2.9)

$$\rho_n = \frac{1}{D_n - \rho_{n+1}}. \tag{2.19}$$

To find the analytic properties of the solution ρ_n it is convenient to regard (2.19) as a contractive equation in the space $\ell^\infty(S_N)$ of sequences $\{\rho_j\}_{j>N}$ in the norm $\|\rho\|_\infty = \sup_{j>N} |\rho_j|$. Let N be large. The map $J \colon S_N \mapsto S_N$ defined by

$$J(\rho)_n = \frac{1}{D_n - \rho_{n+1}}, \tag{2.20}$$

depends analytically on $p_0 \in \mathbb{H}$ and is Lipschitz continuous of exponent at least $1/2$ if $\Re(p_0) \geq 0$. In addition, if $\|\rho_j\|_\infty \leq 1$ we have for sufficiently large $N = N(p_0, \omega, r)$

$$\|J(\rho)\|_\infty \leq \frac{1}{[2/r(\sqrt{|N\omega| - 1 - |p_0|} - 1) - 1]} < \frac{|r|}{|\omega|^{1/2}} \frac{1}{\sqrt{N}}. \tag{2.21}$$

Similarly,

$$\|J(\rho) - J(\rho')\|_\infty \leq \frac{\|\rho - \rho'\|_\infty}{[2/r(\sqrt{|N\omega| - 1 - |p_0|} - 1) - 1]^2} < \frac{|r|^2}{N|\omega|} \|\rho - \rho'\|_\infty, \tag{2.22}$$

for sufficiently large N which shows that J is contractive in the unit ball in $\ell^\infty(S_N)$. Thus, equation (2.20) has a unique solution in S_N, which depends analytically on $p_0 \in \mathbb{H}$ and is Lipschitz continuous of exponent at least $1/2$ if $\Re(p_0) \geq 0$. This also implies the convergence of (2.11).

Note that given $\mathcal{K}_1 \subset \overline{\mathbb{H}}$ and $\mathcal{K}_2 \subset \mathbb{R}^+$ both compact, N can be chosen the same for all $p_0 \in \mathcal{K}_1$ and $r \in \mathcal{K}_2$.

(ii) From (2.21) it is seen that $|\rho_j| = O(j^{-1/2})$ for large j. Thus, we may write, for large j,

$$\frac{1}{\rho_j^+} = D_j - \frac{1}{D_{j+1} - 1/(D_{j+2} + O(j^{-1/2}))}. \tag{2.23}$$

The estimates (2.12) now follow by a straightforward calculation. Since $\ln v_n^+ = \ln v_N^+ + \sum_{j=N+1}^n \ln \rho_j^+$, the estimates follow from (2.23) and the Euler–Maclaurin summation formula.

(iii) As before, we only need to look at v^+. We take two compact sets $\mathcal{K}_1$ and $\mathcal{K}_2$, and choose N as in the note at the end of the proof of (i). Taking the log in the definition (2.17), the infinite sums are absolutely convergent. By standard measure theory, v_n^+ has the same analyticity properties in the interior of $\mathcal{K}_1 \times \mathcal{K}_2$ and Lipschitz continuity in $\mathcal{K}_1 \times \mathcal{K}_2$ as those of ρ^+, when $n > N$. Now, (2.9) easily implies that the same is true for $n \leq N$ as well.

If f_n and g_n are solutions of (2.9) then $(g_{n+1} + g_{n-1})f_n - (f_{n+1} + f_{n-1})g_n = 0$ and thus $W_n(f,g) = f_n g_{n+1} - g_n f_{n+1} = \text{const}$. Thus, if $W_n(f,g) = 0$ for some n then $W_n \equiv 0$ and $f \equiv \text{const}\, g$. The smoothness properties follow from the proof of (iii). □

PROPOSITION 2.6. *There exists a unique solution of* (2.6) *which is bounded as* $\Im(p) \to \pm\infty$ *in* $\mathbb{H}$. *This solution is analytic in* $p \in \mathbb{H}$, *and* $v(p) = O(p^{-2})$ *as* $\Im(p) \to \pm\infty, p \in \mathbb{H}$.

PROOF. By analyticity and continuity $W(v^+, v^-)$ does not vanish for any $p \in \mathbb{H}$ and $r > 0, \omega > 0$. By Lemma 3.1 the function v defined through $v(p_0 + in\omega) = f_n$, where

$$(2.24) \qquad f_n := W(v^+, v^-)^{-1} \left(v_n^+ \sum_{l=-\infty}^{n-1} v_l^- H_l + v_n^- \sum_{l=n}^{\infty} v_l^+ H_l \right),$$

and

$$(2.25) \qquad H_n = \frac{i\omega \exp((\pi p_n)/(2\omega))}{p_n^2 + \omega^2},$$

has the required properties. Since no solution of the homogeneous equation is bounded on $\mathbb{Z}$, v is the unique solution with the desired properties. □

NOTE 2.7. The link between y and f_n is

$$(2.26) \qquad y(p) = 2(\sqrt{1 - ip} - 1)e^{-(\pi p)/(2\omega)} f_n; \quad \text{for } p = p_0 + in\omega.$$

PROPOSITION 2.8. *The function $y(p)$ is analytic in the right half plane, Lipschitz continuous of exponent at least $1/2$ on the imaginary axis and $\lim_{p \to 0} y(p) = i/2$.*

PROOF. Since W is analytic in $\mathbb{H}$, continuous and nonzero in $\overline{\mathbb{H}}$, W is bounded below in compact sets in $\overline{\mathbb{H}}$. Then, the smoothness properties of y derive easily from those of $q_n := W(v^+, v^-) f_n$ on which we concentrate now.

(a) For $n \geq 2$ we write, using (2.25),

$$(2.27) \quad q_n = v_n^+ \sum_{\substack{l=-\infty \\ l \neq \pm 1}}^{n-1} v_l^- H_l + v_n^- \sum_{l=n}^{\infty} v_l^+ H_l$$
$$+ v_n^+ i\omega e^{(\pi p_0)/(2\omega)} \left(\frac{-iv_{-1}^-}{p_0(p_0 - 2i\omega)} + \frac{iv_1^-}{p_0(p_0 + 2i\omega)} \right).$$

The last term in parenthesis can be rewritten, using also (2.9), as

$$(2.28) \quad \frac{i(v_1^- - v_{-1}^-)}{p_0^2 + 4\omega^2} + \frac{2\omega}{p_0^2 + 4\omega^2} \left(\frac{v_1^- + v_{-1}^-}{p_0} \right)$$
$$= \frac{i(v_1^- - v_{-1}^-)}{p_0^2 + 4\omega^2} + \frac{4\omega}{r(p_0^2 + 4\omega^2)} \frac{\sqrt{1 - ip_0} - 1}{p_0} v_0^-.$$

Thus we see that q_n is continuous as $\Re(p_0) \to 0$ and $\Im(p_0) \in [0, \omega)$ [cf. (2.7)], if $n \geq 2$. A very similar calculation shows the continuity of q_n if $n \leq -1$.

(b) By part (a), $y(p)$ is continuous as $\Re(p) \downarrow 0$ with $\Im(p) \geq 2$ or $\Im(p) < 0$. Now, (2.1) written in the form

$$(2.29) \quad rph_2(p)y(p) = rp(h_1(p + 2i\omega) - h_1(p))$$
$$+ rp(y(p + i\omega) + h_2(p + 2i\omega)y(p + 2i\omega)),$$

shows that $y(p)$ is Lipschitz continuous as $\Re(p) \downarrow 0$ if $\Im(p) > -2$ thus for all $\Im(p)$. The value of $y(0)$ is easily calculated using (2.29). □

PROPOSITION 2.9. $1 + 2i \lim_{x \to \infty} \int_0^x Y(s) \, ds = 0.$

PROOF. Indeed,

$$
\begin{aligned}
(2.30)\quad 2\pi i \int_0^\infty Y(s)\,ds \\
= \lim_{x\to\infty}\lim_{\delta\to 0^+}\left(\int_{-i\infty}^{-i\delta}+\int_{i\delta}^{i\infty}\right)\frac{e^{xp}}{p}\bigl(i/2+\bigl(y(p)-i/2\bigr)\bigr)\,dp = -\pi.
\end{aligned}
$$

□

3. Appendix

LEMMA 3.1. *Equation* (1.6) *has a unique solution* $Y \in L^1_{\mathrm{loc}}(\mathbb{R}^+)$ *and* $|Y(x)| < Ke^{Cx}$ *for some* $K \in \mathbb{R}^+$ *and* $C \in \mathbb{R}$.

PROOF. Consider $L^1_{\mathrm{loc}}[0,A]$ endowed with the norm $\|F\|_\nu := \int_0^A |F(s)| e^{-\nu s}\,ds$, where $\nu > 0$. If f is continuous and $F, G \in L^1_{\mathrm{loc}}[0,A]$, a straightforward calculation shows that

$$(3.1)\qquad \|fF\|_\nu < \|F\|_\nu \sup_{[0,A]}|f|,$$

$$(3.2)\qquad \|F * G\|_\nu < \|F\|_\nu \|G\|_\nu,$$

$$(3.3)\qquad \|F\|_\nu \to 0 \text{ as } \nu \to \infty,$$

where the last relation follows from the Riemann–Lebesgue lemma.

The integral equation (1.6) can be written as

$$(3.4)\qquad Y = r\eta + \mathcal{J}Y \text{ where } \mathcal{J}F := r\eta(2i + M) * F.$$

Since M is locally in L^1 and bounded for large x it is clear that for large enough C_2, and for any A, (1.6) is contractive if $\nu > C_2$. □

References

1. S. Albeverio, F. Gesztesy, R. Høegh-Krohn, and H. Holden, *Solvable models in quantum mechanics*, Texts and Monographs in Physics, Springer-Verlag, New York–Berlin, 1988.
2. A. Buchleitner, D. Delande, and J.-C. Gay, *Microwave ionization of three dimensional hydrogen atoms in a realistic numerical experiment*, J. Opt. Soc. Amer. B Opt. Phys. **12** (1995), 505.
3. S. L. Chin and P. Lambropoulos (eds.), *Multiphoton ionization of atoms*, Academic Press, Toronto, New York, 1984.
4. C. Cohen-Tannoudji, J. Duport-Roc, and G. Arynberg, *Atom-photon interactions: Basic processes and applications*, J. Wiley, New York, 1992.
5. O. Costin, J. L. Lebowitz, and A. Rokhlenko, *Exact results for the ionization of a model quantum system*, J. Phys. A **33** (2000), 1–9.
6. H. L. Cycon, R. G. Froese, W. Kirsch, and B. Simon, *Schrödinger operators*, Springer-Verlag, Berlin–New York, 1987.
7. Yu. N. Demkov and V. N. Ostrovskii, *Zero range potentials and their application in atomic physics*, Plenum, 1988.
8. C. Figueira de Morisson Faria, A. Fring, and R. Schrader, *On the influence of pulse shapes on ionization probability*, J. Phys B **31** (1998), 449–464.
9. _____, *Analytical treatment of stabilization*, Laser Phys. **9** (1999), no. 1, 379–387.
10. A. Fring, V. Kostrykin, and R. Schrader, *On the absence of bound-state stabilization through short ultra-intense fields*, J. Phys. B **29** (1996), 5651–5671.
11. _____, *Ionization probabilities through ultra-intense fields in the extreme limit*, J. Phys. A **30** (1997), 8559–8610.
12. S. Geltman, *Ionization of a model atom by a pulse of coherent radiation*, J. Phys. B **5** (1977), 831–840.
13. M. Holthaus and B. Just, *Generalized π pulses*, Phys. Rev. A **49** (1994), 1950–1960.

14. P. M. Koch and K. A. H. van Leeuwen, *The importance of resonances in microwave "ionization" of excited hydrogen atoms*, Phys. Rep. **255** (1995), 289–403.
15. A. Maquet, I.-C. Shih, and W. P. Reinhardt, *Stark ionization of DC and AC fields: An L^2 complex-coordinate approach*, Phys. Rev. A **27** (1983), 2946–2970.
16. C.-A. Pillet, *Some results on the quantum dynamics of a particle in a Markovian potential*, Comm. Math. Phys. **102** (1985), 237–254.
17. _____, *Asymptotic completeness for a quantum particle in a Markovian short range potential*, Comm. Math. Phys. **105** (1986), 259–280.
18. R. M. Potvliege and R. Shakeshaft, *Multiphoton processes in an intense laser field: Harmonic generation and total ionization rates for atomic hydrogen*, Phys. Rev. A **40** (1989), 3061–3079.
19. A. Rokhlenko and J. L. Lebowitz, *Ionization of a model atom by perturbations of the potential*, J. Math. Phys. **41** (2000), 3511–3522.
20. G. Scharf, K. Sonnenmoser, and W. F. Wreszinski, *Sensitive multiphoto ionization*, Phys. Rev. A **44** (1991), 3250–3265.
21. A. Soffer and M. I. Weinstein, *Nonautonomous Hamiltonians*, J. Statist. Phys. **93** (1998), 359–391.
22. S. M. Susskind, S. C. Cowley, and E. J. Valeo, *Multiphoton ionization in a short range potential: A nonperturbative approach*, Phys. Rev. A **42** (1994), 3090–3106.
23. K. Yajima, *Existence of solutions for Schrödinger evolution equations*, Comm. Math. Phys. **89** (1982), 331–352.

DEPARTMENT OF MATHEMATICS, RUTGERS UNIVERSITY, PISCATAWAY, NJ 08854-8019, USA.
E-mail address, O. Costin: `costin@math.rutgers.edu`

DEPARTMENT OF MATHEMATICS AND DEPARTMENT OF PHYSICS, RUTGERS UNIVERSITY, PISCATAWAY, NJ 08854-8019, USA.
E-mail address, J. L. Lebowitz: `lebowitz@math.rutgers.edu`

DEPARTMENT OF MATHEMATICS, RUTGERS UNIVERSITY, PISCATAWAY, NJ 08854-8019, USA.
E-mail address, A. Rokhlenko: `rokhlenko@math.rutgers.edu`

The sine-Gordon Model at $\beta = 4\pi$

J. Dimock

ABSTRACT. We consider the two dimensional sine-Gordon model with parameters (ζ, β). This is a Euclidean quantum field theory with coupling constant ζ and field strength β. It is also describes the classical statistical mechanics of a Coulomb gas with activity ζ and inverse temperature β. The model is analyzed by a renormalization group technique. For $\beta < 8\pi$ and finite volume one can control the ultraviolet divergences and prove that the correlation functions are analytic in ζ around $\zeta = 0$. Hence perturbation theory converges. This fact is used to prove that the model at $\beta = 4\pi$ is equivalent to a theory of massive free fermions.

1. The sine-Gordon Model

We discuss an application of renormalization group techniques to the sine-Gordon model on $\mathbb{R}^2$. We start by giving a formal definition of the model. Let $\{\phi(x)\}$ be a family of Gaussian random variables indexed by $x \in \mathbb{R}^2$ with mean zero and covariance βv where $\beta > 0$ and

(1.1) $$v(x-y) = (-\Delta)^{-1}(x-y) = -\frac{1}{2\pi}\log|x-y|,$$

is the kernel of the inverse Laplacian. Thus we suppose there is a measure $\mu_{\beta v}$, on functions $\phi(x)$ such that for any choice of points $x_1, \ldots, x_n \in \mathbb{R}^2$ and real numbers $\alpha_1, \ldots, \alpha_n$ we have

(1.2) $$\int \exp\left(i\sum_j \alpha_j \phi(x_j)\right) d\mu_{\beta v}(\phi) = \exp\left(-\frac{1}{2}\sum_{j,k} \alpha_j \alpha_k \beta v(x_j - x_k)\right).$$

This is the massless free field. We add an interaction with a potential

(1.3) $$V(\phi) = \int \cos(\phi(x))\, dx.$$

2000 *Mathematics Subject Classification.* Primary: 81T17; Secondary: 81T08, 81T10, 81B28.

I would like to thank Tom Hurd for a long and rewarding collaboration on the sine-Gordon model.

The research is partially supported by NSF Grant PHY9722045.

This is the final form of the paper.

©2001 American Mathematical Society

The problem is to study the measure $\exp(\zeta V)\,d\mu_{\beta v}$ for ζ small. One is interested in the total mass

$$\text{(1.4)} \qquad Z = \int \exp\bigl(\zeta V(\phi)\bigr)\,d\mu_{\beta v}(\phi),$$

and in correlation functions such as

$$\text{(1.5)} \qquad Z^{-1} \int \exp\!\left(i \sum_j \alpha_j \phi(x_j)\right) \exp\bigl(\zeta V(\phi)\bigr)\,d\mu_{\beta v}(\phi).$$

In fact it is convenient to study the convolution

$$\text{(1.6)} \qquad \bigl(\mu_{\beta v} * \exp(\zeta V)\bigr)(\phi) = \int \exp\bigl(\zeta V(\phi + \phi')\bigr)\,d\mu_{\beta v}(\phi').$$

Specializing to $\phi = 0$ gives Z, and taking functional derivatives of the logarithm gives at least some of the correlation functions.

The model is of interest for two distinct problems in physics. On the one hand it is a quantum field theory on a two dimensional Euclidean spacetime, i.e. time is imaginary. To get a true quantum field theory on Minkowski spacetime it suffices to take each point $x_i = (x_{i,0}, x_{i,1})$ in the correlation functions and analytically continue the first components $x_{i,0}$ to imaginary values corresponding to real time. Then one has vacuum expectation values of quantum field operators $e^{i\alpha\phi(x)}$ where $\phi(x)$ satisfies the massless sine-Gordon equation $\beta^{-1}\Box\phi = \zeta \sin\phi$. In this interpretation ζ is a coupling constant and β is a field strength.

On the other hand the model describes the classical statistical mechanics of a Coulomb gas in two dimensional space. The covariance $v(x-y)$ is the Coulomb potential in two dimensions, and if one expands Z in a power series in ζ and carries out the integrals over ϕ one can identify Z as the partition function in the grand canonical ensemble. In this interpretation β is the inverse temperature and $\zeta/2$ is the activity.

The integrals we have quoted to define the model are not well defined. There is a short distance problem connected with the fact that the covariance $v(x-y)$ is singular at $x = y$. This means the measure is not supported on continuous functions, but only on distributions. Thus $\cos(\phi(x))$ is not well defined. There is also a long distance problem connected with the fact that $v(x-y)$ does not decay as $|x-y| \to \infty$. This means the values of the field at distant points are strongly correlated and hence there is major trouble with the definition of integrals like (1.5).

2. The Renormalization Group

We now explain how these problems can be attacked using the renormalization group as formulated by Wilson. First note that the covariance can be formally written as

$$\text{(2.1)} \qquad v(x-y) = \sum_{n=-\infty}^{\infty} C\bigl((x-y)/L^n\bigr),$$

where

$$\text{(2.2)} \qquad C(x-y) = (2\pi)^{-2} \int \frac{e^{ip(x-y)}}{p^2}\bigl(e^{-p^2} - e^{-L^2 p^2}\bigr)\,dp.$$

Then $C\bigl((x-y)/L^n\bigr)$ is the Fourier transform of $p^{-2}(e^{-L^{2n}p^2} - e^{-L^{2n+2}p^2})$ and so is concentrated on the frequencies $L^{-n-1} < |p| < L^{-n}$. As $n \to \infty$ we have the low frequency (infrared, long distance) modes, and as $n \to -\infty$ we have high frequency (ultraviolet, short distance) modes. The theory can now be regularized by replacing $v(x-y)$ above by

$$(2.3) \qquad v_M^N(x-y) = \sum_{n=-N}^{M} C\big((x-y)/L^n\big).$$

Since C is exponentially decaying and smooth the same is true for v_M^N and hence the model with this covariance is well defined. The problem is now to study the limits $N \to \infty$ and/or $M \to \infty$, or at least prove bounds uniform in these limits.

Now suppose we want to study an integral of the form $\mu_{\beta v_M^N} * \mathcal{Z}$ for some function $\mathcal{Z}$. Because of the decomposition of v_M^N the convolution can be understood as the effect of repeatedly taking the convolution with $\mu_{\beta C}$ and scaling. This operation is called a renormalization group (RG) transformation and it is given by

$$(2.4) \qquad \mathcal{Z}'(\phi) = (\mu_{\beta C} * \mathcal{Z})\big(\phi(\cdot/L)\big).$$

To control the integral we have to control the flow of iterated RG transformations.

The case of interest is $\mathcal{Z} = e^{\zeta V}$. When we apply the RG transformation the dominant effect to lowest order in ζ is just to change the coupling constant. We find formally

$$(2.5) \qquad \mathcal{Z}' = \exp\big(\zeta V' + \mathcal{O}(\zeta^2)\big) = \exp\big(\zeta' V + \mathcal{O}(\zeta^2)\big),$$

where the new coupling constant is

$$(2.6) \qquad \zeta' = L^2 e^{-\beta C(0)/2}\zeta = L^{2-\beta/4\pi}\zeta.$$

Here the L^2 comes from the scaling and the $e^{-\beta C(0)/2}$ comes from the calculation $\mu_{\beta C} * V = e^{-\beta C(0)/2} V$.

If $\beta > 8\pi$ then $L^{2-\beta/4\pi} < 1$ and $\zeta' < \zeta$. When the transformation is iterated the coupling constant is driven to zero. i.e. $\zeta = 0$ is an attractive fixed point. Each RG transformation can be thought of as integrating out some short distance modes so the effect of many such transformations is to isolate a effective interaction for the long distance modes. Since the effective interaction tends to zero the long distance behavior is approximately free. This means that long distance problems are amenable to analysis. One refers to the situation as infrared asymptotic freedom. This regime is treated in [4, 6] and is discussed in the lecture by Tom Hurd.

On the other hand if $\beta < 8\pi$ then $L^{2-\beta/4\pi} > 1$ and $\zeta' > \zeta$. Now $\zeta = 0$ is an repulsive fixed point. In this case it turns out that the short distance behavior of the theory is approximately free and can be analyzed. One refers to this situation as ultraviolet asymptotic freedom. We now discuss this regime in more detail.

3. The Ultraviolet Problem for $\beta < 8\pi$

We focus on the short distance modes. Thus we consider the covariance and the potential

$$(3.1) \qquad v^N(x) = \sum_{n=-N}^{0} C\big((x-y)/L^n\big), \quad V^N(\phi) = \int_{|x| \leq 1} :\cos\big(\phi(x)\big):_{\beta v^N} dx.$$

The potential has been restricted to a finite volume, for simplicity a unit square. Also we have replaced the $\cos\phi(x)$ by the Wick ordered version defined by

$$(3.2) \qquad :\cos\big(\phi(x)\big):_{\beta v^N} = \exp\big(\beta v^N(0)/2\big)\cos\big(\phi(x)\big).$$

The Wick ordered function satisfies $\int :\cos(\phi(x)):_{\beta v^N} d\mu_{\beta v^N}(\phi) = 1$. This renormalization is necessary in order to have a nontrivial limit. Our problem is now to study the function

$$(3.3) \qquad (\mu_{\beta v^N} * e^{\zeta V^N})(\phi),$$

as $N \to \infty$.

The first step is to scale up by a factor L^N. The covariance and the potential become

$$(3.4) \qquad v_N(x) = \sum_{n=0}^{N} C\big((x-y)/L^n\big), \quad V_N(\phi) = \int_{|x| \leq L^N} \cos(\phi(x)) \, dx,$$

and we want to study

$$(3.5) \qquad (\mu_{\beta v_N} * e^{\zeta_N V_N})(\phi(\cdot/L^N)),$$

as $N \to \infty$. The Wick ordering constant has been absorbed into the coupling constant which is now

$$(3.6) \qquad \zeta_N = L^{-2N} \exp(\beta v^N(0)/2)\zeta = L^{-(2-\beta/4\pi)N}\zeta.$$

This scaling changes our ultraviolet problem to an infrared problem. It is a good thing to do since infrared problems are the natural home of the renormalization group. Note the the coupling constant ζ_N is very tiny for $\beta < 8\pi$. This is just what is needed to survive the explosive growth under the RG transformations.

We write the integral as iterated RG transformations. Starting with $\mathcal{Z}_N = e^{\zeta_N V_N}$ we define $\mathcal{Z}_j$ for $j = N, N-1, \ldots, 0$ by

$$(3.7) \qquad \mathcal{Z}_{j-1}(\phi) = (\mu_{\beta C} * \mathcal{Z}_j)(\phi(\cdot/L)).$$

We would like to claim that for all j we have something like

$$(3.8) \qquad \mathcal{Z}_j(\phi) = \exp\big(\zeta_j V_j(\phi) + \mathcal{O}(\zeta_j^2) L^{2j}\big),$$

where the term $\mathcal{O}(\zeta_j^2)$ is bounded uniformly in N. The quantity of interest is $\mathcal{Z}_0(\phi)$ and this result would show that it is uniformly bounded away from 0 and ∞.

However this bound is too simple and it will not iterate. One must have more control over the structure of the remainder $\mathcal{O}(\zeta_j^2)$. In particular one must keep track of possible growth in ϕ and especially its local structure. We do this by writing a "polymer expansion" for it, following the general framework of Brydges and Yau [1].

A polymer X is defined to be a union of unit squares, not necessarily connected. A polymer activity $K(X, \phi)$ is a function on polymers X and fields ϕ such that $K(X, \phi)$ depends only on the restriction of ϕ to X. For example $V(X, \phi) = \int_X \cos(\phi(x)) \, dx$ is a polymer activity.

Then one can prove the following:

THEOREM 3.1. *For $\beta < 8\pi$, $|\zeta|$ sufficiently small, and $j = N, N-1, \ldots, 0$ there exist polymer activities $K_j(X, \phi)$ such that*

$$(3.9) \qquad \mathcal{Z}_j(\phi) = e^{\mathcal{E}_j} \sum_{\{X_i\}} \prod_i K_j(X_i, \phi),$$

where the sum is over collections of disjoint polymers $\{X_i\}$ contained in the square $|x| \leq L^j$, and where $K_j(X, \phi)$ has the form

$$(3.10) \qquad K_j(X, \phi) = \left\{ \begin{array}{ll} \zeta_j V(X, \phi) & \text{if } |X| = 1 \\ 0 & \text{if } |X| > 1 \end{array} \right\} + \widetilde{K}_j(X, \phi).$$

The remainder $\widetilde{K}_j(X, \phi)$ is analytic in ζ in a neighborhood of the origin and is $\mathcal{O}(|\zeta_j|^2)$ uniformly in N. Finally the constant $\mathcal{E}_j$ is bounded in N for $\beta < 4\pi$ and is unbounded in N for $\beta \geq 4\pi$.

This result is due to Dimock and Hurd [5, 6].[1] The second reference contains the better proof and corrects some errors in the first reference. The statement of the theorem here is still not completely precise. The bound on $\widetilde{K}_j(X, \phi)$ must also control the decay in X when X is large or widely separated, and it must control possible the growth in ϕ. Then one can give an inductive proof of the theorem starting at $j = N$ and working down to $j = 0$.

The above is the main technical result. As a consequence one can prove uniform bounds and/or limits for the correlation functions as $N \to \infty$. One can also characterize the short distance behavior of correlation functions. Finally one can show that the correlation functions are analytic in ζ in a neighborhood of zero and hence that the perturbation expansion for the these functions converges. For details about these results see [3, 5, 8].

4. Boson–Fermion Equivalence

As an application one can show that the sine-Gordon model at $\beta = 4\pi$ is equivalent to a model of massive free fermions. This has been widely conjectured, but only recently proved [3].

We begin with a discussion of boson-fermion equivalence for massless free fields. The original insight is due to Coleman [2]. The result is roughly that in two dimensions the massless free fermi field ψ is equivalent to the massless free boson field ϕ if we make the identifications $\overline{\psi}\gamma_5\psi \leftrightarrow \sin\phi$ and $\overline{\psi}\psi \leftrightarrow \cos\phi$. More precisely let $v = (-\Delta)^{-1}$ as before and let $v_1 = (-\Delta + 1)^{-1}$. Then there is a constant γ such that for noncoinciding points

$$(4.1) \quad \int \prod_{j=1}^{n} :\sin\phi(x_j):_{4\pi v_1} \prod_{k=1}^{m} :\cos\phi(y_k):_{4\pi v_1} d\mu_{4\pi v}(\phi)$$
$$= (-i\gamma)^n \gamma^m \int \prod_{j=1}^{n} (\overline{\psi}\gamma_5\psi)(x_j) \prod_{k=1}^{m} (\overline{\psi}\psi)(y_k) \, d\mu_{(\not\partial)^{-1}}(\psi).$$

On the left note the mismatch between the Wick ordering and the covariance. On the right we have a fermion Gaussian integral of the type discussed by Feldman in these proceedings. The operator $\not\partial$ is the usual Dirac operator. The proof of this formula is an explicit computation based on the identity

$$(4.2) \qquad\qquad e^{4\pi v(x-y)} = |x - y|^{-2}.$$

This leads to the following result for the sine-Gordon model.

---

[1]The problem is formulated somewhat differently in [5, 6]. Instead of a square in $\mathbb{R}^2$ the model is defined on a unit torus. Also instead of $\sum_{n=-N}^{0} C\big((x-y)/L^n\big)$ as the covariance, all the infrared modes are included and one takes $\sum_{n=-N}^{\infty} C\big((x-y)/L^n\big)$.

THEOREM 4.1. *For any $M > 0$ and $|\zeta|$ sufficiently small,*

$$(4.3) \quad \int \prod_{j=1}^n :\!\sin\phi(x_j)\!:_{4\pi v_1} \exp\left(\zeta \int_{|y|\leq M} :\!\cos\phi(y)\!:_{4\pi v_1} dy\right) d\mu_{4\pi v}(\phi)/(n=0)$$

$$= (-i\gamma)^n \int \prod_{j=1}^n (\overline{\psi}\gamma_5\psi)(x_j) \exp\left(\gamma\zeta \int_{|y|\leq M} (\overline{\psi}\psi)(y)\,dy\right) d\mu_{(\slashed{\partial})^{-1}}(\psi)/(n=0).$$

The proof is given in [**3**]. There it is shown that both sides of the equation are well-defined and analytic in ζ in a neighborhood of the origin. For the left side this is a consequence of Theorem 3.1. The free field result is used to show that both sides have the same power series in ζ. Hence they are equal.

There is a complimentary result for sine-Gordon for $\beta < 4\pi$ due to Fröhlich and Seiler [**7**] . They prove equivalence with a Thirring–Schwinger model of interacting fermions.

References

1. D. C. Brydges and H. T. Yau, *Grad ϕ perturbations of massless Gaussian fields*, Comm. Math. Phys. **129** (1990), 351–392.
2. S. Coleman, *Quantum sine-Gordon equation as the massive Thirring model*, Phys. Rev. D **11** (1975), 2088–2097.
3. J. Dimock, *Bosonization of massive fermions*, Comm. Math. Phys. **198** (1998), 247–281.
4. J. Dimock and T. R. Hurd, *A renormalization group analysis of the Kosterlitz–Thouless phase*, Comm. Math. Phys. **137** (1991), 263–287.
5. _____, *Construction of the two-dimensional sine-Gordon model for $\beta < 8\pi$*, Comm. Math. Phys. **156** (1993), 547–580.
6. _____, *sine-Gordon revisited*, Ann. Henri Poincaré **1** (2000), no. 3, 499–541.
7. J. Fröhlich and E. Seiler, *The massive Thirring–Schwinger model (QED_2): convergence of perturbation theory and particle structure*, Helv. Phys. Acta **49** (1976), 889–924.
8. T. R. Hurd, *Charge correlations for the two-dimensional Coulomb gas*, Constructive Physics (Palaiseau, 1994) (V. Rivasseau, ed.), Lecture Notes in Phys., vol. 446, Springer, Berlin, 1995, pp. 311–326.

DEPARTMENT OF MATHEMATICS, SUNY AT BUFFALO, BUFFALO, NY 14260, USA
E-mail address: dimock@acsu.buffalo.edu

Ground States of Supersymmetric Matrix Models

G. M. Graf

ABSTRACT. We consider supersymmetric matrix Hamiltonians. The existence of a zero-energy bound states, in particular for the $d = 9$ model, is of interest in M-theory. While we do not quite prove its existence, we show that the decay at infinity such a state would have is compatible with normalizability (and hence existence) in $d = 9$. Moreover, it would be unique. Other values of d, where the situation is somewhat different, shall also be addressed. The analysis is based on a Born-Oppenheimer approximation. This seminar is based on joint work with J. Fröhlich, D. Hasler, J. Hoppe and S.-T. Yau.

1. Bosonic Matrix Models

Matrix models are Schrödinger operators which first arose [7,8] in the early 80's as an approximation by finitely many degrees of freedom of relativistic membranes. More recently, supersymmetric matrix models have been interpreted [16] to describe a collection of particles with noncommutative coordinates (D0-branes). In this interpretation, some of the models have been conjectured [2] to describe a strong coupling limit of string theory and, as a result, are believed to have a bound state with energy at the bottom of the essential spectrum.

For expository purposes, let us postpone the definition of supersymmetric matrix models and begin with the simpler *bosonic* matrix models instead. They depend on two integers N, $d \geq 2$ and are as follows. The configuration space is $\mathcal{X} = \left[\mathrm{i} \cdot \mathrm{su}(N)\right]^d$, i.e., each configuration is a d-tuple of symmetric, traceless, $N \times N$ matrices:

$$X = (X_1, \ldots, X_d), \quad X_s \in \mathrm{i} \cdot \mathrm{su}(N), \quad s = 1, \ldots, d.$$

Coordinates $x_{sA} \in \mathbb{R}$ can be introduced through $X_s = T_A x_{sA}$ (with sum over A), where T_A, $A = 1, \ldots, N^2 - 1$ are generators of $\mathrm{i} \cdot \mathrm{su}(N)$ with $\mathrm{tr}(T_A T_B) = \delta_{AB}$. The corresponding momenta are then given as $P_s = -\mathrm{i} T_A \partial/\partial x_{sA}$. The Hamiltonian, acting in the Hilbert space $\mathrm{L}^2(\mathcal{X})$, is

$$(1.1) \qquad H = \sum_{s=1}^d \mathrm{tr}(P_s^2) + \sum_{s<t} \mathrm{tr}\left(\left(\mathrm{i}[X_s, X_t]\right)^2\right),$$

2000 *Mathematics Subject Classification.* Primary: 81T60; Secondary: 81S99.

Reprinted with permission from the Séminaire "Équations aux Dérivées Partielles" of the École Polytechnique, Palaiseau.

where the trace is w.r.t. $\mathrm{su}(N)$. The group $\mathrm{SU}(N) \times \mathrm{SO}(d) \ni (U, R)$ acts as a group of symmetries of the Hamiltonian: $X_s \mapsto U^* X_s U$, $X_s \mapsto R_{st} X_t$.

Note that the potential in (1.1) is large at infinity in $\mathcal{X}$ except for some narrowing "valleys" along its submanifold where all X_s commute. Nevertheless, the quantum mechanical motion of a particle in that potential is confined, because of the increasing zero-point energy associated with the motion transverse to the valley. Indeed, it has been shown [**12**] that the spectrum of H is purely discrete.

Before indicating the supersymmetric extension of the model we shall illustrate a physical motivation for the bosonic matrix models. We sketch their original derivation [**7,8**] as an approximation to 2-dimensional membranes. More generally, but temporarily, consider an M-dimensional closed surface in space $\mathbb{R}^{d+1}$ and view it as an $M+1$-dimensional world sheet in space-time $\mathbb{R}^{d+2}$ parametrized as

$$(1.2) \qquad x^\mu = x^\mu(\lambda_0, \ldots, \lambda_M), \quad (\mu = 0, \ldots, d+1).$$

The dynamics of the surface is governed by the action

$$(1.3) \qquad S = \int d\lambda_0, \ldots, d\lambda_M \sqrt{|G|}, \quad G = \det\left(\frac{\partial x^\mu}{\partial \lambda_a} \frac{\partial x_\mu}{\partial \lambda_b}\right),$$

which represents the volume of the world sheet induced by the Minkowski metric $x_0 = x^0$, $x_i = -x^i$, $(i = 1, \ldots, d+1)$ of space-time. We anticipate that, as result of the invariance of the action (1.3) under a reparametrization of (1.2), one will obtain the Hamiltonian (1.1) *restricted* to $\mathrm{SU}(N)$-invariant states. Moreover, the matrices X_s will be traceless because the membrane will be described in its center of mass frame.

A Hamiltonian description of this model is obtained by passing to light cone coordinates

$$\lambda_0 := \frac{1}{2}(x^0 + x^{d+1}), \quad \xi = \frac{1}{2}(x^0 - x^{d+1}), \quad \vec{x} = (x^1, \ldots, x^d).$$

The coordinate λ_0 is taken as (fictitious) time; at fixed λ_0 the configuration ξ, $\vec{x}$ is a field in the variables $\lambda = (\lambda_1, \ldots, \lambda_M)$. Denoting by $\pi = \pi(\lambda)$, $\vec{p} = \vec{p}(\lambda)$ the canonically conjugate fields, one finds the Hamiltonian

$$(1.4) \qquad H(\vec{x}, \vec{p}; \xi, \pi) = \int \frac{d^M \lambda}{\pi}(\vec{p}^{\,2} + g), \quad g = \det\left(\frac{\partial \vec{x}}{\partial \lambda_a} \cdot \frac{\partial \vec{x}}{\partial \lambda_b}\right),$$

with constraints

$$p \cdot \frac{\partial \vec{x}}{\partial \lambda_a} + \pi \frac{\partial \xi}{\partial \lambda_a} = 0, \quad (a = 1, \ldots, M),$$

resulting from reparametrization invariance. As $\pi = \pi(\lambda)$ is a cyclic coordinate, one is left with a field $\vec{x}(\lambda)$ with d components. The Gram determinant g can be expressed through Lagrange's identity as a sum over $M \times M$ submatrices of $(\partial x_i / \partial \lambda_a)_{i=1,\ldots d; a=1,\ldots M}$:

$$g = \sum_{i_1 < \cdots < i_M} \left(\det \frac{\partial x_{i_a}}{\partial \lambda_b}\right)^2.$$

This expression is particularly simple in the case of membranes $M = 2$:

$$g = \sum_{s<t} \left(\frac{\partial x_s}{\partial \lambda_1} \frac{\partial x_t}{\partial \lambda_2} - \frac{\partial x_t}{\partial \lambda_1} \frac{\partial x_s}{\partial \lambda_2}\right)^2 \equiv \sum_{s<t} \{x_s, x_t\}^2.$$

It should at this point appear plausible that an approximation of the membrane by means of finitely many degrees of freedom results in the replacement of the Hamiltonian (1.4) by (1.1).

2. Supersymmetric Matrix Models

Let us now indicate the supersymmetric extension [3, 15] of the model (1.1). Consider first Clifford generators $\gamma^i, (i = 1, \ldots, d)$, i.e., $\{\gamma^s, \gamma^t\} = 2\delta^{st}$, realized as *real* matrices
$$\gamma^i = (\gamma^i_{\alpha\beta})_{\alpha,\beta=1,\ldots,s_d},$$
where s_d is the dimension of the irreducible (real) representation. Furthermore, consider Clifford generators $\Theta_{\alpha A}(\alpha = 1, \ldots, s_d; A = 1, \ldots, N^2 - 1)$ (with relations $\{\Theta_{\alpha A}, \Theta_{\beta B}\} = \delta_{\alpha\beta}\delta_{AB}$) irreducibly realized on some representation space $\mathcal{C}$. We shall set $\Theta_\alpha = T_A \Theta_{\alpha A}$, a Clifford algebra valued su(N)-matrix.

The Hilbert space is now $L^2(\mathcal{X}, \mathcal{C})$ (instead of $L^2(\mathcal{X})$) and a further term is added to the Hamiltonian (1.1):

$$(2.1) \qquad H = \sum_{s=1}^d \text{tr}(P_s^2) + \sum_{s<t} \text{tr}\big((\text{i}[X_s, X_t])^2\big) - \text{tr}\big(\Theta_\alpha \gamma^s_{\alpha\beta}[X_s, \Theta_\beta]\big).$$

The symmetry group of the model is as in the bosonic case, except that SO(d) is replaced by its covering group Spin(d). In addition the model admits supersymmetry [1] in dimensions $d = 2, 3, 5, 9$: On SU(N) invariant states,
$$\{Q_\alpha, Q_\beta\} = \delta_{\alpha\beta} H,$$
where the Q_α's are the supercharges
$$Q_\alpha = \gamma^s_{\alpha\beta} \text{tr}(\Theta_\beta P_s) - \frac{\text{i}}{4}[\gamma^s, \gamma^t]_{\alpha\beta} \text{tr}(\Theta_\beta[X_s, X_t]).$$

It should be noted that the spectrum of (2.1) is no longer discrete. In fact [14], $\sigma(H) = [0, \infty)$.

According to recent developments in string theory and M-theory [2, 16] the $d = 9$ model is conjectured to describe N D0-branes. In line with this conjecture the following question about the existence of zero-modes is expected to be answered in the affirmative:

Is 0 an eigenvalue of H? More precisely: Does there exist $\psi \in L^2(\mathcal{X}, \mathcal{C})$ with $H\psi = 0$, which is $SU(N) \times Spin(d)$ invariant? Is it unique?

Among the various works predating ours we mention for $N = 2$ [11, 17], indicating the existence of such states, and [9], where an asymptotic analysis of such states (related to ours) is made. See also [10] for general N. On the other hand, the expected answer is no for $d = 2, 3, 5$ (see [5] for $d = N = 2$).

3. A Simple Model

Before stating our result concerning (2.1), let us illustrate the issue and the method by discussing a simpler model. Consider the Hamiltonian

$$(3.1) \qquad H = p_x^2 + p_y^2 + x^2 y^2 + x\sigma_3 + y\sigma_1,$$

acting on $L^2(\mathbb{R}^2, \mathbb{C}^2)$, with σ_i, $(i = 1, 2, 3)$ being the Pauli matrices. The potential $x^2 y^2$ exhibits 4 valleys, one of which, indicated by an equipotential line, is seen here:

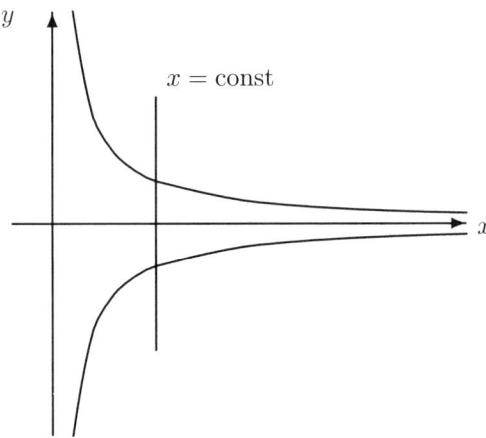

FIGURE 1. The valley.

The Hamiltonian can be written as

$$H = p_x^2 + \int^{\oplus} dx\, H(x),$$

with $H(x)$ acting on the fiber $F = L^2(\mathbb{R}, \mathbb{C}^2)$. Treating $y\sigma_1$ as a perturbation, the ground state of the latter is

$$\varphi(x; y) = \pi^{-1/4} |x|^{1/4} e^{-xy^2/2} \left[\begin{pmatrix} 0 \\ 1 \end{pmatrix} - \frac{y}{4x} \begin{pmatrix} 1 \\ 0 \end{pmatrix} + \cdots \right],$$

with ground state energy

$$E(x) = x - x - \frac{1}{8x^2} + \cdots.$$

Note that the contribution x of the harmonic oscillator $p_y^2 + x^2 y^2$ is canceled by an equal and opposite contribution from $x\sigma_3$. This parallels the fact that bosonic matrix models have purely discrete spectrum, whereas their supersymmetric counterparts have only essential spectrum. It is now natural to postulate that the ground state of H is given, asymptotically along the valley, by a Born-Oppenheimer ansatz:

(3.2) $$\psi(x, y) = f(x) \varphi(x; y).$$

The effective Hamiltonian for f is computed as

$$h = p_x^2 + E(x) + \|A\varphi(x)\|_F^2 + \cdots = p_x^2 - \frac{1}{8x^2} + \frac{1}{8x^2} + \cdots = p_x^2 + O(x^{-5}).$$

Besides of $E(x)$ a term of the same order arises because the infinitesimal translation p_x is dilating the state along the valley ($A - (p_y y + y p_y)/2$ is the generator of dilations in the fiber). A possible zero-mode, i.e., a state satisfying $hf = 0$, should then behave as $f(x) = 1$ or $f(x) = x$ at infinity (actually: the second solution is spurious) and is thus not expected to occur, as it would not be square integrable. Let us mention that this approach, being purely asymptotic, does rule out zero-modes of the form (3.2) with $\varphi(x; \cdot)$ an excited state of $H(x)$.

The Hamiltonian (3.1) is supersymmetric:

$$H = Q^2, \quad Q = p_x \sigma_3 - p_y \sigma_1 - xy \sigma_2.$$

One may thus just look for zero-modes of Q, which we again analyze asymptotically. We anticipate the length scale $x^{-1/2}$ of the ground state by a change of variable, $\widetilde{y} = x^{1/2}y$, as this quantity is effectively of order 1. Then

$$Q = Q_0 x^{1/2} + \widehat{Q}\frac{d}{dx} + Q_1 x^{-1},$$

where the coefficients are operators on F:

$$Q_0 = i\sigma_1 \frac{\partial}{\partial \widetilde{y}}, \quad \widehat{Q} = -i\sigma_3, \quad Q_1 = -(i/2)\sigma_3 \widetilde{y}\frac{\partial}{\partial \widetilde{y}}.$$

The equation $Q\psi = 0$ is therefore an ordinary differential equation in x for $\psi(x,\cdot) \in F$, with $x = \infty$ being a singular point of the second kind [**4**]. The generalized power series ansatz corresponding to the eigenvalue 0 of Q_0 is

$$\psi(x,\widetilde{y}) = x^{-\kappa}\sum_{k=0}^{\infty} x^{-3/2k}\psi_k(\widetilde{y}),$$

which yields

$$Q_0\psi_0 = 0$$
$$\kappa\widehat{Q}\psi_0 = Q_1\psi_0 + Q_0\psi_1$$
$$\cdots.$$

The solution of the first equation is $\psi_0(\widetilde{y}) = e^{-\widetilde{y}^2/2}\binom{0}{1}$, and the projection of the second onto ψ_0 is $\kappa(\psi_0, \widehat{Q}\psi_0)_F = (\psi_0, Q_1\psi_0)_F$, implying $\kappa = -1/4$ (corresponding to $f(x) = 1$ above).

4. $N = 2$ Supersymmetric Matrix Models

The above analysis can be carried over to $N = 2$ supersymmetric matrix models. Writing $X \in i\,\mathrm{su}(N=2)$ as $X = \vec{q}\cdot\vec{\sigma}, \vec{q} \in \mathbb{R}^3$, the configuration spaces becomes $\mathcal{X} = \mathbb{R}^{3d}$, and the Hamiltonian

$$(4.1) \qquad H = \sum_{s=1}^{9} \vec{p}_s^{\,2} + \sum_{s<t}(\vec{q}_s \times \vec{q}_t)^2 + i\vec{q}_s \cdot (\vec{\Theta}_\alpha \times \vec{\Theta}_\beta)\gamma_{\alpha\beta}^s.$$

The potential $\sum_{s<t}(\vec{q}_s \times \vec{q}_t)^2$ vanishes on the $(d+2)$-dimensional manifold

$$\{q = (\vec{q}_1,\ldots,\vec{q}_d) \mid \vec{q}_s \parallel \vec{q}_t \text{ for all } s,t\}.$$

Its points can thus be expressed as $\vec{q}_s = r\vec{e}E_s$ with $r > 0$ and $\vec{e} \in S^{3\ 1}, E \in S^{d\ 1}$; points in a conical neighborhood of the manifold can be expressed in terms of tubular coordinates

$$\vec{q}_s = r\vec{e}E_s + r^{-1/2}\vec{y}_s,$$

with $\vec{y}_s \cdot \vec{e} = 0, \vec{y}_s E_s = 0$. A prefactor has been put explicitly in front of the transversal coordinates $\vec{y}_s$, so as to account for the length scale $r^{-1/2}$ of the ground state. Also note that the change

$$(4.2) \qquad (\vec{e}, E, y) \mapsto (-\vec{e}, -E, y),$$

does not affect $\vec{q}_s$. Hence only states which are even under the antipode map (4.2) lift to $\mathcal{X}$.

We can now describe the structure of a putative ground state.

THEOREM 4.1. *Consider the equations $Q_\beta \psi = 0$ for a formal power series solution near $r = \infty$ of the form*

$$\psi = r^{-\kappa} \sum_{k=0}^{\infty} r^{-3/2k} \psi_k, \qquad (4.3)$$

where

- $\psi_k = \psi_k(\vec{e}, E, y)$ *is square integrable w.r.t.* $\mathrm{d}\vec{e}\,\mathrm{d}E\,\mathrm{d}y$; *where*
- ψ_k *is* $\mathrm{SU}(2) \times \mathrm{Spin}(d)$ *invariant; where*
- $\psi_0 \neq 0$.

Then, up to linear combinations,

$d = 9$: *The solution is unique, and $\kappa = 6$.*
$d = 5$: *There are three solutions with $\kappa = -1$ and one with $\kappa = 3$.*
$d = 3$: *There are two solutions with $\kappa = 0$.*
$d = 2$: *There are no solutions.*

All solutions are even under the antipode map (4.2),

$$\psi_k(\vec{e}, E, y) = \psi_k(-\vec{e}, -E, y),$$

except for the state $d = 5$, $\kappa = 3$, which is odd.

The integration measure is

$$\mathrm{d}q = \mathrm{d}r \cdot r^2 \mathrm{d}e \cdot r^{d-1}\mathrm{d}E \cdot r^{-1/2 \cdot 2(d-1)}\mathrm{d}y = r^2 \mathrm{d}r\,\mathrm{d}e\,\mathrm{d}E\,\mathrm{d}y.$$

The wave function (4.3) is square integrable at infinity if $\int^\infty \mathrm{d}r r^2 (r^{-\kappa})^2 < \infty$, i.e., if $\kappa > 3/2$. The theorem is consistent with the statement according to which only for $d = 9$ a (unique) normalizable ground state for (4.1) (which would have to be even) is possible.

We refer to [6] for the proof of the theorem. Here we merely sketch the argument in the $d = 9$ case for uniqueness of the $\mathrm{SU}(2) \times \mathrm{Spin}(9)$ invariant ground state. As in the simple model described above, the equation to be solved at lowest order is $Q_\alpha^0 \psi_0 = 0$. Ignoring invariance, this equation admits a large space of solutions, namely

$$\psi_0(\vec{e}, E, y) = \mathrm{e}^{-\sum_s \vec{y}_s^2/2} |F(E, \vec{e}\,)\rangle,$$

with $|F(E, \vec{e}\,)\rangle \in N(E, \vec{e}\,)$, a 2^8-dimensional subspace of $\mathcal{C}$. While $\mathrm{SU}(2)$ acts trivially on these solutions, $\mathrm{Spin}(9) \ni R$ does not:

$$(\mathcal{R}(R)\psi_0)(\vec{e}, E, y) = \mathrm{e}^{-\sum_s \vec{y}_s^2/2} \mathcal{R}_\mathrm{F}(R) |F(R^{-1}E, \vec{e}\,)\rangle,$$

where $\mathcal{R}_\mathrm{F}$ denotes the "fermionic part" of the representation, i.e., it acts on $\mathcal{C}$ only. Invariant states, $\mathcal{R}(R)\psi_0 = \psi_0$ are thus in bijective correspondence to states invariant under the "little group" $\mathrm{Spin}(8)$, i.e., to states $|F(E, \vec{e}\,)\rangle \in N(E, \vec{e}\,)$ satisfying

$$\mathcal{R}_F(R)|F(E, \vec{e}\,)\rangle = |F(E, \vec{e}\,)\rangle,$$

for some arbitrary but fixed E and all R with $RE = E$. Infinitesimally, such rotations take place in a plane (with vectors $U, V \in \mathbb{R}^9$) orthogonal to E: $U_s E_s = V_s E_s = 0$. The generators of the little group are represented on $N(E, \vec{e}\,)$ as $M_{st}^\parallel U_s V_t$ with

$$M_{st}^\parallel = -(\mathrm{i}/4)(\vec{\Theta}_\alpha \cdot \vec{e}\,)\gamma_{\alpha\beta}^{st}(\vec{\Theta}_\beta \cdot \vec{e}\,), \qquad (4.4)$$

where $\gamma^{st} = (1/2)(\gamma^s\gamma^t - \gamma^t\gamma^s)$. We need to decompose the $\mathrm{Spin}(8)$ representation on $N(E, \vec{e}\,)$ into irreducibles. To begin with, the Clifford algebra of the operators

$\vec{\Theta}_\alpha \cdot \vec{e}, \alpha = 1, \ldots, s_9 = 16$ acts irreducibly on $N(E, \vec{e}\,)$, but the representation decomposes (see e.g. [**13**]) by passing to the subalgebra of even elements, resp. to Spin(16):
$$N(E, \vec{e}\,) = 128_- \oplus 128_+.$$
The further branching under the embedding Spin(16) $\hookleftarrow$ Spin(9) given by (4.4) is
$$N(E, \vec{e}\,) = (44 \oplus 84) \oplus 128,$$
followed by Spin(9) $\hookleftarrow$ Spin(8):
$$N(E, \vec{e}\,) = (1 \oplus 8_v \oplus 35_v) \oplus (28 \oplus 56_v) \oplus (8_s \oplus 8_c \oplus 56_s \oplus 56_c).$$
This shows that exactly one 1-dimensional representation occurs.

References

1. M. Baake, P. Reinicke, and V. Rittenberg, *Fierz identities for real Clifford algebras and the number of supercharges*, J. Math. Phys. **26** (1985), 1070–1071.
2. T. Banks, W. Fischler, S. H. Shenker, and L. Susskind, *M theory as a matrix model: a conjecture*, Phys. Rev. D **55** (1997), 5112–5128, `hep-th/9610043`.
3. M. Claudson and M. Halpern, *Supersymmetric ground state wave functions*, Nuclear Phys. B **250** (1985), 689–715.
4. E. A. Coddington and N. Levinson, *Theory of ordinary differential equations*, McGraw-Hill Book Company, Inc., New York-Toronto-London, 1955.
5. J. Fröhlich and J. Hoppe, *On zero-mass ground states in super-membrane matrix models*, Comm. Math. Phys. **191** (1998), 613–626, `hep-th/9701119`.
6. J. Fröhlich, G. M. Graf, D. Hasler, J. Hoppe, and S.-T. Yau, *Asymptotic form of zero energy wave functions in supersymmetric matrix models*, Nuclear Phys. B **567** (2000), no. 1-2, 231–248, `hep-th/9904182`.
7. J. Goldstone, unpublished.
8. J. Hoppe, *Quantum theory of a massless relativistic surface*, Constraint's Theory and Relativistic Dynamics (Florence, 1986), World Sci. Publishing, Singapore, 1987, pp. 267–276; MIT Ph.D. Thesis (1982)
9. M. B. Halpern and C. Schwartz, *Asymptotic search for ground states of* SU(2) *matrix theory*, Internat. J. Modern Phys. A **13**, 4367-4408 (1998), `hep-th/9712133`.
10. A. Konechny, *On asymptotic Hamiltonian for* SU(N) *matrix theory*, J. High Energy Phys. (1998), no. 10, Paper 18, `hep-th/9805046`.
11. S. Sethi and M. Stern, *D-Brane bound state redux*, Comm. Math. Phys. **194** (1998), 675–705, `hep-th/9705046`.
12. B. Simon, *Some quantum operators with discrete spectrum but classically continuous spectrum*, Ann. Physics **146** (1983), 209–220.
13. _____, *Representations of finite and compact groups*, Graduate Studies in Mathematics, vol. 10, Amer. Math. Soc., Providence, RI, 1996.
14. B. de Wit, M. Lüscher, and H. Nicolai, *The supermembrane is unstable*, Nuclear Phys. B **320** (1989), 135–159.
15. B. de Wit, J. Hoppe, and H. Nicolai, *On the quantum mechanics of supermembranes*, Nuclear Phys. B **305** (1988), 545–581.
16. E. Witten, *Bound states of strings and p-branes*, Nuclear Phys. B **460** (1996), 335–350, `hep-th/9510135`.
17. P. Yi, *Witten index and threshold bound states of D-branes*, Nuclear Phys. B **505** (1997), 307–318, `hep-th/9704098`.

THEORETISCHE PHYSIK, ETH-HÖNGGERBERG, CH–8093 ZÜRICH
E-mail address: `gmgraf@itp.phys.ethz.ch`

Some Mathematical Problems in the Ginzburg–Landau Theory of Superconductivity

Stephen J. Gustafson

ABSTRACT. This is a brief review of some mathematical aspects of the Ginzburg–Landau theory of superconductivity. We describe a few recent results, focusing especially on vortices and their stability.

1. The Ginzburg–Landau Theory

Some materials, when cooled below a certain temperature (known as the *critical temperature*, T_c) become superconductors. This state is characterized by a negligible resistance to current flow, and by the expulsion of the magnetic field from the bulk of the material (the *Meissner effect*). The Ginzburg–Landau theory is a mathematical model which describes these phenomena.

Ginzburg and Landau proposed their phenomenological theory of superconductivity in 1950 [**17**]. It is based on the general Landau theory, the cornerstone of the theory of phase transitions. The Ginzburg–Landau theory introduces a complex-valued *order parameter*

$$\psi \colon \Omega \to \mathbb{C},$$

(here Ω is the region of space containing the superconductor), whose role it is to indicate locally in space whether the material is in a superconducting or normal (non-superconducting) state. The magnetic field, B, is expressed, as usual, in terms of the vector potential

$$A \colon \Omega \to \mathbb{R}^3,$$

via $B = \operatorname{curl} A$. The difference in the actual free energy density and the free energy density of the normal state is then written as an expansion in powers of ψ and its derivatives. The result (in normalized units) is an energy functional of the form

(1.1) $$\mathcal{E}(\psi, A) = \frac{1}{2} \int_\Omega \left\{ |\nabla_A \psi|^2 + |\operatorname{curl} A|^2 + V(|\psi|) \right\}.$$

To describe the two phases (the normal phase with $|\psi| = 0$ and the superconducting phase with $|\psi| = 1$), V is taken to be

$$V(s) = \frac{\lambda}{4}(1 - s^2)^2,$$

2000 *Mathematics Subject Classification.* Primary: 35Q; Secondary: 82D55.
This is the final form of the paper.

©2001 American Mathematical Society

where $\lambda > 0$ is a material parameter ($\sqrt{\lambda/2}$ is known as the *Ginzburg–Landau parameter*). The coupling of ψ to A occurs through the covariant derivative $\nabla_A = \nabla - iA$, which leads also to the gauge invariance of $\mathcal{E}$:

$$\mathcal{E}(e^{i\gamma}\psi, A + \nabla\gamma) = \mathcal{E}(\psi, A), \tag{1.2}$$

for $\gamma: \mathbb{R}^3 \to \mathbb{R}$. Physical quantities such as $|\psi|$ and $B = \operatorname{curl} A$ are, of course, gauge invariant.

The equilibrium state of the superconducting material is described by configurations (ψ, A) which are stationary points of the functional $\mathcal{E}$. Such critical points satisfy the Ginzburg–Landau (GL) equations

$$-\Delta_A \psi + \frac{\lambda}{2}(|\psi|^2 - 1)\psi = 0, \tag{1.3}$$

$$\operatorname{curl}^2 A + \operatorname{Im}(\overline{\psi}\nabla_A \psi) = 0. \tag{1.4}$$

In physical terms, the quantity $j \equiv \operatorname{Im}(\overline{\psi}\nabla\psi)$ is the *supercurrent*.

The currently accepted microscopic theory of superconductivity was proposed by Bardeen, Cooper, and Schrieffer in 1957. Two years later, Gor'kov [18] proved that the Ginzburg–Landau theory occurs as a limiting case of the (generalized) BCS theory, confirming its fundamental nature. In the BCS theory, electrons form superconducting bound pairs (Cooper pairs). The squared modulus of the order parameter in the Ginzburg–Landau theory, $|\psi|^2$, represents the local density of Cooper pairs.

This paper contains a brief review of some mathematical aspects of the Ginzburg–Landau theory, and describes a few recent results. Many more details may be found in the books [15, 26, 40, 41], for example, and the review articles [13, 16], and [33].

2. Domains and Boundary Conditions

A typical physical problem involves a finite superconductor ($\Omega \subset \mathbb{R}^3$ bounded) placed in an external magnetic field $H : \mathbb{R}^3 \to \mathbb{R}^3$. In this case the energy functional is modified to become

$$\mathcal{E}_H(\psi, A) = \frac{1}{2}\int_\Omega \left\{ |\nabla_A \psi|^2 + |\operatorname{curl} A - H|^2 + \frac{\lambda}{4}(1 - |\psi|^2)^2 \right\}.$$

Often, the external field is a constant, $H = hv_0$, where $h \geq 0$ and v_0 is a unit vector (this will be the case from now on when an external field is present). In this case, the cross term $\int \operatorname{curl} A \cdot H$ in $\mathcal{E}_H$ makes no contribution to the variational equations, so the GL equations are unchanged. Typical boundary conditions are

$$\nabla_A \psi \cdot \widehat{n}|_{\partial\Omega} = 0 \quad (\operatorname{curl} A - H) \times \widehat{n}|_{\partial\Omega} = 0,$$

where $\widehat{n}$ denotes the unit normal to $\partial\Omega$.

A common situation is when the fields are approximately constant in one direction (say the x_3 direction), and the external field points in that same direction ($v_0 = (0, 0, 1)$). In this case, a two dimensional version of the GL equations is studied. That is, $\Omega \subset \mathbb{R}^2$, $\psi: \Omega \to \mathbb{C}$, $A: \Omega \to \mathbb{R}^2$, and $\operatorname{curl} A = \partial_1 A_2 - \partial_2 A_1$ is a scalar. The energy functional and the GL equations then retain the same form, with the proviso that for a scalar B, $\operatorname{curl} B = (-\partial_2 B, \partial_1 B)$ is a vector. This leads us to the functional

$$\mathcal{E}_h(\psi, A) = \frac{1}{2}\int_{\Omega \subset \mathbb{R}^2} \left\{ |\nabla_A \psi|^2 + (\operatorname{curl} A - h)^2 + \frac{\lambda}{4}(1 - |\psi|^2)^2 \right\}, \tag{2.1}$$

whose critical points satisfy the GL equations (1.3)–(1.4). This is the setting for the remainder of this paper.

3. Flux Quantization and Topology

One immediate consequence of the Ginzburg–Landau theory is *flux quantization*. Let the total flux of the magnetic field through $\Omega \subset \mathbb{R}^2$ be

$$\Phi = \int_\Omega \operatorname{curl} A.$$

Using Stokes' theorem, we may rewrite this as a line integral around the boundary:

(3.1) $$\Phi = \int_{\partial\Omega} A \cdot dl.$$

Now we suppose that the supercurrent $j = \operatorname{Im}(\overline{\psi} \nabla_A \psi)$ vanishes on $\partial\Omega$ (this will be the case, for example, when we study finite energy solutions on the plane $\mathbb{R}^2$). We assume also that the material is superconducting at the boundary ($|\psi|$ bounded away from zero on $\partial\Omega$). Writing $\psi = |\psi|e^{i\phi}$ with ϕ real, we have $j = |\psi|^2(\nabla\phi - A) = 0$ and hence $(\nabla\phi - A)|_{\partial\Omega} = 0$. By (3.1), then,

$$\Phi = \int_{\partial\Omega} \nabla\phi.$$

Since ϕ is the phase, and must change by an integral multiple of 2π as $\partial\Omega$ is traversed, we have

(3.2) $$\Phi = 2\pi n,$$

where $n \in \mathbb{Z}$ is the topological winding number (or *degree*) of ψ around $\partial\Omega$. This is flux quantization. There is a more general notion of "fluxoid" quantization in the case j does not vanish on the boundary (see, e.g. [40]).

4. Length Scales

A more subtle feature of the GL theory is the appearance of length scales. We consider a finite energy solution of GL on the whole plane $\mathbb{R}^2$, without external field ($h \equiv 0$). We would like to determine the rate of decay of the (gauge-invariant) quantities $w \equiv 1 - |\psi|$ and $B = \operatorname{curl} A$ as $|x| \to \infty$ (these quantities must decay for the energy to be finite). Writing $\psi = |\psi|e^{i\phi}$ we obtain from (1.3)

(4.1) $$(-\Delta + \lambda)w = |A - \nabla\phi|^2(w-1) + \frac{3}{2}\lambda w^2 - \frac{\lambda}{2}w^3.$$

Taking the curl of (1.4) we obtain

(4.2) $$\left(-\Delta + (1-w)^2\right)B = -2(1-w)(\operatorname{curl} w) \cdot (\nabla\phi - A).$$

We study these equations as $|x| \to \infty$, and hence $w, B \to 0$. We remark that

$$|\nabla_A \psi|^2 = |\nabla w|^2 + (1-w)^2|\nabla\phi - A|^2,$$

and so $\nabla\phi - A \to 0$ as $|x| \to \infty$. Since $\operatorname{curl}(\nabla\phi - A) = -B$ we expect $\nabla\phi - A$ to decay with the same rate as B. In the leading order as $|x| \to \infty$, (4.2) becomes

$$(-\Delta + 1)B = -2(\operatorname{curl} w) \cdot (\nabla A - \phi).$$

We recall that the Greens function for $-\Delta + 1$ (the integral kernel of the operator $(-\Delta + 1)^{-1}$) decays like $e^{-|x-y|}$ for $|x-y|$ large. Since the right side decays at

least as quickly as B itself (by the above comments), the solution of this equation decays like $e^{-|x|}$. Turning now to (4.1), we obtain
$$(-\Delta + \lambda)w = -|A - \nabla\phi|^2,$$
in the leading order as $|x| \to \infty$. The right side decays like $e^{-2|x|}$, so the solution of this equation decays as $e^{-\min(\sqrt{\lambda},2)|x|}$. To summarize, our formal considerations predict decay of the form
$$(4.3) \qquad w \sim ce^{-m_L|x|}, \quad B \sim \tilde{c}e^{-m_T|x|},$$
as $|x| \to \infty$, where $m_T = 1$ and $m_L = \min(\sqrt{\lambda}, 2)$. Physicists associate corresponding length scales to these rates of exponential decay. The *penetration depth*, $\eta = 1/m_T = 1$, is the typical length scale for variations in the magnetic field, B. The *coherence length* is the typical length scale over which the modulus of the order parameter, $|\psi|$ varies, and is taken in the physics literature to be $\xi = 1/\sqrt{\lambda}$. The Ginzburg–Landau parameter, $\kappa = \sqrt{\lambda/2}$ is essentially the ratio of the penetration depth to the coherence length: $\kappa = \eta/\sqrt{2}\xi$.

Rigorous results are available. In [**24**] it is shown that for any finite energy solution, and any $\varepsilon > 0$, there is a constant $M(\varepsilon)$ such that
$$|w| \leq M(\varepsilon)e^{-[\min(\sqrt{\lambda},2)-\varepsilon]|x|}, \quad |B| \leq M(\varepsilon)e^{-[1-\varepsilon]|x|},$$
(so that the decay is at least as fast as predicted above). Their proof uses maximum principle arguments applied to the equations (4.1)–(4.2).

5. Normal and Superconducting Solutions

There are two trivial solutions to the GL equations. The first, $u_n = (\psi_n, A_n)$ where
$$\psi_n \equiv 0, \quad \operatorname{curl} A_n \equiv H,$$
corresponds to the normal state (with constant external field $H = hv_0$). The second, $u_s = (\psi_s, A_s)$ where
$$\psi_s \equiv 1, \quad A_s \equiv 0,$$
represents the purely superconducting state (which exhibits the Meissner effect: $\operatorname{curl} A_s \equiv 0$). The respective energies of these solutions are easily computed:
$$\mathcal{E}_h(u_n) = \frac{\lambda}{8}|\Omega|, \quad \mathcal{E}_h(u_s) = \frac{1}{2}h^2|\Omega|.$$
The value $h_c = \sqrt{\lambda}/2$ of h for which $\mathcal{E}_H(u_n) = \mathcal{E}_H(u_s)$ is given a name: it is known as the *critical magnetic field*.

A glance at the functional (2.1) suggests that when the external field h is sufficiently large, the global energy minimizer of the energy $\mathcal{E}$ is the normal solution, u_n. When h is sufficiently small, the superconducting solution, u_s is the minimizer. A basic question is how the phase transition from the normal state to the superconducting state occurs as h is lowered. To gain some insight into this question, we linearize GL around the normal solution u_n, and seek nontrivial solutions. In order to incorporate the external parameter h into the GL equations, we rescale, setting $\tilde{A} = A/h$. The GL equations now read
$$(5.1) \qquad -\Delta_{h\tilde{A}}\psi + \frac{\lambda}{2}(|\psi|^2 - 1)\psi = 0,$$

$$h \cdot \mathrm{curl}^2 \tilde{A} + \mathrm{Im}(\overline{\psi}\nabla_{h\tilde{A}}\psi) = 0.$$

The normal solution $\tilde{A}_n = A_n/h$ satisfies $\mathrm{curl}\,\tilde{A}_n = 1$. We can choose $\tilde{A}_n(x_1,x_2) = (0,x_1)$, for example. Writing $\psi = \sqrt{\varepsilon}\xi$ and $\tilde{A} = \tilde{A}_n + \varepsilon\alpha$, and retaining only the leading order terms in (5.1) as $\varepsilon \to 0$, we obtain

(5.2) $$\left(-\Delta_{h\tilde{A}_n} - \frac{\lambda}{2}\right)\xi = 0,$$

(5.3) $$\mathrm{curl}^2\,\alpha + \mathrm{Im}(\bar{\xi}\nabla_{h\tilde{A}_n}\xi) = 0.$$

Equation (5.3) determines α once ξ is known. Equation (5.2) has a very simple form: it is a Schrödinger eigenvalue equation for a charged particle in a constant magnetic field. The solutions are well-known. The ground state energy (bottom of the spectrum) of $-\Delta_{h\tilde{A}_n}$ is h. Thus the highest value of h for which nontrivial solutions exist is $h_{c_2} = \lambda/2$ (h_{c_2} is known as the *upper critical field*). The corresponding family of generalized eigenfunctions is

(5.4) $$\xi_k(x_1,x_2) = e^{ikx_2}e^{-h/2(x_1-k/h)^2}.$$

This computation suggests that a superconducting solution can bifurcate from the normal solution at $h = h_{c_2}$. We remark that for $\lambda < 1$, $h_{c_2} = \lambda/2 < h_c = \sqrt{\lambda}/2$. As the field is lowered, then, by the time a bifurcation from the normal solution is possible, the purely superconducting solution is energetically favorable. In this case, one expects an abrupt phase transition (a phase transition of the *first kind*) to the purely superconducting state. Conversely, if $\lambda > 1$, a gradual transition for $h_c < h < h_{c_2}$ through a "mixed state" is possible. We therefore expect superconductors with $\lambda < 1$ (known as *type I superconductors*), to exhibit qualitatively different properties than those with $\lambda > 1$ (*type II superconductors*).

6. The Normal-Superconducting Interface

In [17], Ginzburg and Landau calculated the "surface energy" of an interface separating a normal from a superconducting region. The sign of this quantity was found (numerically) to be the same as the sign of $1-\lambda$. In other words, the normal-superconducting surface energy is another feature which distinguishes type I superconductors (positive surface energy) from type II superconductors (negative surface energy). This calculation is described briefly here.

We consider a one-dimensional solution of the GL equations describing a transition between the normal and superconducting states. The fields A (real-valued now) and ψ depend on a single variable, x. By a gauge transformation, we can assume ψ is real-valued. To the far left, we impose the condition that the material is in the normal state with an external field strength of $h_c = \sqrt{\lambda}/2$:

$$\psi(x) \to 0, \quad A(x) \to h_c x,$$

as $x \to -\infty$. To the far right, we impose the normal state:

$$\psi(x) \to 1, \quad A(x) \to 0,$$

as $x \to \infty$. Let $u_{\text{trans}} = (\psi_{\text{trans}}, A_{\text{trans}})$ denote a solution of the GL equations, which now take the form

$$-\psi'' + A^2\psi + \frac{\lambda}{2}(\psi^2-1)\psi = 0, \quad A'' = \psi^2 A,$$

satisfying the above boundary conditions (such a solution is sometimes called a *domain wall*). Since $h = h_c$, the energy densities of the normal and purely superconducting states are the same (equal to the constant $\lambda/8$). If $e(u_{\text{trans}})$ denotes the free energy density of the solution u_{trans}, then $e(u_{\text{trans}}) - \lambda/8$ vanishes as $x \to \pm\infty$. Set $\delta = \int_{-\infty}^{\infty} \{e(u_{\text{trans}}) - \lambda/8\}$. Thus

$$\delta = \int_{-\infty}^{\infty} \left\{ (\psi')^2 + A^2\psi^2 + (A' - h_c)^2 + \frac{\lambda}{2}(\psi^2 - 1)^2 - \frac{\lambda}{8} \right\}.$$

We are interested in the sign of the quantity δ as an indicator of the energy cost (or profit) associated with the formation of a normal-superconducting interface. Originally, (see, e.g. [40]), heuristic considerations were used to determine the approximate value of δ in the limiting cases $\lambda \ll 1$ ($\delta > 0$) and $\lambda \gg 1$ ($\delta < 0$), and numerical computations suggested that δ changes sign exactly at $\lambda = 1$. More recently, it has been rigorously established in [11] that, δ does, in fact, have the same sign as $1 - \lambda$.

7. Abrikosov Vortices

The nature of the mixed state in a superconductor of the second kind ($\lambda > 1$) was studied by Abrikosov in 1957 [1]. Beginning with the idea that a normal-superconducting interface carries a positive surface energy for a type II superconductor, he predicted the existence of a "mixed state" in the form of a periodic array of "flux tubes," for h below h_{c_2}.

In studying the nature of nucleation of the normal state from the purely superconducting one as a small external field is increased, Abrikosov predicted an initially sparse flux lattice, and was led to the problem of an isolated flux cell in an infinite domain. This motivated his introduction of a family of radially-symmetric (or *equivariant*) solutions of GL on $\mathbb{R}^2$. These have the form

(7.1) $$\psi^{(n)}(x) = f_n(r)e^{in\theta}, \quad A^{(n)}(x) = n\frac{a_n(r)}{r}\hat{x}^{\perp},$$

where n is an integer, (r, θ) are polar coordinates on $\mathbb{R}^2$, $\hat{x}^{\perp} = (-x_2, x_1)/r$, and $f_n, a_n \colon [0, \infty) \to [0, 1]$. Such a solution is known as an n-vortex, and is easily seen to have degree n (i.e., quantized flux $2\pi n$—see (3.2)). One can think of the n-vortex as a "tube" (in three-dimensions) of magnetic flux localized at the origin (it is localized because the vortex profiles $f_n(r)$ and $a_n(r)$ tend to 1 at an exponential rate as $r \to \infty$). Away from the origin, the solution is superconducting ($|\psi| \sim 1$). The n-vortices will be studied in more detail in the next section.

For h just below, h_{c_2}, the Abrikosov flux lattice is described by a superposition of the solutions (see (5.4)) of the linearized GL equations:

(7.2) $$\psi(x_1, x_2) = \sum_{n=-\infty}^{\infty} C_n \exp\left\{iknx_2 - \frac{h}{2}(x_1 - nk/h)^2\right\},$$

which is periodic in x_2 (with period $2\pi/k$). If $C_{n+N} = C_n$ for some integer $N \geq 1$, then (7.2) is also periodic in x_1 (with period kN/h). It turns out that N is the winding number of ψ around the fundamental cell. The choice $C_n = C$ for all n is known (for $k = \sqrt{2\pi h}$) as Abrikosov's square lattice. Each cell contains a single vortex of degree one. The choice $C_{2n} = C$, $C_{2n+1} = \pm iC$ is known (for $k = \sqrt{\pi\sqrt{3}h}$) as the triangular lattice. Here $N = 2$, and each cell contains two vortices in such a way that the vortices form a triangular array. The lowest energy lattice was determined numerically in [25] to be the triangular one.

Some rigorous work on critical fields, bifurcations from the normal solution, and the role of vortices has been done, mostly in the large λ limit. In [29] and [28], the bifurcation at h_{c_2} is studied. For results on critical fields and on the vortex structure of solutions (for large λ primarily) see [10, 34, 36, 37], and references therein. In [5], the bifurcation from the normal state has been rigorously studied for the case of a disk, and λ large. Here, boundary effects play a role. At a critical field $h_{c_3} > h_{c_2}$, a high-degree vortex appears in the center of the disk, and the superconducting region is limited to a neighborhood of the boundary. Sandier and Serfaty [35] have recently obtained related results.

Some results on vortex lattices are also available. Odeh [29] proved the existence of periodic solutions to the full GL equations. Almog [4] and Chapman [12] establish the existence of many periodic solutions of the linearized equations, and investigate the stability (with respect to periodic perturbations) of some of them.

8. Vortex Stability

The existence of solutions of GL of the form (7.1) was established in [31] (see also [6]) using variational techniques. It was proved in [2] that if $\lambda \geq 4n^2$, there is a unique solution of this form.

The *vortex profiles* f_n and a_n increase monotonically from 0 to 1 as $r \to \infty$. Plohr [32] has determined the asymptotic behaviour of the profiles. As $r \to \infty$,

$$1 - |\psi^{(n)}|^2 = 1 - f_n^2(r) \sim ce^{-\min(\sqrt{\lambda},2)r}, \quad \operatorname{curl} A^{(n)} = na'_n(r)/r \sim \tilde{c}e^{-r}.$$

This agrees with the computation (4.3) above.

In the GL theory, the n-vortex plays the role of a basic building block for solutions. Thus the question of its stability is a fundamental one. The stability question was first addressed by Bogomolnyi [9] who provided a nonrigorous argument suggesting that the n-vortex with $|n| \geq 2$ is unstable (not a minimizer) when $\lambda > 1$—i.e., a type II superconductor (see [3] for a rigorous result in the large λ limit). Bogomolnyi's work also emphasized the special role played by the transitional value $\lambda = 1$.

The solutions of the GL equations are well-understood in the case of *critical coupling*, $\lambda = 1$. In this case, the *Bogomolnyi method* [9] gives a pair of first-order equations whose solutions are global minimizers of $\mathcal{E}(\psi, A)$ among fields of fixed degree, and hence solve the GL equations. Taubes [38, 39] has shown that all solutions of GL with $\lambda = 1$ are solutions of these first-order equations, and that for a given degree n, the gauge-inequivalent solutions form a $2|n|$-parameter family. One can think of these as "multi-vortex" solutions, where the $2|n|$ parameters describe the vortex locations (i.e., the locations of the zeros of the order parameter ψ). This is discussed in more detail in [24]. We remark that for $\lambda = 1$, the n-vortex solution (7.1) corresponds to the case when all $|n|$ zeros of ψ lie at the origin.

There is a somewhat vague general picture concerning inter-vortex interactions. Consider a finite energy configuration $u = (\psi, A)$ for which ψ has isolated zeros with well-defined local winding numbers. We think of these zeros as localized vortices. The general picture suggests that any pair of vortices experiences an inter-vortex force whose sign is determined as follows:

(1) if the winding numbers have different signs, the force is attractive,

(2) if the winding numbers have the same sign, the force is,
   (a) attractive if $\lambda < 1$,
   (b) zero if $\lambda = 1$,
   (c) repulsive if $\lambda > 1$.

An attractive (resp. repulsive) force means the energy is lowered by decreasing (resp. increasing) the vortex separation. There is some minimal evidence supporting this picture. Jacobs and Rebbi [23] have performed a numerical calculation in which the energy is first minimized among configurations containing two vortices, each of degree $+1$, a fixed distance R apart. As R is increased, the energy of the minimizer is found to increase when $\lambda < 1$, remain unchanged for $\lambda = 1$, and decrease for $\lambda > 1$, as is predicted by the "rule" described above. The Taubes solution for $\lambda = 1$ is also consistent with this picture, as it allows arbitrary placement of vortex centres with no energy variation. The repulsive nature of intervortex forces has been well-established in the limit $\lambda \to \infty$. In this limit, the essential feature of a field configuration is the location of its vortices (see, e.g. [7, 8, 14, 27, 30]).

Indeed, this general picture underlies the following old conjecture, which is clearly formulated by Jaffe and Taubes in [24]: for $\lambda < 1$, the n-vortex is stable for all n, whereas for $\lambda > 1$, only the $0, \pm 1$-vortices are stable, while higher-degree ($|n| \geq 2$) vortices are unstable. The idea is that for $\lambda > 1$, a high-degree vortex can lower its energy by splitting into several vortices of lower degree.

Recent work [20, 21] settles this conjecture. To state the strongest result, is is necessary to introduce dynamic versions of the Ginzburg–Landau equations.

The simplest dynamic GL equation is the gradient flow GF for the functional $\mathcal{E}(u)$, where $u = (\psi, A)$:

(GF) $$-\dot{u}(t) = \partial_u \mathcal{E}(u(t)).$$

This equation arises in the study of vortex dynamics in superconductors ([19]).

The GL equations are also studied in the context of gauge field theories (see [24]). In this case, the appropriate dynamic equation consists of the Maxwell equations for the electric field $E = \partial_t A - V$ and the magnetic field $B = \nabla \times A$ coupled to a nonlinear wave equation for ψ (we have introduced an additional field, V, the electric potential). The result, known as the Maxwell–Higgs (MH) (or Abelian–Higgs) equations is

(MH) $$\begin{cases} -(\partial_t - iV)^2 \psi = -\Delta_A \psi + \frac{\lambda}{2}(|\psi|^2 - 1)\psi, \\ \partial_t(\dot{A} - \nabla V) = \operatorname{curl}^2 A + \operatorname{Im}(\overline{\psi}\nabla_A \psi), \\ \nabla \cdot (\partial_t A - \nabla V) = \operatorname{Im}(\overline{\psi}(\partial_t - iV)\psi), \end{cases}$$

(see, e.g. [24]). The (MH) is also studied as a model in superconductivity ([22]).

We consider the appropriate Cauchy problem for (GF) or (MH), and study the *orbital stability* of the n-vortex, which is a static solution of either of these equations. To explain this notion, we recall that that the Ginzburg–Landau energy functional (and hence the GL equations) is invariant with respect to gauge transformations (1.2), as well as coordinate translations. Denote the (infinite-dimensional) group composed of these symmetry transformations by G_{sym}. We will say that the n-vortex $u^{(n)} = (\psi^{(n)}, A^{(n)})$ is *orbitally stable* if given any $\varepsilon > 0$, there is a $\delta > 0$ such that for any solution $u(t)$ of the dynamic equation,

$$\operatorname{dist}(u(0), G_{\text{sym}} u^{(n)}) < \delta \implies \operatorname{dist}(u(t), G_{\text{sym}} u^{(n)}) < \varepsilon \quad \forall t > 0.$$

Otherwise, we will say it is *unstable*. The "distance" function here is defined by

$$\text{dist}(u, G_{\text{sym}} u^{(n)}) \equiv \inf_{g \in G_{\text{sym}}} \|u - g u^{(n)}\|_{H^1}.$$

Given these preliminaries, we can now state the main result.

THEOREM 8.1 ([**20, 21**]). *As a solution of* (GF) *or* (MH), *the n-vortex is orbitally stable if either* $n = \pm 1$ *or* $\lambda < 1$. *The n-vortex is unstable if* $\lambda > 1$ *and* $|n| \geq 2$.

More details, and the proof of this result, can be found in [**21**] and [**20**]. The main ingredient is an involved spectral analysis of the linearized operator $\mathcal{E}''(u^{(n)})$, which is carried out in [**21**]. Roughly speaking, this analysis has two basic themes. The first can be described as a quantitative characterization of the symmetry breaking of the vortex. Combined with ideas of Perron–Frobenius theory, this enables us to locate the bottom of the spectrum of the linearized operator. The second theme is exploitation of the degeneracy at the critical value $\lambda = 1$. In particular, an idea of Bogomolnyi [**9**] lies at the heart of the proof of instability of high-degree vortices for $\lambda > 1$.

References

1. A. A. Abrikosov, *On the magnetic properties of superconductors of the second group*, Soviet Physics JETP **5** (1957), no. 6, 1174–1182.
2. S. Alama, L. Bronsard, and T. Giorgi, *Uniqueness of symmetric vortex solutions in the Ginzburg–Landau model of superconductivity*, J. Funct. Anal. **167** (1999), no. 2, 399–424.
3. L. Almeida, F. Bethuel, and Y. Guo, *A remark on the instability of symmetric vortices with large coupling constant*, Comm. Pure Appl. Math. **50** (1997), 1295–1300.
4. Y. Almog, *On the bifurcation and stability of periodic solutions of the Ginzburg–Landau equations in the plane*, Preprint, 1998.
5. P. Bauman, D. Phillips, and T. Qi, *Stable nucleation for the Ginzburg–Landau system with an applied magnetic field*, Arch. Rational Mech. Anal. **142** (1998), 1–43.
6. M. S. Berger and Y. Y. Chen, *Symmetric vortices for the nonlinear Ginzburg–Landau equations of superconductivity, and the nonlinear desingularization phenomenon*, J. Funct. Anal. **82** (1989), 259–295.
7. F. Bethuel, H. Brezis, and F. Hélein, *Ginzburg–Landau vortices*, Birkhäuser, 1994.
8. F. Bethuel and T. Riviere, *Vortices for a variational problem related to superconductivity*, Ann. Inst. Henri Poincaré Théor. **12** (1995), 243–303.
9. E. B. Bogomol'nyi, *The stability of classical solutions*, Soviet J. Nuclear Phys. **24** (1976), no. 4, 449–454.
10. C. Bolley and B. Helffer, *Rigorous results on Ginzburg–Landau models in a film submitted to an exterior parallel magnetic field. I*, Nonlinear Stud. **3** (1996), no. 1, 1–29.
11. S. J. Chapman, *Asymptotic analysis of the Ginzburg–Landau model of superconductivity: reduction to a free-boundary problem*, Quart. Appl. Math. **53** (1995), 601–627.
12. _____, *Nucleation of superconductivity in decreasing fields. I*, Europhys. J. Appl. Math. **5** (1994), 449–468.
13. S. J. Chapman, S. D. Howison, and J. R. Ockendon, *Macroscopic models for superconductivity*, SIAM Rev. **34** (1992), no. 4, 529–560.
14. J. Colliander and R. Jerrard, *Vortex dynamics for the Ginzburg–Landau Schrödinger equation*, Internat. Math. Res. Notices **7** (1998), 333–358.
15. P. de Gennes, *Superconductivity of metals and alloys*, Benjamin, New York, 1966.
16. Q. Du, M. Gunzburger, and J. Peterson, *Analysis and approximation of the Ginzburg–Landau model of superconductivity*, SIAM Rev. **34** (1992), no. 1, 54–81.
17. V. L. Ginzburg and L. D. Landau, *On the theory of superconductivity*, J. Experiment. Theoret. Phys. **20** (1950), 1064–1082.
18. L. Gork'ov, *Macroscopic derivation of the Ginzburg–Landau equations in the theory of superconductivity*, Soviet Physics JETP **9** (1959), 1364–1367.

19. L. P. Gork'ov and G. M. Eliashberg, *Generalization of the Ginzburg–Landau equations for non-stationary problems in the case of alloys with paramagnetic impurities*, Soviet Physics JETP **27** (1968), no. 2, 328–334.
20. S. Gustafson, *Dynamic stability of magnetic vortices*, (in preparation).
21. S. Gustafson and I. M. Sigal, *The stability of magnetic vortices*, Comm. Math. Phys. **212** (2000), no. 2, 257–275.
22. R. P. Huebener, *Magnetic flux structures in superconductors*, Springer, Berlin, 1979.
23. L. Jacobs and C. Rebbi, *Interaction of superconducting vortices*, Phys. Rev. B **19** (1979), 4486–4494.
24. A. Jaffe and C. Taubes, *Vortices and monopoles*, Birkhauser, Boston, 1980.
25. W. H. Kleiner, L. M. Roth, and S. H. Autler, *Bulk solution of Ginzburg–Landau equations for type. II. Superconductors*, Phys. Rev. **133** (1964), 1226–1227.
26. L. D. Landau and E. M. Lifshitz, *Statistical physics*, Pergamon, 1980.
27. F.-H. Lin and J. Xin, *On the incompressible fluid limit and the vortex motion law of the nonlinear Schrödinger equation*, Comm. Math. Phys. **200** (1999), 249–274.
28. M. H. Milman and J. B. Keller, *Perturbation theory of nonlinear boundary-value problems*, J. Math. Phys. **10** (1969), 342–361.
29. F. Odeh, *Existence and bifurcation theorems for the Ginzburg–Landau equations*, J. Math. Phys. **8** (1967), 2351–2356.
30. Y. Ovchinnikov and I. M. Sigal, *Long-time behaviour of Ginzburg–Landau vortices*, Nonlinearity **11** (1998), 1295–1309.
31. B. Plohr, *The existence, regularity, and behavior at infinity of isotropic solutions of classical gauge field theories*, Ph.D. thesis, Princeton, 1980.
32. _____, *The behaviour at infinity of isotropic vortices and monopoles*, J. Math. Phys. **22** (1981), no. 10, 2184–2190.
33. J. Rubinstein, *Six lectures on superconductivity*, Boundaries, Interfaces, and Transitions (Banff, 1995), CRM Proc. and Lecture Notes, vol. 13, Amer. Math. Soc., Providence, RI, 1998, pp. 163–184.
34. E. Sandier and S. Serfaty, *Global minimizers for the Ginzburg–Landau functional below the first critical magnetic field*, Ann. Inst. H. Poincaré Anal. Non Linéaire **17** (2000), no. 1, 119–145.
35. _____, *Vorticity in the Ginzburg–Landau model of superconductors in a magnetic field*, CRM Workshop on Nonlinear Dynamics and Renormalization Group (Montréal, 1999), (to appear).
36. S. Serfaty, *Local minimizers for the Ginzburg–Landau energy near critical magnetic field. I*, Commun. Contemp. Math. **1** (1999), 213–254.
37. _____, *Local minimizers for the Ginzburg–Landau energy near critical magnetic field. II*, Commun. Contemp. Math. **1** (1999), 295–333.
38. C. Taubes, *Arbitrary n-vortex solutions to the first order Ginzburg–Landau equations*, Comm. Math. Phys. **72** (1980), 277–292.
39. _____, *On the equivalence of the first and second order equations for gauge theories*, Comm. Math. Phys. **75** (1980), 207–227.
40. M. Tinkham, *Introduction to superconductivity*, McGraw-Hill, New York, 1975.
41. T. Tsuneto, *Superconductivity and superfluidity*, Cambridge, 1988.

COURANT INST. OF MATHEMATICAL SCIENCES, 251 MERCER STREET, NEW YORK, NY10012, USA.

E-mail address: gustaf@cims.nyu.edu

Renormalization in Radiation Reaction: New Developments in Classical Electron Theory

Michael K.-H. Kiessling

ABSTRACT. Recently the notorious radiation reaction problem of classical electrodynamics has received its first clean treatment in several rigorous works, by Dürr, Spohn, and collaborators, on the semi-relativistic theory of a rigid 'Abraham electron' with finite bare mass but without spin. In an effort to extend the treatment to the relativistic regime including classical spin, W. Appel and the author have recently succeeded in the construction of the first consistent relativistic theory of a deformable 'Lorentz electron' with spin and positive bare mass, in which the renormalized particles behave dynamically as spinning solitons, their characteristic data can be matched to the physical electron data, and for which a renormalization flow toward the limit of vanishing bare rest mass can be studied. The paper summarizes these developments.

1. Introduction

The classical electron theory (CET) of the late 19th and early 20th century was an early, pre-relativistic and pre-quantum attempt to construct a consistent microscopic dynamical theory of electrons and their electromagnetic field. To overcome the problem of infinite radiation reactions and self energies of point charges, the theory postulated stable, extended 'atoms of electricity' (electrons) that generated the electromagnetic fields, Maxwell's equations in 'the free aether' for the dynamics of these fields, and Newton's equation of motion for the dynamics of these electrons, equipped with the Lorentz force averaged over an electron. Our modern notion of *mass renormalization* has its roots in CET in form of the concept of the purely electromagnetic electron of Abraham [1] and Lorentz [22], and in this lecture I will take us 'back to our roots.'

The idea of the purely electromagnetic classical electron was a very happy one at first, combined with the idea of the deformable electron, leading Lorentz [21] to special relativity. Ironically, special relativity and renormalization should later

---

2000 *Mathematics Subject Classification*. Primary: 83A05; Secondary: 83C50.

Joint work with Walter Appel, Laboratoire de Physique, Unité de recherche 1325 associée au CNRS, École normale supérieure de Lyon, 46 allée d'Italie, 69364 Lyon Cedex 07, France.

My sincere thanks go to the organizers of this workshop, Michael Israel Sigal and Catherine Sulem. Work of M.K. funded in parts by NSF Grant DMS 9623220.

This is the final form of the paper.

contribute to the demise of CET more than quantum mechanics,[1] namely when in the 1920's Lorentz concluded [**29**, p. 35] that, in order to account for the discovered magnetic moment of the physical electron the purely electromagnetic electron had to rotate at its equator faster than the speed of light.[2]

Judged from our modern perspective on renormalization [**7, 11**], however, the purely electromagnetic calculations by Lorentz and his contemporaries do not qualify as proper renormalization analysis. Very recently now, W. Appel from the ENS Lyon and I succeeded [**2**] in constructing the first consistent relativistic model for the dynamics of a spinning classical charged particle and its electromagnetic self-fields which allows one to conduct such a proper renormalization analysis.

To set things into perspective, let me list some minimal requirements that any such classical model of an electron has to satisfy *beside* being relativistically invariant (and which our model does satisfy):

(1) As a physical evolution model the dynamical equations should be well posed as a Cauchy problem;
(2) To guarantee that, poetically speaking, 'an electron remains an electron', the dynamics of the renormalized particle must be solitonic;
(3) As a classical model of an electron, it must be possible to choose the parameters of the bare particle such that charge, magnetic moment, and renormalized mass of the stationary particle can be matched to the physical electron data in the limit of vanishing bare rest mass.

It is not so obvious that it is possible to satisfy all these requirements. Indeed, an earlier relativistically invariant dynamical model with finite bare rest mass, by Nodvik [**24**], fails to satisfy points (1) and (3), and we don't know whether (2) could hold, then. Our model, which owes much to the monumental relativistic work of Nodvik [**24**], removes the deficiencies of the latter through a single but nontrivial modification. To our delight, the modified model satisfies all points (1), (2), and (3).

Furthermore, we were able to prove global existence and uniqueness and long-time approach to a stationary state through radiation damping for the special case of purely spin dynamics. The relation to the BMT equation has yet to be worked out. Global existence and uniqueness for the full dynamics needs yet to be established (under reasonable assumptions) and effective dynamical equations extracted.

---

[1] Of course, the discovery of quantum mechanics removed classical electron theory from the list of candidates for a truly fundamental theory of electrodynamics, but that alone would not have 'killed' CET. (You would not be unhappy with a consistent classical microscopic theory of electrodynamics that would be valid in the classical regime, and to which a consistent quantum electrodynamics would—presumably—reduce in some classical limit, or would you?) However, the absurd conclusions reached on basis of relativity and mass renormalization convinced people that it was impossible to use CET to model even a single stationary electron in otherwise empty space.

[2] A naive calculation with a rotating charged sphere of radius equal to the classical electron radius produces equatorial speeds of about $200c$ to account for the observed magnetic moment. However, the particle's magnetic energy then exceeds its electrostatic energy by several orders of magnitude, while only the latter one was equated with the physical mass to compute the radius in the first place. A consistent calculation that matches the total electromagnetic mass to the physical mass and the magnetic moment to its physical value still gives superluminal speeds, only this time the quantitative result is not as dramatic: about ten times the speed of light.

Meanwhile we have carried out a preliminary renormalization flow analysis for a stationary rotating charged particle with its charge *and* positive bare rest mass distributed uniformly over a sphere. The bare rest mass, the size, and the rotation frequency of the bare electron (the charge is fixed) for matched physical electron data behave qualitatively simply like this: a reduction of the positive bare rest mass leads to a shrinking of the particle and a simultaneous speed-up at the equator. This procedure can, in principle, be continued until the equatorial speed of the model electron reaches the speed of light. Clearly, if the superluminal results of Lorentz and others are bearers of bad tidings, then one should expect that the speed of light will be reached before the bare rest mass has vanished. However, the analysis leads to a most remarkable result:[3] The model electron reaches equatorial speed of light *precisely* when its bare rest mass vanishes, which happens when its radial size is of the order of the Compton radius of the physical electron, *not* the classical electron radius! The renormalized mass of the limit particle consists of the traditional electromagnetic contribution *plus* a 'photonic' contribution associated with its spin: like a photon having a mass but no rest mass, so also our model particle has a mass associated with its luminal rotational motion while its bare rest mass is zero. In this renormalized sense, the purely electromagnetic classical electron of Lorentz exists, after all!

A future project is to study the renormalization flow toward vanishing bare rest mass of the dynamics of the model. Also the many-particle version should be studied; in particular, what is its relation to the relativistic Vlasov–Maxwell equations [8–10, 31] of relativistic statistical mechanics? Finally, it should be studied whether the results have any bearing on the renormalization program in quantum electrodynamics, and on radiation reaction in general relativity.

In the remaining pages I now outline the mathematical structure of the model by Appel and myself, the full story of which will appear elsewhere [2]. Our work was prompted by the recent rigorous papers [4, 17–19, 27] on the semi-relativistic theory of Abraham. Hence I begin with a brief review of the semi-relativistic theory and then turn to the relativistic theory.

2. Semi-Relativistic CET

In the semi-relativistic formulation of CET, the charge ($-e$ for an electron) is distributed around the instantaneous location $\boldsymbol{q}(t) \in \mathbb{R}^3$ of the particle by a charge density f_e with SO(3) symmetry and compact support, satisfying $\int_{\mathbb{R}^3} f_e(\boldsymbol{x}) \mathrm{d}^3 x = -e$. The charge density f_e is rigidly carried along by the particle with linear velocity $\dot{\boldsymbol{q}}(t)$, a point advocated in particular by Abraham [1], and rotating rigidly with angular velocity $\boldsymbol{\omega}(t)$. Hence, charge and current densities, $\rho(\boldsymbol{x},t)$ and $\boldsymbol{j}(\boldsymbol{x},t)$, of a single particle are given by the Abraham–Lorentz expressions

(2.1) $$\rho(\boldsymbol{x},t) = f_e(\boldsymbol{x} - \boldsymbol{q}(t)),$$
(2.2) $$\boldsymbol{j}(\boldsymbol{x},t) = f_e(\boldsymbol{x} - \boldsymbol{q}(t))\boldsymbol{v}(\boldsymbol{x},t),$$

with

(2.3) $$\boldsymbol{v}(\boldsymbol{x},t) = \dot{\boldsymbol{q}}(t) + \boldsymbol{\omega}(t) \times (\boldsymbol{x} - \boldsymbol{q}(t)).$$

---

[3]These results were finalized shortly after the workshop in Montreal. They are included here, for the benefit of the reader.

Considering only one particle interacting with the electromagnetic field, (2.1) and (2.2) with (2.3) are the source terms in the classical Maxwell–Lorentz equations

$$\frac{1}{c}\frac{\partial}{\partial t}\boldsymbol{B}(\boldsymbol{x},t) + \nabla \times \boldsymbol{E}(\boldsymbol{x},t) = 0, \tag{2.4}$$

$$-\frac{1}{c}\frac{\partial}{\partial t}\boldsymbol{E}(\boldsymbol{x},t) + \nabla \times \boldsymbol{B}(\boldsymbol{x},t) = 4\pi\frac{1}{c}\boldsymbol{j}(\boldsymbol{x},t), \tag{2.5}$$

$$\nabla \cdot \boldsymbol{B}(\boldsymbol{x},t) = 0, \tag{2.6}$$

$$\nabla \cdot \boldsymbol{E}(\boldsymbol{x},t) = 4\pi\rho(\boldsymbol{x},t), \tag{2.7}$$

where $\boldsymbol{E}(\boldsymbol{x},t) \in \mathbb{R}^3$ is the electric field and $\boldsymbol{B}(\boldsymbol{x},t) \in \mathbb{R}^3$ the magnetic field at the point $\boldsymbol{x} \in \mathbb{R}^3$ at time $t \in \mathbb{R}$. The evolution equations for the dynamical variables of the particle, i.e. position $\boldsymbol{q}(t)$, linear velocity $\dot{\boldsymbol{q}}(t)$, and angular velocity $\boldsymbol{\omega}(t)$, are Newton's and Euler's equations of motion, equipped with the Abraham–Lorentz expressions for the volume-averaged Lorentz force and torque [20] felt by the 'electron'. Newton's equation thus reads

$$\frac{\mathrm{d}\boldsymbol{p}_\mathrm{b}}{\mathrm{d}t} = \int_{\mathbb{R}^3} \left[\boldsymbol{E}(\boldsymbol{x},t) + \frac{1}{c}\boldsymbol{v}(\boldsymbol{x},t) \times \boldsymbol{B}(\boldsymbol{x},t)\right] f_\mathrm{e}(\boldsymbol{x} - \boldsymbol{q}(t))\, \mathrm{d}^3 x, \tag{2.8}$$

where

$$\boldsymbol{p}_\mathrm{b} = \begin{cases} m_\mathrm{b}\dot{\boldsymbol{q}} & \text{Newton,} \\ \frac{m_\mathrm{b}\dot{\boldsymbol{q}}}{\sqrt{1-|\dot{\boldsymbol{q}}|^2/c^2}} & \text{Einstein,} \end{cases} \tag{2.9}$$

is the linear particle momentum, with m_b the bare mass ('material mass' in [22]). Euler's equation here reads

$$\frac{\mathrm{d}\boldsymbol{s}_\mathrm{b}}{\mathrm{d}t} = \int_{\mathbb{R}^3} (\boldsymbol{x} - \boldsymbol{q}(t)) \times \left[\boldsymbol{E}(\boldsymbol{x},t) + \frac{1}{c}\boldsymbol{v}(\boldsymbol{x},t) \times \boldsymbol{B}(\boldsymbol{x},t)\right] f_\mathrm{e}(\boldsymbol{x} - \boldsymbol{q}(t))\, \mathrm{d}^3 x, \tag{2.10}$$

where

$$\boldsymbol{s}_\mathrm{b} = I_\mathrm{b}\boldsymbol{\omega}, \tag{2.11}$$

is the classical particle spin associated with the bare moment of inertia I_b. When the bare inertias $m_\mathrm{b} \neq 0$ and $I_\mathrm{b} \neq 0$, the semi-relativistic equations listed above pose a Cauchy problem for the following initial data, posed at time $t = t_0$: for the mechanical variables of the particles, the data are $\boldsymbol{q}(t_0)$, $\dot{\boldsymbol{q}}(t_0)$, and $\boldsymbol{\omega}(t_0)$; and for the fields, $\boldsymbol{B}(\boldsymbol{x},t_0)$ satisfying (2.6), and $\boldsymbol{E}(\boldsymbol{x},t_0)$ satisfying (2.7) at $t = t_0$. Notice that (2.6) and (2.7) are merely initial constraints, which then remain satisfied by the fields $\boldsymbol{B}(\boldsymbol{x},t)$ and $\boldsymbol{E}(\boldsymbol{x},t)$ in the ensuing evolution. For (2.6) this is seen by taking the divergence of (2.4). For (2.7) this is seen by taking the divergence of (2.5) and the time-derivative of (2.7), then using the continuity equation for the charge, which is a consequence of (2.1), (2.2), and (2.3) alone.

Various special cases are scattered in the literature. In particular, while in earlier versions with bare inertias of course the Newtonian momentum with $m_\mathrm{b} > 0$ and the Eulerian angular momentum with $I_\mathrm{b} > 0$ occur, in the later works of Abraham [1] and Lorentz [21,22] just before relativity the equations occur without bare inertias, i.e. $m_\mathrm{b} = 0 = I_\mathrm{b}$ in (2.9) and (2.11), and correspondingly the left sides in (2.8) and (2.10) are replaced by $\boldsymbol{0}$. Actually, it seems that only Abraham wrote down (2.8) *and* (2.10) explicitly, while Lorentz left it with some descriptive remarks as to the spinning motion.

Interestingly, most works after Abraham [1] consider only the translational equations of motion of the particle, assuming $\boldsymbol{\omega} = \mathbf{0}$ identically, see [12, 25, 26, 32]. This simpler case was recently treated rigorously in the papers [4, 17–19, 27], where for technical reasons the Einsteinian momentum with $m_b \neq 0$ is used instead of the Newtonian one (see also [15, 16] for a simpler scalar theory). In [4, 17], the global existence and uniqueness for the Cauchy problem of a particle without spin and $m_b \neq 0$ was proven. The papers [17–19, 27] address the long time asymptotics of the motion with $m_b > 0$, and derive effective equations of motion in slowly varying external fields using center manifold theory. These effective equations were originally obtained by Landau–Lifshitz by a physically correct but heuristical closure argument for the (in)famous Abraham–Lorentz–Dirac equation [5], whose ominous triple derivative of $\boldsymbol{q}$ had given rise to countless speculations as to its meaning. In the rigorous work of Spohn et al. the Landau–Lifshitz closure for the ALD equation is given its first clean derivation.

Moreover, in [4] it was shown that the motion is stable if $m_b > 0$ and unstable if $m_b < 0$. Their work thus shows that the bare mass should not be negative even in a theory without spin, contrary to what was advocated by Dirac [5].

Similar studies for the full set of semi-relativistic equations including spin have not yet been carried out. In this general format the equations are discussed in [14], where it was shown that all the conventional conservation laws are satisfied. It was also shown that the conventional angular momentum conservation does not hold if the 'classical particle spin' is omitted from the dynamics.[4]

After this preparation, we now turn to the relativistic theory.

3. Relativistic CET

The material of this section is taken from [2] which, by and large, follows the notational conventions of the book by C. W. Misner, K. Thorne and J. A. Wheeler [23]. However, since the model is essentially a flat space model, instead of forms, four-vectors and tensors formed from them are employed in noncomponent notation to facilitate the comparison with the semi-relativistic theory. Thus, points in Minkowski space-time $\mathbb{R}^{1,3}$ are given by ordered real 4-tuples $x = (x^0, x^1, x^2, x^3)$, where $x^\mu = x \cdot e_\mu$ for $\mu = 1, 2, 3$, and $x^0 = -x \cdot e_0$, are the components with respect to some Lorentz frame $\{e_0, e_1, e_2, e_3\}$, satisfying the elementary inner product rules

$$(3.1) \qquad e_\mu \cdot e_\nu = \begin{cases} -1 & \text{for } \mu = \nu = 0 \\ 1 & \text{for } \mu = \nu > 0 \\ 0 & \text{for } \mu \neq \nu \end{cases},$$

whence $x = x^\mu e_\mu$ with Einstein summation convention. Using the elementary inner product rules (3.1), the inner product of two four-vectors reads $x \cdot y = -x^0 y^0 + x^1 y^1 + x^2 y^2 + x^3 y^3$. For convenience, $x \cdot x$ will sometimes be abbreviated thus, $x \cdot x = \|x\|^2$, where $\|x\|$ is the Minkowski norm of x, defined as the principal value of $(x \cdot x)^{1/2}$. Notice that $\|x\|^2$ is negative for timelike vectors. The tensor product between any two four-vectors a and b is a tensor of rank two, denoted by $a \otimes b$, and defined by its inner-product action on four-vectors thus, $(a \otimes b) \cdot x \stackrel{\text{def}}{=} a(b \cdot x)$ and $x \cdot (a \otimes b) \stackrel{\text{def}}{=} (a \cdot x)b$. Any tensor of rank two, T, can be uniquely written as

---

[4]The model is invariant under rotations also without spin, but the conserved quantity associated with this invariance does not qualify as physical angular momentum in the conventional sense.

$\mathsf{T} = T^{\mu\nu} e_\mu \otimes e_\nu$. In particular, the *metric tensor* reads $\mathsf{g} = g^{\mu\nu} e_\mu \otimes e_\nu$ where $g^{\mu\nu} = e_\mu \cdot e_\nu$ as given in (3.1). Notice that $\mathsf{g} \cdot x = x$, i.e. g acts as identity on four-vectors. The exterior product between two four-vectors defines an anti-symmetric tensor of rank two, denoted by a wedge product $a \wedge b$, and given by $a \wedge b \stackrel{\text{def}}{=} a \otimes b - b \otimes a$. Notice that $(a \wedge b) \cdot x = -x \cdot (a \wedge b)$. The four-trace, or contraction, of a tensor of rank two is given by sum over the diagonal after multiplying the time-component by negative one. Thus, the four-trace of an antisymmetric tensor vanishes, and $\mathrm{Tr}\, \mathsf{g} = 4$. For a differentiable function $f(x)$, its four-gradient is denoted by $\partial_x f$, where

$$(3.2) \qquad \partial_x = \left(-\frac{\partial}{\partial x^0}, \frac{\partial}{\partial x^1}, \frac{\partial}{\partial x^2}, \frac{\partial}{\partial x^3} \right),$$

is the four-gradient operator in local coordinates. The four-Laplacian is simply the (negative of the) d'Alembertian, or wave operator, and given by $\partial_x \cdot \partial_x = -\Box$. This fixes the general notation. We now come to the model-specific setup.

Kinematically, the motion of the particle will be described by one timelike oriented space-time curve $\tau \mapsto q(\tau) \in \mathbb{R}^{1,3}$ and one spacelike oriented space-time curve $\tau \mapsto w_\mathrm{E}(\tau) \in \mathbb{R}^{1,3}$, where $\tau \in \mathbb{R}$ is the proper time of the particle, i.e. $c^2 (\mathrm{d}\tau)^2 \stackrel{\text{def}}{=} -\mathrm{d}q \cdot \mathrm{d}q$, and c is the speed of light. We shall use units such that $c = 1$. The map $q\colon \tau \mapsto q(\tau)$ describes the 'travel history' of the particle, while the map $w_\mathrm{E}\colon \tau \mapsto w_\mathrm{E}(\tau)$ describes the 'history of the particle's inertial rotation'. By $u(\tau) \stackrel{\text{def}}{=} \mathrm{d}q/\mathrm{d}\tau$ we denote the four-velocity of the particle, which with our convention of units is a dimensionless timelike unit vector, i.e., $u \cdot u = -1$. We will sometimes write $u = \mathring{q} = (\gamma, \gamma \dot{q})$, where $\dot{q}$ is the usual three-velocity, and where $\gamma \stackrel{\text{def}}{=} 1/\sqrt{1 - |\dot{q}|^2}$. By $\boldsymbol{\Omega}_\mathrm{T} = \mathring{u} \wedge u$ we denote the antisymmetric tensor of the celebrated *Thomas precession* [28] in Minkowski space-time ($\boldsymbol{\Omega}_\mathrm{FW}$ in [23]). By $\boldsymbol{\Omega}_\mathrm{E}$ we denote the antisymmetric tensor of the *Euler rotation*, which is the inertial rotation of the particle w.r.t. the co-moving particle frame $\{\bar{e}_\mu\}_{\mu=0,\ldots,3}$, with $\bar{e}_0 = u$, that satisfies the law of *Fermi–Walker transport* [23]

$$(3.3) \qquad \frac{\mathrm{d}}{\mathrm{d}\tau} \bar{e}_\mu = -\boldsymbol{\Omega}_\mathrm{T} \cdot \bar{e}_\mu.$$

Our four-vector w_E that describes the particle's history of inertial rotation is the dual to the tensor $\boldsymbol{\Omega}_\mathrm{E}$, whence $\boldsymbol{\Omega}_\mathrm{E} \cdot w_\mathrm{E} = 0$, and since $\boldsymbol{\Omega}_\mathrm{E} \cdot u = 0$, also $w_\mathrm{E} \cdot u = 0$, i.e. w_E is spacelike.

Electrically, the 'classical electron' carries a negative unit of the elementary charge e. This charge is distributed, in any of the particle's rest frames, by a spherically symmetric electrical charge density $f_e(|\,.\,|)\colon \mathbb{R}^3 \to \mathbb{R}^-$, satisfying

$$(3.4) \qquad \int_{\mathbb{R}^3} f_e(|\boldsymbol{x}|)\, \mathrm{d}^3 \boldsymbol{x} = -e.$$

In general, $f_e \colon \mathbb{R}^+ \to \mathbb{R}^-$ is a measure. The relativistic analog of (2.1), (2.2), (2.3) for such a charge distribution is the four-current density computed by Nodvik [24],

$$(3.5) \qquad j(x) = \int_{-\infty}^{+\infty} \bigl(u - \boldsymbol{\Omega} \cdot (x - q) \bigr) f_e\bigl(\|x - q\| \bigr) \delta\bigl(u \cdot (x - q) \bigr) \mathrm{d}\tau,$$

where

$$(3.6) \qquad \boldsymbol{\Omega} = \boldsymbol{\Omega}_\mathrm{T} + \boldsymbol{\Omega}_\mathrm{E}.$$

Considering, as before, only one particle, the four-current density (3.5) is the source term in the inhomogeneous Maxwell–Lorentz equations

$$\partial_x \cdot \mathsf{F}(x) = 4\pi j(x), \tag{3.7}$$

supplemented by the homogeneous Maxwell equations

$$\partial_x \cdot {}^*\mathsf{F}(x) = 0. \tag{3.8}$$

Here, F is the standard anti-symmetric, rank-two, electromagnetic field strength tensor, and ${}^*\mathsf{F}$ is the star dual to F.

Different from the semi-relativistic theory, we now also need a compactly supported spherical mass density $f_\mathrm{m}(|\,.\,|) : \mathbb{R}^3 \to \mathbb{R}^+$, where $f_\mathrm{m} : \mathbb{R}^+ \to \mathbb{R}^+$ is in general a measure giving the particle a positive *bare rest mass*

$$m_\mathrm{b} = \int_{\mathbb{R}^3} f_\mathrm{m}(|\boldsymbol{x}|)\, \mathrm{d}^3\boldsymbol{x}, \tag{3.9}$$

where $r = |\boldsymbol{x}|$, and positive nonrelativistic moment of inertia[5]

$$I_\mathrm{b} = \frac{2}{3} \int_{\mathbb{R}^3} |\boldsymbol{x}|^2 f_\mathrm{m}(|\boldsymbol{x}|)\, \mathrm{d}^3\boldsymbol{x}. \tag{3.10}$$

Through its inertial Euler rotation the bare particle acquires a *rotational relativistic mass*, which is simply the sum of its bare rest mass and its rotational kinetic energy,

$$\mathcal{M}(\omega_\mathrm{E}) = \int \frac{1}{\sqrt{1 - \|\boldsymbol{\Omega}_\mathrm{E} \cdot x\|^2}} f_\mathrm{m}(\|x\|) \delta(u \cdot x)\, \mathrm{d}^4 x. \tag{3.11}$$

Contact with the semi-relativistic CET of the previous section is made in the limit of small rotation frequency,

$$\mathcal{M}(\omega_\mathrm{E}) = m_\mathrm{b} + \frac{1}{2} I_\mathrm{b} \omega_\mathrm{E}^2 + O(\omega_\mathrm{E}^4), \tag{3.12}$$

where $\omega_\mathrm{E} = \sqrt{-\operatorname{Tr} \boldsymbol{\Omega}_\mathrm{E} \cdot \boldsymbol{\Omega}_\mathrm{E}/2}$ is the inertial Euler rotation frequency.

Our relativistic dynamical equations can be written in a format closely reminiscent of the semi-relativistic CET of the previous section. For the translational degrees of freedom, we have

$$\frac{\mathrm{d}}{\mathrm{d}\tau} p_\mathrm{M} = \int \mathsf{F}(q+x) \cdot (u - \boldsymbol{\Omega} \cdot x) f_\mathrm{e}(\|x\|) \delta(u \cdot x)\, \mathrm{d}^4 x, \tag{3.13}$$

where

$$p_\mathrm{M}(\tau) = \mathsf{M}_\mathrm{M}(\tau) \cdot u(\tau), \tag{3.14}$$

---

[5]It is here where Nodvik's relativistic model differs from ours. Nodvik assumes that the inertia is concentrated at the center of the particle, i.e. $f_\mathrm{m}(|\,.\,|) = m_\mathrm{b} \delta(\,.\,)$ in our setting. This implies that the rotational moment of bare inertia vanishes, and with it vanish several terms in the relativistic equations of motion. As a consequence the mathematical structure of a Cauchy problem is lost for the spin variable. Incidentally, this also removes the possibility of matching the physical electron data to Nodvik's model.

is the *Minkowski four-momentum*, with

$$(3.15) \quad \mathsf{M}_\mathrm{M} = \int \frac{1}{\sqrt{1 - \|\mathbf{\Omega}_\mathrm{E} \cdot x\|^2}} f_\mathrm{m}(\|x\|) \delta(u \cdot x) \, \mathrm{d}^4 x \, \mathsf{g}$$

$$- \int \big[x \otimes x, [\mathsf{F}(q+x), \mathbf{\Omega}_\mathrm{E}]_+\big]_+ f_\mathrm{e}(\|x\|) \delta(u \cdot x) \, \mathrm{d}^4 x$$

the *symmetric Minkowski mass tensor*. The first term is simply the rotational bare mass of the particle. The second term originates in an electromagnetic spin-orbit coupling and describes a generally anisotropic translational inertia of the spinning charged particle. For the rotational history, we have

$$(3.16) \quad \frac{\mathrm{d}}{\mathrm{d}\tau} \mathsf{S}_\mathrm{b} + [\mathbf{\Omega}, \mathsf{S}_\mathrm{b}]_-$$

$$= \int x \wedge (\mathsf{g} + u \otimes u) \cdot \mathsf{F}(q+x) \cdot (u - \mathbf{\Omega} \cdot x) f_\mathrm{e}(\|x\|) \, \delta(u \cdot x) \, \mathrm{d}^4 x,$$

where

$$(3.17) \quad \mathsf{S}_\mathrm{b}(\tau) = -\int x \wedge \frac{\mathbf{\Omega}_\mathrm{E}(\tau) \cdot x}{\sqrt{1 - \|\mathbf{\Omega}_\mathrm{E}(\tau) \cdot x\|^2}} f_\mathrm{m}(\|x\|) \delta\big(u(\tau) \cdot x\big) \, \mathrm{d}^4 x$$

is the *spin angular momentum* (about q) of the bare particle. Notice that

$$(3.18) \quad \mathsf{S}_\mathrm{b} \cdot u = 0,$$

so that S_b is of space-space type w.r.t. u.

The equations for the histories $\tau \mapsto q(\tau)$, $\tau \mapsto \mathbf{\Omega}_\mathrm{E}(\tau)$, have to be supplemented by the Cauchy data $q(\tau_1) = q_1$, $u(\tau_1) = u_1$, with u_1 satisfying $\|u_1\|^2 = -1$, furthermore $\mathbf{\Omega}_\mathrm{E}(\tau_1) = \mathbf{\Omega}_{\mathrm{E}1}$ and $\mathsf{F}(\tau_1) = \mathsf{F}_1$.

So far, under the choice $f_\mathrm{m}(|\,.\,|) = m_\mathrm{b} \delta(|\,.\,| - R)$, so that the bare rest mass is concentrated on a sphere, and similarly for the charge, we showed that our dynamical equations for the variables q, u, $\mathbf{\Omega}_\mathrm{E}$, F are locally well-posedness as a Cauchy problem if the initial data for the fields are sufficiently small. For the special case that only the spin degree of freedom is involved in the dynamics, we also proved global existence and uniqueness. Global well-posedness as Cauchy problem of our set of equations for the variables q, u, $\mathbf{\Omega}_\mathrm{E}$, F is an important issue that we will have to be addressed in future work.

Given global existence, our equations satisfy the conservation laws analogous to the semi-relativistic CET with spin. In addition, there is a further conservation law for the magnitude of the renormalized spin.

Since our relativistic CET is superficially reminiscent of the semi-relativistic CET, let me conclude with pointing out some peculiar points about it. The four-momentum p_M is not scalarly proportional to the four-velocity but rather related by a linear tensor proportion. Hence, in general we will have $u \cdot f \neq 0$ for f given in the right side of (3.13). Another significant difference to the semi-relativistic Abraham model of a spinning charge is the occurrence of $\dot{u}$, through $\mathbf{\Omega}$, in the Minkowski force. This term has its origin in the Thomas precession. Therefore the equation of motion is, in its current form, rather implicit in the highest derivative. This makes their control more subtle than in the semi-relativistic case.

References

1. M. Abraham, *Theorie der Elektrizität*, II, Teubner, Leipzig, 1905; 1923.
2. W. Appel and M. K.-H. Kiessling, *Mass and spin renormalization in Lorentz electrodynamics*, Ann. Physics (2001) (to appear).
3. V. Bargmann, L. Michel, and V. L. Telegdi, *Precession of the polarization of particles moving in a homogeneous electromagnetic field*, Phys. Rev. Lett. **2** (1955), 435–436.
4. G. Bauer and D. Dürr, *The Maxwell–Lorentz system of a rigid charge distribution*, Preprint, LMU München (1999).
5. P. A. M. Dirac, *Classical theory of radiating electrons*, Proc. Roy. Soc. London. Ser. A **167** (1938), 148–169.
6. E. Fermi, *Über einen Widerspruch der elektrodynamischen und der relativistischen Theorie der elektromagnetischen Masse*, Physik. Zeitschr. **23** (1922), 340–344.
7. R. Fernandez, J. Fröhlich, and A. Sokal, *Random walks, critical phenomena, and triviality in quantum field theory*, Springer, New York, 1992.
8. R. Glassey and J. Schaeffer, *Global existence for the relativistic Vlasov–Maxwell system with nearly neutral initial data*, Comm. Math. Phys. **119** (1988), 353–384.
9. R. Glassey and W. Strauss, *Absence of shocks in an initially dilute collisionless plasma*, Comm. Math. Phys. **113** (1987), 191–208.
10. _____, *Singularity formation in a collisionless plasma could occur only at high velocities*, Arch. Rational Mech. Anal. **92** (1986), 59–90.
11. J. Glimm and A. Jaffe, *Quantum physics*, 2nd. ed., Springer, New York, 1980.
12. J. D. Jackson, *Classical electrodynamics*, 3rd ed., Wiley, New York, 1999.
13. J. M. Jauch and F. Rohrlich, *The theory of photons and electrons*, Springer, New York, 1976.
14. M. K.-H. Kiessling, *Classical electron theory and conservation laws*, Phys. Lett. A **258** (1999), 197–204.
15. A. Komech, M. Kunze, and H. Spohn, *Long-time asymptotics for a classical particle interacting with a scalar wave field*, Comm. Partial Differential Equations **22** (1997), 307–335.
16. _____, *Effective dynamics for a mechanical particle coupled to a wave field*, Comm. Math. Phys. **203** (1999), 1–19.
17. A. Komech and H. Spohn, *Long-time asymptotics for the coupled Maxwell–Lorentz equations*, Comm. Partial Differential Equations **25** (2000), no. 3-4, 559–584.
18. M. Kunze and H. Spohn, *Adiabatic limit of the Maxwell–Lorentz equations*, Ann. Inst. H. Poincaré, Phys. Théor. (in press).
19. _____, *Radiation reaction and center manifolds*, SIAM J. Math. Anal. (in press).
20. H. A. Lorentz, *La théorie électromagnétique de Maxwell et son application aux corps mouvants*, Archives néerl. Sciences exactes et naturelles **25** (1892), 363–552.
21. _____, *Electromagnetic phenomena in a system moving with any velocity less than that of light*, Proc. K. Akad. Wet. **6** (1904), Amsterdam.
22. _____, *The theory of electrons and its applications to the phenomena of light and radiant heat*, 2nd ed., 1915; reprinted by Dover, New York, 1952.
23. C. W. Misner, K. S. Thorne, and J. A. Wheeler, *Gravitation*, W. H. Freeman Co., New York, 1973.
24. J. S. Nodvik, *A covariant formulation of classical electrodynamics for charges of finite extension*, Ann. Phys. **28** (1964), 225–319.
25. F. Rohrlich, *Classical charged particles*, Addison Wesley, Redwood City, 1990.
26. A. Sommerfeld, *Electrodynamics*, Academic Press, New York, 1952.
27. H. Spohn, *Runaway charged particles and center manifold theory*, Preprint, TU Munich, 1998.
28. L. H. Thomas, *The motion of the spinning electron*, Nature **117** (1926), 514; *On the kinematics of an electron with an axis*, Phil. Mag. **3** (1927), 1–22.
29. S. Tomonaga, *The story of spin*, Univ. Chicago Press, 1997.
30. G. E. Uhlenbeck and S. A. Goudsmit, *Spinning electrons and the structure of spectra*, Nature **117** (1926), 264–265.
31. A. A. Vlasov, *Many-particle theory and its application to plasma*, Trans. Russian Gordon and Breach Science Publishers, Inc., New York, 1961.
32. A. D. Yaghjian, *Relativistic dynamics of a charged sphere*, Lect. Notes Phys. **11**, Springer, Berlin, 1992.

Department of Mathematics, Rutgers University, 110 Frelinghuysen Rd., Piscataway, NJ 08854, USA.

E-mail address: `miki@math.rutgers.edu`

Singular Limit of the Modified Nonlinear Schrödinger Equation

Chi-Kun Lin

Dedicated to Professor Fon-Che Liu on his sixtieth birthday

ABSTRACT. We study the semiclassical limit of the general modified nonlinear Schrödinger equation for initial data with Sobolev regularity, before shocks appear in the limit system. The strict hyperbolicity and genuine nonlinearity are proved for the dispersion limit of the cubic nonlinear case. The limiting transition from MNLS equation to the NLS equation is also discussed.

1. Introduction

In this paper we consider the Cauchy problem with rapidly oscillating initial data for the general modified nonlinear Schrödinger equation:

$$i\hbar\partial_t\psi^\hbar + \frac{\hbar^2}{2}\partial_x^2\psi^\hbar + i\frac{\hbar}{2}\partial_x\big(\Phi'(|\psi^\hbar|^2)\psi^\hbar\big) - U'(|\psi^\hbar|^2)\psi^\hbar = 0, \tag{1.1}$$

$$\psi^\hbar(0,x) = \psi_0^\hbar(x) = A_0^\hbar(x)\exp\Big(\frac{i}{\hbar}S_0(x)\Big), \tag{1.2}$$

where $\Phi, U \in C^\infty(\mathbb{R}^+;\mathbb{R})$, $S_0 \in H^s(\mathbb{R})$ for s large enough, and $A_0^\hbar$ is a function, polynomial in $\hbar$, with coefficients of Sobolev regularity in x. We will study the behavior of solutions of the problem (1.1), (1.2) as $\hbar \to +0$ and $-\infty < x < \infty$, $0 \le t \le T$, i.e., within an arbitrary finite time T. In the sequel we will refer to this system as MNLS equation. When $\Phi' = 0$, (1.1) is the nonlinear Schrödinger equation (NLS equation for short). If $U' = 0$ then MNLS equation (1.1) reduces to the derivative nonlinear Schrödinger equation (DNLS equation for short). Compared with the NLS equation, the present MNLS equation (1.1) includes an additional term, the space derivative nonlinear term $i\hbar/2\partial_x\big(\Phi'(|\psi|^2)\psi\big)$, which plays an important role when the characteristic length of the variation of the envelope and the wave length of the carrier wave become of the same order. Thus

---

2000 *Mathematics Subject Classification.* Primary: 35Q40, 35Q53; Secondary: 76Y07.

This work is partial supported by the National Science Council of Taiwan under the grant NSC 89-2914-1-006-032-A1. The author thanks Professors Catherine Sulem and Michael Sigal for their invitation and the Centre de recherches mathématiques in Montreal, Canada for their hospitality.

This is the final form of the paper.

©2001 American Mathematical Society

the features of the steepening of the envelope and the modulational instabilities are affected significantly by the additional term except at an early stage of the time evolution.

The semiclassical limit of the MNLS equation (1.1)–(1.2) is to determine the limiting dynamics of any function of the field $\psi^\hbar$ as $\hbar \to 0$. However, it is not clear directly from (1.1) what form such a dynamics might take. Insight into this question can be gained by considering the conservation laws associated with the MNLS equation. Recall that the NLS equation, also well known as the Gross–Pitaevskiy equation, appear in the phenomenological description of superfluidity of an almost ideal Bose gas. In this case, the squared modulus of the wave function $\bar\psi\psi$ is interpreted as the particle number density in the condensate state, while the gradient of the phase is proportional to the superfluid velocity $\mathbf{u} = \nabla \arg \psi$. The similar hydrodynamical interpretation for the decomposition of quantum mechanical wave function was considered by Madelung in 1927, which applies to MNLS equation (1.1), too. By decomposing the wave function $\psi \in \mathbf{C}$, $\psi = A\exp(iS/\hbar)$, in terms of A, S, the model (1.1) can be represented as the Madelung fluid

$$(1.3) \qquad \partial_t A + \frac{1}{2}(AS_{xx} + 2A_x S_x) + \frac{1}{2}\partial_x\big(\Phi'(|A|^2)A\big) = 0,$$

$$(1.4) \qquad \partial_t S + \frac{1}{2}(S_x)^2 + \frac{1}{2}\Phi'(|A|^2)S_x + U'(|A|^2) = \frac{\hbar^2}{2}\frac{A_{xx}}{A},$$

which are equivalent to the MNLS equation (1.1). Equation (1.4) is the Hamilton–Jacobi equation and (1.3) turns out to be the continuity equation for the "quantum fluid". Introducing two new functions $\rho = |A|^2 = |\psi|^2$, $u = S_x$, we have the system

$$(1.5) \qquad \partial_t \rho + \partial_x\left(\rho u + \rho\Phi'(\rho) - \frac{1}{2}\Phi(\rho)\right) = 0,$$

$$(1.6) \qquad \partial_t u + u\partial_x u + \partial_x\left(\frac{1}{2}\Phi'(\rho)u\right) + \partial_x\big(U'(\rho)\big) = \frac{\hbar^2}{2}\partial_x\left(\frac{\partial_x^2\sqrt{\rho}}{\sqrt{\rho}}\right).$$

It follows immediately from (1.5)–(1.6) that the momentum $\mu \equiv \rho u$ satisfies differential equation

$$(1.7) \quad \partial_t \mu + \partial_x\left(\frac{\mu^2}{\rho} + \frac{1}{2}\mu\Phi'(\rho)\right) + \rho U''(\rho)\partial_x\rho + \mu\Phi''(\rho)\partial_x\rho = \frac{\hbar^2}{4}\partial_x(\rho\partial_x^2\log\rho),$$

which is not a conservation law except when $\Phi''(\rho) = 0$. This is due to the *derivative loss* caused by the space derivative nonlinear term, $i\hbar/2\partial_x\big(\Phi'(|\psi|^2)\psi\big)$. Like the DNLS equation, the MNLS equation (1.1) does not possess parity and Galilean invariance and therefore the canonical momentum is not conservative although it is translation invariant with respect to space variable x. To obtain the conservation law of the momentum, we have to introduce the *noncanonical momentum* which is indeed *the* conservative quantity of the MNLS equation. Indeed, the transport equation (1.5) also gives

$$(1.8) \qquad \partial_t \Phi(\rho) + \partial_x\big(\Phi(\rho)u\big) + \big(\rho\Phi'(\rho) - \Phi(\rho)\big)\partial_x u + \partial_x\left(\frac{1}{2}\rho\big(\Phi'(\rho)\big)^2\right) = 0.$$

As in [3], we need to introduce the noncanonical momentum

$$(1.9) \qquad M \equiv \mu + \Phi(\rho) = \rho u + \Phi(\rho),$$

it means that even if the fluid velocity vanishes, i.e., $u = 0$, the flow has background momentum with characteristic speed. This is the point that solitons of DNLS equation have nontrivial static limit. In the field theoretical language we can say that the spectrum of excitations has always the gap (like in superfluidity). We have the local conservation laws associated with the MNLS equation (1.1):

$$\partial_t \rho + \partial_x \left(M + \rho \Phi' - \frac{3}{2}\Phi \right) = 0, \tag{1.10}$$

$$\partial_t M + \partial_x \left[\frac{M}{\rho} \left(M + \rho \Phi' - \frac{3}{2}\Phi + \frac{1}{2}(\rho\Phi' - \Phi) \right) \right] \tag{1.11}$$
$$+ \partial_x \left[\frac{1}{\rho}(\rho\Phi' - \Phi)\left(\frac{1}{2}\rho\Phi' - \Phi \right) \right] + \partial_x P(\rho) = \frac{\hbar^2}{4}\partial_x(\rho \partial_x^2 \log \rho),$$

where $P(\rho) = \rho U'(\rho) - U(\rho)$ is the canonical pressure. Equation (1.11) is derived by adding (1.7) and (1.8) together. This equation also tells us that the space derivative nonlinear term, $i\hbar/2 \partial_x(\Phi'(|\psi|^2)\psi)$, not only affects the momentum but also the pressure from the point of view of hydrodynamics. Equations (1.10) and (1.11) correspond to the mass and momentum conservation laws respectively. These densities are related to the wave function $\psi = \psi(t,x)$ through

$$\rho = |\psi|^2, \quad M = \frac{i\hbar}{2}(\psi\bar{\psi}_x - \psi_x\bar{\psi}) + \Phi(|\psi|^2), \tag{1.12}$$

where the bar denotes the complex conjugate. The initial conserved densities are then

$$\rho(0,x) = |A_0(x)|^2, \quad M(0,x) = |A_0(x)|^2 \partial_x S_0(x) + \Phi(|A_0(x)|^2). \tag{1.13}$$

Equations (1.10)–(1.11) comprise a closed system governing ρ and M which have the form of a perturbation of the modified Euler equations. Arguing formally, the modified Euler system that describe the formal semiclassical limit reduces to (formally letting $\hbar \to 0$ in (1.10)–(1.11))

$$\partial_t \rho + \partial_x \left(M + \rho\Phi' - \frac{3}{2}\Phi \right) = 0, \tag{1.14}$$

$$\partial_t M + \partial_x \left[\frac{M}{\rho}\left(M + \rho\Phi' - \frac{3}{2}\Phi + \frac{1}{2}(\rho\Phi' - \Phi) \right) \right] \tag{1.15}$$
$$+ \partial_x \left[\frac{1}{\rho}(\rho\Phi' - \Phi)\left(12\rho\Phi' - \Phi \right) \right] + \partial_x P(\rho) = 0,$$

with initial data

$$\rho(0,x) = \rho_0(x), \quad M(0,x) = M_0(x). \tag{1.16}$$

The rest of the paper is organized as follows. In Section 2, we represent the MNLS equation as a linear dispersive perturbation of a symmetric quasilinear hyperbolic system, and prove the semiclassical limit by employing the classical theory of quasilinear hyperbolic systems. The Section 3 is devoted to the cubic nonlinear case. We apply Lax's theory to prove the strict hyperbolicity and genuine nonlinearity for the dispersion limit of the MNLS equation. In Section 4, we study the semiclassical limit of the MNLS equation. We also compare this result with the case of the NLS equation and discuss the transition from MNLS equation to NLS equation. This explain that the effect of the space derivative nonlinear term in MNLS equation can be neglected if the characteristic length of the variation of envelope is much larger than the wave length of the carrier wave. Therefore the MNLS equation reduces to the NLS equation.

2. Semiclassical Limit and WKB Expansion

The conservation laws (1.10)–(1.11) give the insight into the semiclassical limit of the MNLS equation (1.1). However, due to the nonlinear dispersive term, $\hbar^2/4\partial_x(\rho\partial_x^2 \log \rho)$, it is still difficult to treat this problem directly from (1.10)–(1.11). As suggested by Grenier [4] (see also [3]), the modified Madelung's transformation can be utilized in the study of semi-classical limit. More precisely, we will look for solutions $\psi^\hbar$ of the form

$$(2.1) \qquad \psi^\hbar(t,x) = A^\hbar(t,x) \exp\left(\frac{i}{\hbar} S^\hbar(t,x)\right),$$

where the complex-value function $A^\hbar = a^\hbar + ib^\hbar$ represents the amplitude and the real-valued $S^\hbar$ represents the phase. Here we allow the phase function $S^\hbar$ to depend on $\hbar$.

Now insert (2.1) in MNLS equation (1.1), we obtain

$$(2.2) \quad iA_t^\hbar - \frac{1}{\hbar}A^\hbar S_t^\hbar + \frac{\hbar}{2}A_{xx}^\hbar - \frac{1}{2\hbar}A^\hbar(S_x^\hbar)^2 + \frac{i}{2}(A^\hbar S_{xx}^\hbar + 2A_x^\hbar S_x^\hbar)$$
$$- \frac{1}{\hbar}U'A^\hbar - \frac{1}{2\hbar}\Phi' A^\hbar S_x^\hbar + \frac{i}{2}\Phi' A_x^\hbar + i\frac{1}{2}A^\hbar \Phi'' \partial_x |A^\hbar|^2 = 0,$$

where U', Φ' and Φ'' denote $U'(|A^\hbar|^2)$, $\Phi'(|A^\hbar|^2)$ and $\Phi''(|A^\hbar|^2)$ respectively. We split (2.2) into two parts, of order $O(1/\hbar)$ and $O(1)$ respectively;

$$(2.3) \qquad S_t^\hbar + \frac{1}{2}(S_x^\hbar)^2 + \frac{1}{2}\Phi' S_x^\hbar + U' = 0,$$

$$(2.4) \qquad iA_t^\hbar + \frac{i}{2}(A^\hbar S_{xx}^\hbar + 2A_x^\hbar S_x^\hbar) + \frac{i}{2}\Phi' A_x^\hbar + \frac{i}{2}A^\hbar \Phi'' \partial_x |A^\hbar|^2 + \frac{\hbar}{2}A_{xx}^\hbar = 0.$$

The expression (2.3)–(2.4) is not the same as (1.3)–(1.4) where we split into the real and imaginary parts. The second derivative term (the dispersive term) is highly nonlinear in (1.4) but it is *linear* in (2.4). Hence the classical quasilinear hyperbolic theory can be applied to the semi-classical limit of Schrödinger type equation. Considering the change of variable $w^\hbar = S_x^\hbar$ and using the fact that $A^\hbar = a^\hbar + ib^\hbar$ we have

$$(2.5) \quad a_t^\hbar + \left(w^\hbar + \frac{1}{2}\Phi'\right)a_x^\hbar + \frac{1}{2}\Phi'' a^\hbar \left[(a^\hbar)^2 + (b^\hbar)^2\right]_x + \frac{1}{2}a^\hbar w_x^\hbar = -\frac{\hbar}{2}b_{xx}^\hbar$$

$$(2.6) \quad b_t^\hbar + \left(w^\hbar + \frac{1}{2}\Phi'\right)b_x^\hbar + \frac{1}{2}\Phi'' b^\hbar \left[(a^\hbar)^2 + (b^\hbar)^2\right]_x + \frac{1}{2}b^\hbar w_x^\hbar = \frac{\hbar}{2}a_{xx}^\hbar$$

$$(2.7) \quad w_t^\hbar + \left(w^\hbar + 12\Phi'\right)w_x^\hbar + \left(\frac{1}{2}\Phi'' w^\hbar + U''\right)\left[(a^\hbar)^2 + (b^\hbar)^2\right]_x = 0.$$

This system can be written in the form

$$(2.8) \qquad V_t^\hbar + \mathcal{A}(V^\hbar) V_x^\hbar = \frac{\hbar}{2}\mathcal{L}(V^\hbar), \quad V^\hbar = (a^\hbar, b^\hbar, w^\hbar)^t,$$

where

$$(2.9) \quad \mathcal{A}(V^\hbar) \equiv \begin{pmatrix} w^\hbar + \frac{1}{2}\Phi' + (a^\hbar)^2 \Phi'' & a^\hbar b^\hbar \Phi'' & \frac{1}{2}a^\hbar \\ a^\hbar b^\hbar \Phi'' & w^\hbar + \frac{1}{2}\Phi' + (b^\hbar)^2 \Phi'' & \frac{1}{2}b^\hbar \\ a^\hbar(w^\hbar \Phi'' + 2U'') & b^\hbar(w^\hbar \Phi'' + 2U'') & w^\hbar + \frac{1}{2}\Phi' \end{pmatrix}$$

and

$$\mathcal{L}(V^\hbar) = \begin{pmatrix} 0 & -\partial_x^2 & 0 \\ \partial_x^2 & 0 & 0 \\ 0 & 0 & 0 \end{pmatrix} \begin{pmatrix} a^\hbar \\ b^\hbar \\ w^\hbar \end{pmatrix} = \begin{pmatrix} -b^\hbar_{xx} \\ a^\hbar_{xx} \\ 0 \end{pmatrix}, \tag{2.10}$$

is an antisymmetric matrix. The matrix $\mathcal{A}(V^\hbar)$ can be symmetrized by

$$\mathcal{S}(V^\hbar) = \begin{pmatrix} 2w^\hbar \Phi'' + 4U'' & 0 & 0 \\ 0 & 2w^\hbar \Phi'' + 4U'' & 0 \\ 0 & 0 & 1 \end{pmatrix} \tag{2.11}$$

which is symmetric and positive if $w^\hbar \Phi'' + 2U'' > 0$, for all $V^\hbar$. Thus, we write MNLS equation (1.1) as a *linear dispersive perturbation of a quasilinear symmetric hyperbolic system*:

$$\mathcal{S}(V)V_t + \tilde{\mathcal{A}}(V)V_x = \frac{\hbar}{2}\tilde{\mathcal{L}}(V), \tag{2.12}$$

where $\tilde{\mathcal{A}} = \mathcal{S}\mathcal{A}$ is a symmetric matrix (we omit $\hbar$ for convenience). The antisymmetric operator $\hbar/2\tilde{\mathcal{L}} = \hbar/2\mathcal{S}\mathcal{L}$ reflects the dispersive nature of the equations. Moreover, the classical energy estimate shows that this term contributes nothing to the estimate, i.e., the singular perturbation does not create energy. Therefore, the existence of the classical solutions and its semiclassical limit proceed along the lines of the classical theory of quasilinear symmetric hyperbolic systems (see [**7,15**]) with modifications. Indeed, we have the following theorems.

THEOREM 2.1. *Let $\Phi, U \in C^\infty(\mathbb{R}^+, \mathbb{R})$ with $w^\hbar \Phi'' + 2U'' > 0$. Let $s > 2 + 1/2$, for $A_0^\hbar$ and $\partial_x S_0$ uniformly bounded in H^s, solutions $\psi^\hbar$ to equation (1.1) exists on a small time interval $[0, T]$, T independent of $\hbar$. Moreover, $\psi^\hbar(t, x) = A^\hbar(t, x) \exp(i/\hbar S^\hbar(t, x))$ with $A^\hbar$ and $S_x^\hbar$ bounded in $L^\infty([0, T]; H^s)$ uniformly in $\hbar$ and limit points of $(A^\hbar, S_x^\hbar)$ are solutions of the modified Euler equation (1.14)–(1.16).*

THEOREM 2.2. *Let (ρ, M) be a solution of the quasilinear hyperbolic system (1.14)–(1.16) (the modified Euler equations) for $0 \leq t \leq T$ with initial condition*

$$\rho_0(x) = \rho(0, x) = |A_0(x)|^2,$$
$$M_0(x) = M(0, x) = |A_0(x)|^2 \partial_x S_0(x) + \Phi(|A_0(x)|^2).$$

Then there exists a critical value of $\hbar$, $\hbar_c$, and a constant $C > 0$ such that under the hypothesis

  (1) $A_0^\hbar(x)$ *converges strongly to* A_0 *in* H^s *as $\hbar$ tends to 0,*
  (2) $\|\rho_0\|_{H^s} < \infty$, $\|M_0\|_{H^s} < \infty$, $s \geq 3$,
  (3) $0 < \hbar < \hbar_c$,

the initial value problem for the MNLS equation (1.1) has a unique classical solution of the form $\psi^\hbar(t,x) = A^\hbar(t,x)\exp(i/\hbar S^\hbar(t,x))$ on $[0,T]$. Moreover, $A^\hbar$ and $S_x^\hbar$ are bounded in $L^\infty([0,T]; H^s)$ uniformly in $\hbar$.

The proofs of the above theorems is nearly identical to those given in the semiclassical limit of the DNLS equation [**3**]; we therefore omit the details.

REMARK 2.3. The hypothesis $w^\hbar \Phi'' + 2U'' > 0$ will guarantee the hyperbolicity of the *Euler part* of (1.10)–(1.11). The theorem is true even if U is not a strictly convex function provided $\Phi'' \neq 0$. The above theorems also apply to other Schrödinger type equations. In the case of NLS equation, $\Phi' = 0$, in this case we replace the hypothesis $w^\hbar \Phi'' + 2U'' > 0$ by $U'' > 0$ (see [**4**] and [**5**]). This means that U must be a strictly convex function of ρ and corresponds to the *defocusing* NLS equation. When $U' = 0$, the DNLS equation, we require $w^\hbar \Phi'' > 0$. Hence, Φ is a strictly convex or concave function of ρ depending on the sign of fluid velocity $S_x^\hbar = w^\hbar$ (see [**3**]).

THEOREM 2.4 (WKB expansion). *Under the assumption of Theorem 2.1, suppose the initial amplitude $A_0^\hbar(x)$ admits the following expansion:*

$$(2.13) \qquad A_0^\hbar(x) = \sum_{\kappa=0}^{N} A_0^{(\kappa)}(x)\hbar^\kappa + R_N^{in}(x,\hbar)\hbar^N,$$

where

$$(2.14) \qquad \lim_{\hbar \to 0} \|R_N^{in}(x,\hbar)\|_{H^\sigma} = 0$$

for $N \in \mathbf{N}$ and $\sigma 2N + 2 + 1/2$, then the solution of the MNLS equation (1.1)–(1.2) can be represented as

$$(2.15) \quad \psi^\hbar(t,x) = A^\hbar(t,x) \exp\left(\frac{i}{\hbar} S^\hbar(t,x)\right)$$

$$= \sum_{\kappa=0}^{N} A^{(\kappa)}(t,x)\hbar^\kappa \exp\left(\frac{i}{\hbar} S(t,x)\right) + R_N(t,x,\hbar)\hbar^N$$

and where

$$(2.16) \qquad \lim_{\hbar \to 0} \|R_N(t,x,\hbar)\|_{C([0,T];H^{\sigma-2N-\varepsilon})} = 0, \quad \forall \varepsilon > 0.$$

3. Hyperbolic Nature of the Semiclassical Limit

In this section we will consider the MNLS equation in the following form

$$(3.1) \qquad i\hbar \partial_t \psi^\hbar + \frac{\hbar^2}{2} \partial_x^2 \psi^\hbar + \frac{i\hbar}{2} \partial_x(\alpha |\psi^\hbar|^2 \psi^\hbar) + (1 - |\psi^\hbar|^2)\psi^\hbar = 0,$$

where the term with the real parameter α governs the effect of the Kerr nonlinearity dispersion. For brevity we suppress the superscript $\hbar$ of $\rho^\hbar$, $\mu^\hbar$ and $M^\hbar$. In this case

$$(3.2) \qquad \Phi(\rho) = \alpha\rho^2, \quad M = \mu + \alpha\rho^2, \quad U(\rho) = \frac{1}{2}(\rho - 1)^2$$

and (1.14)–(1.15) become

$$(3.3) \qquad \partial_t \rho + \partial_x\left(M + \frac{\alpha}{2}\rho^2\right) = 0,$$

$$(3.4) \qquad \partial_t M + \partial_x\left(\frac{M^2}{\rho} + \alpha\rho M + \frac{\rho^2}{2}\right) = \frac{\hbar^2}{4}\partial_x(\rho \partial_x^2 \log \rho).$$

The energy is decomposed into classical, noncanonical and quantum parts respectively. Each part propagates according to

$$(3.5) \quad \partial_t\left(\frac{M^2}{2\rho} + \frac{1}{2}(\rho-1)^2\right) + \partial_x\left[\left(\frac{M^2}{2\rho} + \frac{1}{2}(\rho-1)^2 + \frac{1}{2}(\rho^2-1)\right) \cdot \frac{M}{\rho}\right]$$
$$+ \alpha\partial_x\left(\frac{M^2}{2} + \frac{\rho^3}{3} - \frac{\rho^2}{2}\right) + \alpha\frac{M^2}{2\rho}\partial_x\rho = \frac{\hbar^2}{4}\frac{M}{\rho}\partial_x\left(\rho\partial_x^2 \log\rho\right),$$

$$(3.6) \quad \partial_t(\rho M) + \partial_x\left(\frac{3}{2}M^2 + \alpha\rho^2 M + \frac{\rho^3}{3}\right) - \frac{M^2}{\rho}\partial_x\rho = \frac{\hbar^2}{4}\rho\partial_x\left(\rho\partial_x^2\log\rho\right)$$

and

$$(3.7) \quad \partial_t\left(\frac{\rho_x^2}{2\rho}\right) + \partial_x\left(\frac{\rho_x^2}{2\rho}\cdot\frac{M}{\rho}\right) + \frac{\alpha}{2}\partial_x\left(\frac{5}{2}\rho_x^2 - \rho\rho_{xx}\right)$$
$$= \partial_x\left(\frac{\rho_{xx}M}{\rho} - \frac{\rho_x M_x}{\rho}\right) - \frac{M}{\rho}\partial_x(\rho\partial_x^2\log\rho) - \frac{\alpha}{2}\rho\partial_x(\rho\partial_x^2\log\rho),$$

respectively. Therefore the local conservation law of energy is derived by (3.5) + $\alpha 2 \cdot (3.6) + \hbar^2/4 \cdot (3.7)$;

$$(3.8) \quad \partial_t E + \partial_x\left(\frac{M}{\rho}(E_1 + P)\right) + \alpha\partial_x\left(\frac{5}{4}M^2 + \frac{\alpha}{2}\rho^2 M + \frac{\rho^3}{2} - \frac{\rho^2}{2}\right)$$
$$+ \frac{\alpha\hbar^2}{8}\partial_x\left(\frac{5}{2}\rho_x^2 - \rho\rho_{xx}\right) = \frac{\hbar^2}{4}\partial_x\left(\frac{\rho_{xx}M}{\rho} - \frac{\rho_x M_x}{\rho}\right),$$

where $P(\rho) = 1/2(\rho^2 - 1)$ and

$$(3.9) \quad E = E_1 + E_2 + E_3 \equiv \left(\frac{M^2}{2\rho} + \frac{1}{2}(\rho-1)^2\right) + \frac{\alpha}{2}\rho M + \frac{\hbar^2}{4}\cdot\frac{\rho_x^2}{2\rho}.$$

The crossing term $E_2 = \alpha/2\rho M$ comes from the first order space derivative nonlinear term. Arguing formally, the modified Euler system that describe the formal semiclassical limit reduces to (formally letting $\hbar \to 0$ in (3.3)–(3.4))

$$(3.10) \quad \partial_t\rho + \partial_x\left(M + \frac{\alpha}{2}\rho^2\right) = 0,$$

$$(3.11) \quad \partial_t M + \partial_x\left(\frac{M^2}{\rho} + \alpha\rho M + \frac{\rho^2}{2}\right) = 0,$$

with the initial condition

$$\rho(0,x) = A^2(x), \quad M(0,x) = A^2(x)S_x(x) + A^4(x).$$

The energy equation (3) becomes

$$(3.12) \quad \partial_t\left(\frac{M^2}{2\rho} + \frac{\alpha}{2}\rho M + \frac{1}{2}(\rho-1)^2\right)$$
$$+ \partial_x\left(\left(\frac{M^2}{2\rho} + \frac{1}{2}(\rho-1)^2 + \frac{1}{2}(\rho^2-1)\right)\cdot\frac{M}{\rho}\right)$$
$$+ \alpha\partial_x\left(\frac{5}{4}M^2 + \frac{\alpha}{2}\rho^2 M + \frac{\rho^3}{3} - \frac{\rho^2}{2}\right) = 0.$$

Riemann invariants for the modified Euler system (3.10)–(3.11) are given by

$$R_\pm = \sqrt{\frac{\alpha M}{\rho} + 1} \pm \alpha\sqrt{\rho},$$

and the system can be placed in the Riemann invariant form

(3.13) $$\partial_t R_+ + \frac{1}{\alpha}\left(\frac{3}{4}R_+^2 + \frac{1}{4}R_-^2 - 1\right)\partial_x R_+ = 0,$$

(3.14) $$\partial_t R_- + \frac{1}{\alpha}\left(\frac{1}{4}R_+^2 + \frac{3}{4}R_-^2 - 1\right)\partial_x R_- = 0.$$

THEOREM 3.1. *The break-time t_b for (3.10)–(3.11) can be estimated in the following way*:

$$t_b = \min\{t_{+,b}, t_{-,b}\},$$

where

$$t_{\pm,b} \leq \inf_{x_0 \in \Omega_\pm}\{t\colon G_\pm(t, x_0) = 0\}, \quad \Omega_\pm = \{x_0\colon \operatorname{sgn}\alpha \cdot \partial_x R_\pm(x_0) \leq 0\},$$

with

$$G_\pm = 1 + \alpha\frac{3}{2}\frac{\partial_x R_\pm(x)}{\sqrt{R_+^2(x) - R_-^2(x)}}$$

$$\times \int_0^t R_\pm\bigl(\tau, x_\pm(\tau)\bigr)\sqrt{R_+^2\bigl(\tau, x_\pm(\tau)\bigr) - R_-^2\bigl(\tau, x_\pm(\tau)\bigr)}\,d\tau$$

and particle path $x_\pm = x_\pm(t)$ satisfies the differential equation

$$\frac{dx_\pm}{dt} = \frac{1}{2\alpha}R_\pm^2(t, x_\pm) + \frac{1}{4\alpha}\bigl[R_+^2(t, x_\pm) + R_-^2(t, x_\pm) - 1\bigr], \quad x_\pm(0) = x.$$

PROOF. The break-time t_b can be estimated by Lax's recipe (see [**9**] or [**15**]). We start from the Riemann invariant form (3.13);

(3.15) $$R'_+ = \partial_t R_+ + \frac{1}{\alpha}\left(\frac{3}{4}R_+^2 + \frac{1}{4}R_-^2 - 1\right)\partial_x R_+ = 0,$$

where $'$ denotes the differentiation in the characteristic direction

$$\frac{dx_+}{dt}(t) = \lambda_+\bigl(t, x_+(t)\bigr) = \frac{1}{\alpha}\left(\frac{3}{4}R_+^2 + \frac{1}{4}R_-^2 - 1\right)(t, x_+(t)).$$

Similarly,

(3.16) $$\dot{R}_- = \partial_t R_- + \frac{1}{\alpha}\left(\frac{1}{4}R_+^2 + \frac{3}{4}R_-^2 - 1\right)\partial_x R_- = 0,$$

where the dot "$\cdot$" denotes the differentiation in the characteristic direction

$$\frac{dx_-}{dt}(t) = \lambda_-\bigl(t, x_-(t)\bigr) = \frac{1}{\alpha}\left(\frac{1}{4}R_+^2 + \frac{3}{4}R_-^2 - 1\right)(t, x_-(t)).$$

Differentiate (3.15) with respect to x and set $Z_+ \equiv \partial_x R_+$ we have

(3.17) $$\partial_t Z_+ + \frac{1}{\alpha}\left(\frac{3}{2}Z_+^2 R_+ + \frac{1}{2}Z_+ R_- \partial_x R_-\right) + \frac{1}{\alpha}\left(\frac{3}{4}R_+^2 + \frac{1}{4}R_-^2 - 1\right)\partial_x Z_+ = 0.$$

From (3.15) we deduce that
$$R'_- = \partial_t R_- + \frac{1}{\alpha}\left(\frac{3}{4}R_+^2 + \frac{1}{4}R_-^2 - 1\right)\partial_x R_-,$$
hence
$$(3.18) \qquad \partial_x R_- = \frac{2\alpha R'_-}{R_+^2 - R_-^2}.$$

Substituting this relation into (3.17) we obtain
$$(3.19) \qquad Z'_+ + \frac{R_- R'_-}{R_+^2 - R_-^2} Z_+ + \frac{3\alpha}{2} R_+ Z_+^2 = 0.$$

Let $h = h(R_+, R_-)$ satisfying $h' = (R_- R'_-)/(R_+^2 - R_-^2)$, $h = 1/2 \log 1/(R_+^2 - R_-^2)$, for example. Then multiplying (3.19) by $e^h = 1/(\sqrt{R_+^2 - R_-^2})$ and using the abbreviation
$$(3.20) \qquad q_+ \equiv e^h Z_+ = \frac{\partial_x R_+}{\sqrt{R_+^2 - R_-^2}}, \quad k_+ \equiv \frac{3}{2} R_+ e^{-h} = \frac{3}{2} R_+ \sqrt{R_+^2 - R_-^2},$$

we obtain the standard Riccati equation
$$(3.21) \qquad q'_+ + \alpha k_+ q_+^2 = 0.$$

The solution of (3.21) is given by
$$(3.22) \qquad q_+(t, x) = \frac{q_+^0}{1 + \alpha q_+^0 K_+(t)}, \quad q_+^0 = q_+(0, x(0)),$$

where
$$(3.23) \quad \begin{aligned} K_+(t) &= \int_0^t k_+(\tau, x_+(\tau))\, d\tau \\ &= \frac{3}{2} \int_0^t R_+(\tau, x_+(\tau))\sqrt{R_+^2(\tau, x_+(\tau)) - R_-^2(\tau, x_+(\tau))}\, d\tau, \end{aligned}$$

the integration along the λ_+-characteristic. The sign of the function $\alpha q_0 K(t)$ is important for the formation of singularity. For $\alpha > 0$, at initial time if $\partial_x R_+(0, x(0)) = \partial_x R_+(x_0) < 0$, i.e., $q_0 < 0$ then $q_+(t, x)$ must become unbounded in finite time. Similarly, for $\alpha < 0$ if $\partial_x R_+(0, x(0)) = \partial_x R_+(x_0) > 0$, i.e., $q_0 > 0$ then $q_+(t, x)$ must become unbounded in finite time. This means that $q_+(t, x)$ must blow up at some later time t, where $1 + \alpha q_0 K(t) = 0$. Therefore, the break t_b can be estimated by the following rules. Let $t_{+,b}$ satisfy
$$t_{+,b} \leq \inf_{x_0 \in \Omega} \{t \colon G_+(t, x_0) - 0\}, \quad \Omega = \{x_0 \colon \operatorname{sgn}\alpha \cdot \partial_x R_+(x_0) \leq 0\},$$

where
$$G_+ = 1 + \alpha \frac{3}{2} \frac{\partial_x R_+(x)}{\sqrt{R_+^2(x) - R_-^2(x)}}$$
$$\times \int_0^t R_+(\tau, x_+(\tau))\sqrt{R_+^2(\tau, x_+(\tau)) - R_-^2(\tau, x_+(\tau))}\, d\tau.$$

The particle path $x_+ = x_+(t)$ satisfies $x_+(0) = x$ and
$$\frac{dx_+}{dt}(t) = \frac{1}{2\alpha}R_+^2(t, x_+(t)) + \frac{1}{4\alpha}\left[R_+^2(t, x_+(t)) + R_-^2(t, x_+(t)) - 1\right].$$
Similarly, we can also estimate $t_{-,b}$ by considering the λ_--characteristic. $\square$

The limit conserved densities can be described by the Riemann invariants.

PROPOSITION 3.2. *For $0 \le t < t_b$, the limiting densities are given by*
$$\rho = w\text{-}L^1 \lim_{\hbar \to 0} |\psi^\hbar|^2,$$
$$M = w\text{-}L^1 \lim_{\hbar \to 0} \left(\frac{i\hbar}{2}(\psi^\hbar \bar{\psi}_x^\hbar - \psi_x^\hbar \bar{\psi}^\hbar)\right) = \frac{1}{16\alpha^3}\left[(R_+ + R_-)^2 - 1\right](R_+ - R_-)^2,$$
uniformly on compact subsets of t. Hence, the classical momentum is given by
$$\mu = w\text{-}L^1 \lim_{\hbar \to 0} \frac{i\hbar}{2}(\psi^\hbar \bar{\psi}_x^\hbar - \psi_x^\hbar \bar{\psi}^\hbar) = \frac{1}{16\alpha^3}(R_+ - R_-)^2(4R_+ R_- - 1).$$

4. The Transition to NLS Equation

In this section, we consider the semiclassical limit of the MNLS equation (3.1). For convenience, we relabel it as

(4.1) $$i\hbar \partial_t \psi^\hbar + \frac{\hbar^2}{2}\partial_x^2 \psi^\hbar + \frac{i\hbar}{2}\partial_x(\alpha|\psi^\hbar|^2 \psi^\hbar) + (1 - |\psi^\hbar|^2)\psi^\hbar = 0.$$

When $\alpha = 0$, the NLS equation, this problem has been studied by Colin and Soyeur [1] in the case when there are no vortices (uniformly bounded energy as $\hbar \to 0$). In two space dimensions, when there are vortices, it is treated by Lin and Xin [14].

The conserved densities of MNLS equation (3.1) are related to the wave function $\psi^\hbar$ through

(4.2) $$\rho^\hbar = |\psi^\hbar|^2 = \psi^\hbar \bar{\psi}^\hbar,$$

(4.3) $$M^\hbar = \mu^\hbar + \alpha|\psi^\hbar|^2 = \frac{i\hbar}{2}(\psi^\hbar \bar{\psi}_x^\hbar - \psi_x^\hbar \bar{\psi}^\hbar) + \alpha|\psi^\hbar|^2,$$

(4.4) $$E^\hbar = \frac{|M^\hbar|^2}{2\rho^\hbar} + \frac{1}{2}(\rho^\hbar - 1)^2 + \frac{\alpha}{2}\rho^\hbar M^\hbar + \frac{\hbar^2}{4} \cdot \frac{(\rho_x^\hbar)^2}{2\rho^\hbar}.$$

The mass and noncanonical momentum densities determines the field ψ up to a constant phase. Therefore the energy density can be rewritten as

(4.5) $$E^\hbar = \hbar^2 |\psi_x^\hbar|^2 + \frac{1}{2}(|\psi^\hbar|^2 - 1)^2 + \frac{\alpha^2}{2}|\psi^\hbar|^6$$
$$+ \hbar \frac{3\alpha}{2}|\psi^\hbar|^2 \cdot \frac{i}{2}(\psi^\hbar \bar{\psi}_x^\hbar - \psi_x^\hbar \bar{\psi}^\hbar) + C,$$

where the constant is also taken into account. We will consider the case when $(\rho^\hbar, \mu^\hbar)$ is near the constant state $(1, 0)$. Then the noncanonical momentum, $M^\hbar \approx \alpha$, this means that even the fluid velocity vanishes, the flow still has a background momentum with characteristic speed. Under the above assumption we can choose the initial data such that

(4.6) $$E^\hbar = \hbar^2 |\psi_x^\hbar|^2 + \frac{1}{2}(|\psi^\hbar|^2 - 1)^2 + \frac{\alpha^2}{2}(|\psi^\hbar|^6 - 1)$$
$$+ \hbar \frac{3\alpha}{2}|\psi^\hbar|^2 \cdot \frac{i}{2}(\psi^\hbar \bar{\psi}_x^\hbar - \psi_x^\hbar \bar{\psi}^\hbar).$$

We will assume the initial energy is of order $O(\hbar^2)$ then the conservation of energy can be recast as

$$(4.7) \quad \int \left[|\psi_x^\hbar|^2 + \frac{1}{2}\left(\frac{|\psi^\hbar|^2 - 1}{\hbar}\right)^2 + \frac{\alpha^2}{2} \frac{|\psi^\hbar|^6 - 1}{\hbar^2} \right.$$
$$\left. + \frac{3\alpha}{2}|\psi^\hbar|^2 \cdot \frac{i}{2\hbar}(\psi^\hbar \bar\psi_x^\hbar - \psi_x^\hbar \bar\psi^\hbar) \right] dx = C = \text{constant}.$$

The conservation of mass (1.14) can be recast as

$$(4.8) \quad \frac{\partial}{\partial t}\left(\frac{|\psi^\hbar|^2 - 1}{\hbar}\right) + \frac{\alpha}{2}\frac{\partial}{\partial x}\left(\frac{|\psi^\hbar|^2 - 1}{\hbar}\right) + \frac{\partial}{\partial x}W^\hbar = 0,$$

where

$$(4.9) \quad W^\hbar = \frac{i}{2}(\psi^\hbar \bar\psi_x^\hbar - \psi_x^\hbar \bar\psi^\hbar).$$

Recall the hypothesis of Theorem 2.1, $w^\hbar \Phi'' + U'' > 0$, this implies the nonnegativity of $3\alpha/2|\psi^\hbar|^2 \cdot i/2\hbar(\psi^\hbar \bar\psi_x^\hbar - \psi_x^\hbar \bar\psi^\hbar)$. We can combine this property with the conservation of energy (4.7) to conclude

$$(4.10) \quad \int_{-\infty}^{\infty} \left[|\psi_x^\hbar|^2 + \frac{(|\psi^\hbar|^2 - 1)^2}{2\hbar^2} + \frac{\alpha^2}{2} \cdot \frac{|\psi^\hbar|^6 - 1}{\hbar^2} \right] dx < C.$$

It follows from (4.10) that

(4.11) $\quad \psi^\hbar A \quad$ is bounded in $L^\infty([0,T]; H^1)$,

(4.12) $\quad \psi_t^\hbar \quad$ is bounded in $L^\infty([0,T]; H^{-1})$,

(4.13) $\quad \dfrac{|\psi^\hbar|^2 - 1}{\hbar} \quad$ is bounded in $L^\infty([0,T]; L^2)$,

and

(4.14) $\quad |\psi^\hbar|^2 \to 1$ strongly in $\quad L^2$ and a.e.

Now, by classical compactness arguments, we deduce from (4.11) and (4.12) the existence of a subsequence of $\{\psi^\hbar\}$ (always denoted by $\psi^\hbar$) such that

(4.15) $\quad \psi^\hbar \to \psi \quad$ strongly in $C([0,T]; L^2)$,

(4.16) $\quad \psi^\hbar \rightharpoonup \psi \quad$ weakly in $\quad L^\infty([0,T]; H^1)$.

We introduce the Lagrangian coordinates

$$(4.17) \quad \tau = t, \quad \xi = x - \frac{\alpha}{2}t,$$

then (4.8) becomes

$$(4.18) \quad \frac{\partial}{\partial \tau}\left(\frac{|\psi^\hbar|^2 - 1}{\hbar}\right) + \partial_\xi W^\hbar = 0.$$

Furthermore, from the equation (4.13) and (4.18) we also have

$$(4.19) \quad \frac{|\psi^\hbar|^2 - 1}{\hbar} \rightharpoonup -\int_0^\tau \partial_\xi W^\hbar(\sigma)\,d\sigma$$

in the sense of distributions. The integral is along the characteristic $dx/dt = \alpha.2$. Therefore, by (4.15)–(4.16) and using that fact that $|\psi^\hbar|^2 \to 1$ a.e., as $\hbar$ tends to zero in (4.1) we have

$$i\partial_t \psi + \frac{\alpha}{2} i\partial_x \psi + \psi \int_0^\tau W_\xi(\sigma) \, d\sigma = 0, \tag{4.20}$$

or

$$i\partial_\tau \psi + \psi \int_0^\tau W_\xi(\sigma) \, d\sigma = 0. \tag{4.21}$$

Since $|\psi| = 1$, (4.21) becomes

$$\psi_{\tau\tau} - \psi_{\xi\xi} = \psi\big(|\psi_\xi|^2 - |\psi_\tau|^2\big). \tag{4.22}$$

Using the fact $|\psi| = 1$ again, write $\psi = e^{i\theta}$ shows (by (4.21))

$$-\theta_\tau + \int_0^\tau \theta_{\xi\xi} \, d\tau = 0, \tag{4.23}$$

or

$$\theta_{\tau\tau} - \theta_{\xi\xi} = 0. \tag{4.24}$$

THEOREM 4.1. *We have $\psi^\hbar \to \psi$ strongly in $C\big([0,T]; L^2\big)$, and $\psi^\hbar \rightharpoonup \psi$ weakly in $L^\infty\big([0,T]; H^1\big)$, where ψ satisfies*

$$\psi_{\tau\tau} - \psi_{\xi\xi} = \psi\big(|\psi_\xi|^2 - |\psi_\tau|^2\big),$$

or equivalently $\psi = e^{i\theta}$ with phase function θ satisfying the wave equation

$$\theta_{\tau\tau} - \theta_{\xi\xi} = 0.$$

REMARK 4.2. When $\alpha = 0$, the NLS equation, the uniformly bounded energy is good enough for the semiclassical limit ([**1**]). We don't need to require the well-prepared initial data. However, for GMNLS equation $\alpha \neq 0$, the nonlinear term $\alpha|\psi^\hbar|^2$ also contributes the potential energy. Therefore the interaction between different nonlinear terms (reaction terms) is much more involved and makes the limiting process (depending on the initial conditions) complicated.

REMARK 4.3. In particular $\alpha = O(\hbar)$, say $\alpha = \tilde{\alpha}\hbar$ for example, then the conservation of energy becomes

$$\int \left[|\psi_x^\hbar|^2 + \frac{1}{2}\left(\frac{|\psi^\hbar|^2 - 1}{\hbar}\right)^2 + \frac{\tilde{\alpha}^2}{2}|\psi^\hbar|^6 \right. \tag{4.25}$$
$$\left. + \frac{3\tilde{\alpha}}{2}|\psi^\hbar|^2 \cdot \frac{i}{2}(\psi^\hbar \bar{\psi}_x^\hbar - \psi_x^\hbar \bar{\psi}^\hbar) \right] dx = C = \text{constant}.$$

Since $H^1 \hookrightarrow L^6$, the L^6 bound can be replaced by H^1 bound. Using the nonnegativity of $(3\tilde{\alpha})/2|\psi^\hbar|^2 \cdot i/2\hbar(\psi^\hbar\bar{\psi}_x^\hbar - \psi_x^\hbar\bar{\psi}^\hbar)$ again we can also deduce the same result even for not well-prepared initial data.

References

1. T. Colin and A. Soyeur, *Some singular limits for evolutionary Ginzburg Landau equations*, Asymptot. Anal. **13** (1996), 361–372.
2. B. Desjardins and C.-K. Lin, *On the semiclassical limit of the modified nonlinear Schrödinger equation*, J. Math. Anal. Appl. (1999) (submitted).
3. B. Desjardins, C.-K. Lin, and T.-C. Tso, *Semiclassical limit of the derivative nonlinear Schrödinger equation*, Math. Models Methods Appl. Sci. **10** (2000), no. 2, 261–285.

4. E. Grenier, *Semiclassical limit of the nonlinear Schrödinger equation in small time*, Proc. Amer. Math. Soc. **126** (1998), 523–530.
5. S. Jin, C. D. Levermore, and D. W. McLaughlin, *The semiclassical limit of the defocusing NLS hierarchy*, Comm. Pure Appl. Math. **52** (1999), 613–564.
6. M. Khanna and R. Rajaram, *Evolution of nonlinear Alfvén waves propagating along the magnetic fields in a collisionless plasma*, J. Plasma Phys. **28** (1982), 459–468.
7. S. Klainerman and A. Majda, *Singular limits and quasilinear systems with large parameter and the incompressible limit of compressible fluids*, Comm. Pure Appl. Math. **34** (1981), 481–524.
8. L. D. Landau and E. M. Lifshitz, *Statistical physics*, vol. 2, Pergamon Press, Oxford, 1982.
9. P. D. Lax, *Hyperbolic systems of conservation laws and the mathematical theory of schock waves*, Conference Board of the Mathematical Sciences Regional Conference Series in Applied Mathematics, no. 11, SIAM, Philadelphia, Pa., 1973.
10. J. H. Lee, *Global solvability of the derivative nonlinear Schrödinger equation*, Trans. Amer. Math. Soc. **314** (1989), 107–118.
11. _____, *On the dissipative evolution equations associated with the Zakharov Shabat system with a quadratic spectral parameter*, Trans. Amer. Math. Soc. **316** (1989), 327–336.
12. C.-K. Lin, *On the fluid dynamical analogue of the general nonlinear Schrödinger equation*, Southeast Asia Bull. Math. **22** (1998), no. 1, 45–56.
13. _____, *Remark on the singularity of the nonlinear Schrödinger equation*, Southeast Asia Bull. Math. **22** (1998), no. 2, 161–170.
14. F.-H. Lin and J. X. Xin, *On the incompressible fluid limit and the vortex motion law of the nonlinear Schrödinger equation*, Comm. Math. Phys. **200** (1999), 249–274.
15. A. Majda, *Compressible fluid flow and systems of conservation laws in several space variables*, Appl. Math. Sci., vol. 53, Springer-Verlag, Berlin–New York, 1984.
16. K. Mio, T. Ogino, K. Minamy, and S. Takeda, *Modified nonlinear Schrödinger equation for Alfvén waves propagating along the magnetic field in cold plasma*, J. Phys. Soc. Japan **41** (1976), 265–273.
17. E. Mjolhus, *On the modulational instability of hydromagnetic waves parallel to the magnetic field*, J. Plasma Phys. **16** (1976), 321–334.
18. T. Ozawa, *On the nonlinear Schrödinger equations of derivative type*, Indiana Univ. Math. J. **45** (1996), 137–163.
19. O. K. Pashaev and J.-H. Lee, *Resonance NLS solitons as black holes in Madelung fluid*, 1998, hep-th/9810139v2.
20. O. K. Pashaev, J.-H. Lee, and C.-K. Lin, *Dissipative and envelope solitons: Equivalent relations*, Preprint, no. 1, Academia Sinica, 1998.
21. _____, *Integrable homogeneous perturbations of nonlinear Schrödinger type equations with resonance soliton dynamics*, Preprint, 1999.
22. H. Schochet and M. Weinstein, *The nonlinear Schrödinger limit of the Zakharov equations governing Langmuir turbulence*, Comm. Math. Phys. **106** (1986), 569–580.
23. M. Wadati, K. Konno, and Y.-K. Ichikawa, *A generalization of inverse scattering method*, J. Phys. Soc. Japan **46** (1979), 1965–1966.
24. G. B. Whitham, *Linear and nonlinear waves*, Pure and Applied Mathematics, Wiley-Interscience Publ., John Wiley & Sons, Inc., New York, 1999.

DEPARTMENT OF MATHEMATICS, NATIONAL CHENG KUNG UNIV., TAINAN 70101, TAIWAN
E-mail address: cklin@mail.ncku.edu.tw

Dynamics of Quantum Resonances

Marco Merkli

ABSTRACT. We develop a general theory describing the *time evolution of metastable states* resulting from perturbations of unstable eigenvalues. We apply this theory to *Many Body Quantum Systems* and to the problem of *Quasi-classical Tunneling*.

This theory links the physical picture of a resonance (which is a metastable state, i.e. a time-dependent picture) to its mathematical definition (given in time-independent terms, e.g. as poles of a complex dilated resolvent). In particular, we find that the lifetimes of the metastable states (resonances) are given by the famous *Fermi Golden Rule*.

1. Short History

Quantum resonances have played a central role in physics since the early days of quantum mechanics, see e.g. Weisskopf and Wigner [9]. In its modern form, the mathematical theory of quantum resonances has been laid down by Simon [6] who used the theory of dilation analytic Hamiltonians due to J. Aguilar and J. M. Combes [1] and E. Balslev and J. M. Combes [2]. This approach was further developed by (among others) Simon [7], Sigal [5], Hunziker [3].

So far, the theory developed was a static theory, despite the fact that the physical picture is that of a metastable state. Soffer and Weinstein [8] proposed a new approach to the *time dependent* theory of quantum resonances, inspired by ideas of Weisskopf and Wigner. The approach we take follows Soffer and Weinstein.

2. General Setting

Consider a quantum system with a Hamiltonian H_0 acting on a Hilbert space $\mathcal{H}$ such that H_0 has an eigenvalue E_0 with eigenstate φ_0:

$$H_0 \varphi_0 = E_0 \varphi_0.$$

We perturb H_0 by an operator W and study the Schrödinger evolution governed by the perturbed Hamiltonian

$$H := H_0 + W.$$

2000 *Mathematics Subject Classification.* Primary: 35Q40; Secondary: 35B25.
This is the final form of the paper.

For "small" perturbations W (in a sense to be specified), one expects that the eigenvalue E_0 of H_0 is unstable and consequently, that any solution to the Schrödinger equation
$$i\hbar \partial_t \psi = H\psi,$$
with initial condition ψ_0 spectrally localized "close" to E_0 with respect to H should exhibit *local decay* (Ruelle). More precisely, we expect that $\forall R > 0$,
$$\chi(|A| \leq R)e^{-iHt/\hbar}\psi_0 \to 0, \quad \text{as } t \to \infty,$$
where $\chi(\cdot \leq R)$ is the indicator function of $[-R, R]$ and A is an operator on $\mathcal{H}$, called the *adjoint operator*. Think of A as being the multiplication operator $x \in \mathbb{R}^d$, if $\mathcal{H} = L^2(\mathbb{R}^d)$. Notice that the convergence to zero is a priori in the average sense.

On the other hand, if the perturbation W is "small" and the initial condition ψ_0 is "close" to the unperturbed bound state φ_0, then we expect $e^{-iHt/\hbar}\psi_0$ to behave almost like $e^{-iH_0 t/\hbar}\varphi_0$ which does not decay locally.

The synthesis of these two aspects leads us to expect that $e^{-iHt/\hbar}\psi_0$ does decay locally, but stays localized for a long time: this is the resonance behaviour. Our goal is to understand the dynamics of decay, in particular, we wish to obtain the lifetime of the resonance, i.e. the duration during which it behaves like a bound state.

Let P denote the projector onto the eigenspace $\text{Ker}(H_0 - E_0)$ and introduce the perturbation parameter $\kappa := \|\langle A \rangle^\alpha W P\|$ that measures "smallness" of the perturbation. Here, α is any number > 2 and $\langle \cdot \rangle := (1 + |\cdot|^2)^{1/2}$. The adjoint operator A must satisfy certain criteria and for concrete systems it can be chosen explicitly (e.g. as the dilation generator, see below). Define "closeness" of the initial state ψ_0 to the unperturbed bound state φ_0 as meaning that $\|\langle A \rangle^\alpha W \overline{P} \psi_0\|$ is small, where $\overline{P} := 1 - P$.

Next, we introduce the *Fermi Golden Rule Condition*. This condition says that the positive operator $W\delta(\overline{H} - E_0)\overline{P}W$ is *strictly* positive on Ran P:
$$\Gamma := \pi P W \delta(\overline{H} - E_0)\overline{P} W P \geq C_0 \kappa^2 P,$$
for some $C_0 > 0$. Here, $\overline{H} := \overline{P} H \overline{P}$ is the reduced Hamiltonian. Assuming the Fermi Golden Rule Condition holds (for a discussion, see below), we can state our main (abstract) result (Ref. [**4**, Theorem 2.1] in a simplified version):

THEOREM 2.1. *Let ψ be the solution to the Schrödinger equation with initial condition ψ_0 close to the unperturbed bound state φ_0 and let $0 < \beta < 1/2$. Then for small κ we have*
$$\psi(t) = e^{-i\lambda t} P \psi_0 + \psi_{\text{disp}}(t),$$
where $\|\langle A \rangle^{-\alpha} \psi_{\text{disp}}(t)\| = O(\kappa^{1-4\beta} \langle t \rangle^{-\beta})$ and the imaginary part of λ, $\text{Im}\,\lambda = 1/2i(\lambda - \lambda^*)$, is given by
$$\text{Im}\,\lambda = -\Gamma + O(\kappa^3).$$

We can prove in fact a much more general result, where initial conditions do not have to be close to φ_0, and we give a more detailed picture of the evolution of the solution (see [**4**]).

A comparison between the two parts in the solution $\psi(t)$ given in Theorem 2.1 yields that for all $R > 0$, $\chi(|A| \leq R)\psi(t)$ is close to the "*stationary state*" $\chi(|A| \leq R)e^{-i\lambda t} P \psi_0$ in a time-interval of the order $\kappa^{-2}\ln(\kappa^{-1})$, during which $\chi(|A| \leq R)\psi_{\text{disp}}$ is relatively negligible.

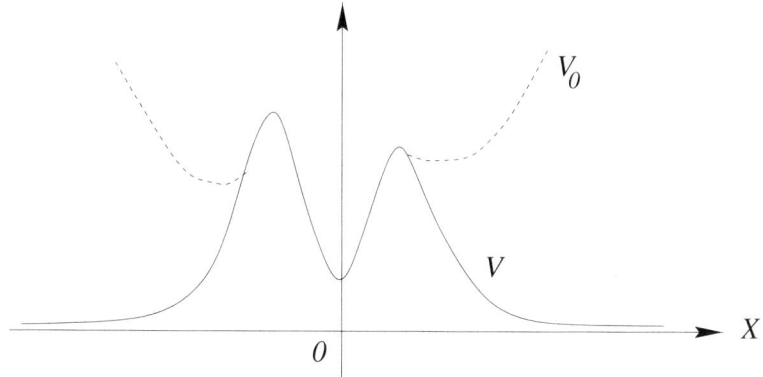

FIGURE 1. Tunneling potential V and confining potential V_0.

The role of the Fermi Golden Rule Condition becomes now clear. If $\Gamma = 0$, then Im λ is at least of order three in κ and in order to obtain a physically interesting picture of the resonance evolution as above, one has to push the procedure leading to Theorem 2.1 to a higher order in κ.

The Fermi Golden Rule Condition is conjectured to hold generically and in that case we say that the resonance behaviour is observable in the second order in the perturbation.

2.1. Many body systems. We apply the abstract theorem to the case where $\mathcal{H} = L^2(X)$ (X is the configuration space) and the unperturbed Hamiltonian is a Schrödinger operator

$$H_0 = -\Delta + V.$$

Resonances in this context emerge from perturbation of embedded eigenvalues of H_0 (which are naturally present due to symmetries of the system, i.e. if H_0 is invariant under e.g. rotation or permutation of particles).

Theorem 2.1 applies with a perturbation of the form $W = \kappa' U$, where U is a potential with $\|\langle A \rangle^\alpha U P\| < \infty$ and for κ' small. Here, A is the dilation generator, $A = (p \cdot x + x \cdot p)/2$, $p = -i\nabla_x$. In particular, if V is an N-body potential, we get:

THEOREM 2.2. *The conclusion of Theorem 2.1 holds provided E_0 is separated from the thresholds of H_0, and if the Fermi Golden Rule Condition is satisfied.*

2.2. Quasiclassical tunneling. We consider the following initial value problem on $L^2(\mathbb{R}^d)$ (for any d):

$$i\hbar \partial_t \psi = H\psi,$$

with initial condition $\psi_0 \in \operatorname{Ran} E_{\hbar\Delta}(H)$. Here, $\hbar > 0$ is considered to be a small parameter ($\hbar \to 0$ is called the *semiclassical limit*) and the Schrödinger operator H is given by

$$H = p^2 + V, \quad p = -i\hbar\nabla,$$

where V is a volcano shaped bounded potential with a local minimum at the origin, decaying to zero as $|x| \to \infty$ (see Fig. 1). Δ is an interval around $V(0)$ of some fixed length (say 1) and $\hbar\Delta := \{\hbar E \mid E \in \Delta\}$.

We introduce a *reference Hamiltonian* $H_0 = p^2 + V_0$, where V_0 is a *confining potential* coinciding with V in a neighborhood of the origin (see Fig. 1). Let us denote the ground state of H_0 by φ_0: $H_0\varphi_0 = E_0\varphi_0$. Setting $W := V - V_0$ yields $H = H_0 + W$.

Notice that the spectrum of H is absolutely continuous, while the spectrum of H_0 is purely discrete. Resonances in this context emerge from a large perturbation of the confining system (the perturbation considered here turns purely discrete spectrum into absolutely continuous spectrum!). The reason why we are able to apply our theory is that even though W is a large perturbation (in particular $|W| \to \infty$ as $|x| \to \infty$), $\kappa = \|\langle A\rangle^\alpha W \varphi_0\|$ is exponentially small in $\hbar \to 0$, i.e. $\kappa \sim e^{-C/\hbar}$ (for some $C > 0$, independent of $\hbar$). This results from the fact that φ_0, being an eigenfunction of H_0, is spatially exponentially localized around the origin (Agmon), while the support of W is by construction away from the origin.

This explains the somewhat surprising fact that perturbation theory to the second degree (the Fermi Golden Rule) yields *exponential* lifetimes for quasiclassical tunneling. The reason is that we expand to order two in κ, but κ is already exponentially small in $\hbar$.

Here is our result for quasiclassical tunneling in the case where the initial condition ψ_0 is close to the ground state φ_0 (for more general initial conditions and a more detailed description of the dynamics, see Ref. [**4**, Theorem 2.5]):

THEOREM 2.3. *Let β as in Theorem 2.1 and suppose that the Fermi Golden Rule Condition holds*:
$$\Gamma = \pi\langle \overline{P}W\varphi_0, \delta(\overline{H} - E_0)\overline{P}W\varphi_0\rangle \geq C_0 \hbar^p \kappa^2,$$
for some $C_0 > 0$ and some $p \geq 0$. Then for $\hbar$ small enough, the solution to the initial value problem has an expansion
$$\psi(t) = a(t)\varphi_1 + \psi_{\mathrm{disp}}(t),$$
where $\varphi_1 = \varphi_0 + O(\kappa/\hbar)$ and $\|\langle A\rangle^{-\alpha}\psi_{\mathrm{disp}}(t)\| = O(\langle t\rangle^{-\beta}\hbar^{-q}\kappa^{1-2\beta})$ and q is some positive integer. Moreover, $a(t) = \langle\varphi_0, \psi(t)\rangle$ satisfies
$$a(t) = e^{-i\lambda t/\hbar}a(0) + O(\langle t\rangle^{-\beta}\hbar^{-q}\kappa^{1-4\beta}), \quad \text{with } \mathrm{Im}\,\lambda = -\Gamma + O(\hbar^{-2}\kappa^3).$$

Taking $\beta = 0$ in this expansion yields

COROLLARY. *For some $q > 0$ and for all $t \geq 0$, we have*
$$|\langle\varphi_0, e^{-iHt/\hbar}\varphi_0\rangle| = e^{-\Gamma t/\hbar} + O(\kappa/\hbar^q).$$

REMARKS. (1) Recall that $\kappa \sim e^{-C/\hbar}$, so $\kappa/\hbar^q \to 0$ as $\hbar \to 0$ (any q).

(2) In physics literature, resonances are often "defined" as states with the decay property $|\langle\varphi_0, e^{-iHt/\hbar}\varphi_0\rangle| = e^{-\Gamma t/\hbar}$.

References

1. J. Aguilar and J. M. Combes, *A class of analytic perturbations for one-body Schrödinger Hamiltonians*, Comm. Math. Phys. **22** (1971), 269–279.
2. E. Balslev and J. M. Combes, *Spectral properties of many-body Schrödinger operators with dilation analytic interactions*, Comm. Math. Phys. **22** (1971), 280–294.
3. W. Hunziker, *Distortion analyticity and molecular resonance curves*, Ann. Inst. H. Poincaré Phys. Théor. **45** (1986), 339–358.
4. M. Merkli and I. M. Sigal, *A time-dependent theory of quantum resonances*, Comm. Math. Phys. **201** (1999), 549–576.
5. I. M. Sigal, *Complex transformation method and resonances in one-body quantum systems*, Ann. Inst. H. Poincaré Phys. Théor. **41** (1984), 103–114.

6. B. Simon, *Resonances in n-body quantum systems with dilation analytic potentials and the foundations of time-dependent perturbation theory*, Ann. of Math. (2) **97** (1973), 247–274.
7. _____, *Resonances and complex scaling: A rigorous overview*, Internat. J. Quantum Chem. **14** (1978), 529–542.
8. A. Soffer and M. I. Weinstein, *Time dependent resonances theory*, Geom. Funct. Anal. **8** (1998), 1086–1128.
9. V. Weisskopf and E. Wigner, *Berechnung der natürlichen Linienbreite auf Grund der Diracschen Lichttheorie*, Z. Phys. C **63** (1930), 54–73.

Current address: Department of Mathematics, ETH Zurich, Switzerland

DEPARTMENT OF MATHEMATICS, UNIVERSITY OF TORONTO, TORONTO, ONTARIO, M5S 3G3, CANADA

E-mail address: merkli@math.toronto.edu; merkli@math.ethz.ch

Nelson Diffusions and Blow-Up Phenomena in Solutions of the Nonlinear Schrödinger Equation With Critical Power

Hayato Nawa

Introduction

We shall concern the following Cauchy problem for the nonlinear Schrödinger equation (NSC):

(NSC) $$2i\frac{\partial \psi}{\partial t} + \triangle \psi + |\psi|^{4/d}\psi = 0, \quad (x,t) \in \mathbb{R}^d \times \mathbb{R}_+.$$

Here $i = \sqrt{-1}$, and $\triangle$ is the Laplace operator on $\mathbb{R}^d$. We associate this equation with initial data from usual Sobolev space $H^1(\mathbb{R}^d)$:

(IV) $$\psi(x,0) = \psi_0(x), \quad x \in \mathbb{R}^d.$$

We summarize here the basic properties of this Cauchy problem (NSC)–(IV) (see, e.g., [**9, 13, 31**]). The unique local existence of solutions is well known: for any $\psi_0 \in H^1(\mathbb{R}^d)$, there exists a unique solution $\psi(x,t)$ in $C([0,T_m); H^1(\mathbb{R}^d))$ for some $T_m \in (0,\infty]$ (maximal existence time); for simplicity, we shall consider the forward problem only), and $\psi(t)$ satisfies the following three conservation laws of L^2, the energy $\mathcal{H}$ and the momenta, in this order:

(0.1) $$\|\psi(t)\| = \|\psi_0\|,$$

(0.2) $$\mathcal{H}(\psi(t)) \equiv \|\nabla \psi(t)\|^2 - \frac{2}{2+4/d}\|\psi(t)\|_{2+4/d}^{2+4/d} = \mathcal{H}(\psi_0),$$

(0.3) $$\Im \int_{\mathbb{R}^d} \overline{\psi(x,t)} \frac{\partial}{\partial x_l}\psi(x,t)\,dx = \Im \int_{\mathbb{R}^d} \overline{\psi_0(x)}\frac{\partial}{\partial x_l}\psi_0(x)\,dx \quad l = 1,2,\ldots,d,$$

2000 *Mathematics Subject Classification.* Primary: 35Q55; Secondary: 60G35.

I would like to express my heartily gratitude to Professors Israel M. Sigal and Catherine Sulem (University of Toronto) for having given me a chance to talk these results at the occasion of this well organized "Workshop on Nonlinear Dynamics and Renormalization Group" held at Montreal in August 22–27, 1999, and for their concern and encouragements.

I am very grateful to Professors Takehiko Morita (Tokyo Institute of Technology) and Takashi Kumagai (Kyoto University) from "Stochastic clan" in Japan for very useful, helpful discussions on the materials in Sections 5 and 6, and for their friendship.

This work was partially supported by the Grant-in-Aid for Scientific Research (#108874033) of the Ministry of Education, Science, Sports and Culture, Japan.

This is the final form of the paper.

©2001 American Mathematical Society

for $t \in [0, T_m)$, where $\|\cdot\|$ and $\|\cdot\|_{2+4/d}$ denotes the L^2 norm and $L^{2+4/d}$ norm respectively. Furthermore we have the following alternatives: $T_m = \infty$ or $T_m < \infty$ and $\lim_{t \to T_m} \|\nabla \psi(t)\| = \infty$ (blow-up).

If we replace the nonlinear term by $|\psi|^{p-1}\psi$, it is known that the exponent $p = p_c \equiv 1 + 4/d$ in dimension d is the critical value for the nonexistence of global solutions: If $p < p_c$, every solution exists globally in time; If $p \geqq p_c$, there is a class of initial data leading to blow-up solutions (see, e.g., [**10, 30, 32, 34**]).

We will make a overview around the blow-up phenomena in solutions of (NSC)–(IV) through the keyword *tightness*: we can safely say that our main mathematical subject is to investigate conditions which imply the tightness of the family of Radon measures $\{|\psi(x,t)|^2 dx\}_{0 \leqq t < T_m}$ defined by a blow-up solution ψ (see Sections 1, 2, and 5). We consider its limiting profiles in Sections 3 and 5. And our final object is the so-called log log law for the rate of blow-up solutions of (NSC)–(IV): we shall cast another light on the study of the limiting and asymptotic profiles of blow-up solutions and of their blow-up rate (see Section 6).

The last two Sections 5 and 6 are based on the author's recent paper [**23**]. For more details, further results, and related facts, see [**22, 23, 31**], and their references.

1. Finite Variance

If, in addition, $|x|\psi_0 \in L^2(\mathbb{R}^d)$, then the solution $\psi(t)$ also enjoys $|x|\psi(\cdot) \in C([0, T_m); L^2(\mathbb{R}^d))$, and satisfies the following virial identity (see, e.g., [**31, 32, 34**]):

$$(1.1) \quad \big\||x-a|\psi(t)\big\|^2 = \big\||x-a|\psi_0\big\|^2 + 2t\Im\langle \psi_0, (x-a)\cdot\nabla\psi_0\rangle + t^2 \mathcal{H}(\psi_0),$$

where we have used the notation:

$$(1.2) \quad \langle f, g \rangle = \int_{\mathbb{R}^d} \overline{f(x)} g(x)\, dx.$$

We recall here the argument of M. Tsutsumi [**32**] (see also Weinstein [**34**]) to prove the existence of blow-up solutions to (NSC) for this case of $\psi_0 \in H^1(\mathbb{R}^d) \cap L^2(|x|^2 dx)$: we are led to a contradiction through the identity (1.1) and

$$(1.3) \quad \int_{|x|>R} |\psi(x,t)|^2 dx \leq \frac{1}{R^2}\big\||x-a|\psi(t)\big\|^2,$$

if we assume $T_m = \infty$ and assume one of the following three conditions:

$$(1.4) \quad \mathcal{H}(\psi_0) < 0,$$
$$(1.5) \quad \mathcal{H}(\psi_0) = 0 \quad \text{and} \quad \exists a \in \mathbb{R}^d;\ \Im\langle u\psi_0, (x-a)\cdot\nabla\psi_0\rangle < 0,$$
(1.6) $\mathcal{H}(\psi_0) > 0$ and
$$\exists a \in \mathbb{R}^d;\quad \Im\langle \psi_0, (x-a)\cdot\nabla\psi_0\rangle \leqq -\big\||x-a|\psi_0\big\|^2 \sqrt{\mathcal{H}(\psi_0)}.$$

Summing up, we have:

PROPOSITION 1.1. *If ψ_0 satisfies one of three conditions (1.4), (1.5), or (1.6), then the corresponding solution ψ of (NSC)–(IV) blows-up in finite time. Furthermore, the family of Radon measures $\{|\psi(x,t)|^2 dx\}_{0 \leq t < T_m}$ is tight.*

The tightness of $\{|\psi(x,t)|^2 dx\}_{0 \leq t < T_m}$ comes from the Chebychev inequality (1.3).

In the sequel of this paper, we will be mainly working in the framework of pure energy space $H^1(\mathbb{R}^d)$ instead of $H^1(\mathbb{R}^d) \cap L^2(|x|^2 dx)$.

2. General Case

We consider the following two cases:
(i) $d = 1$ and $\mathcal{H}(\psi_0) < 0$;
(ii) $d \geq 2$, $\mathcal{H}(\psi_0) < 0$ and ψ_0 being radially symmetric.

In either case, we put
$$-\mathcal{H}^* \equiv \mathcal{H}(\psi_0) < 0.$$

We note here that, if the initial datum $\psi_0(x)$ is radially symmetric, so is the corresponding solution $\psi(x,t)$ of (NSC)–(IV) with respect to $x \in \mathbb{R}^N$ for any $t \in [0, T_m)$.

We shall show that

PROPOSITION 2.1. *Suppose one of the conditions (i) or (ii) above. Then we have $T_m < \infty$, that is, the corresponding solution ψ of (NSC)–(IV) blows-up in finite time T_m. Furthermore the family of Radon measures $\{|\psi(x,t)|^2 dx\}_{0 \leq t < T_m}$ defined by the solution ψ is tight.*

We introduce a $W^{3,\infty}(\mathbb{R})$ odd function, following Ogawa–Y. Tsutsumi [26, 27]:

$$(2.1) \qquad w(\xi) = \begin{cases} \xi, & 0 \leq \xi < 1, \\ \xi - (\xi-1)^3, & 1 \leq \xi < 1 + \frac{1}{\sqrt{3}}, \\ \text{smooth}, (w' \leq 0) & 1 + \frac{1}{\sqrt{3}} \leq \xi < 2, \\ 0, & 2 \leq \xi. \end{cases}$$

We put $r \equiv |x| = \sqrt{\sum_{k=1}^N x_k^2}$ for $x = (x_1, \ldots, x_N)$. This convention will be also applied to one dimensional case. Using $w(\xi)$ defined in (2.1), we define, for $R > 0$,

$$(2.2) \qquad \vec{w}_R(x) = \frac{x}{r} w_R(r) = \frac{x}{r} R w\left(\frac{r}{R}\right),$$

$$(2.3) \qquad W_R(x) = 2 \int_0^r w_R(s)\, ds.$$

For these functions, we have: there exist constants $K_l > 0$ ($l = 0, 1, 2, 3$) such that

$$(2.4) \qquad \left|\frac{d^l}{dr^l} w_R(r)\right| \leq \frac{K_l}{R^{l-1}}, \quad l = 0, 1, 2, 3,$$

$$(2.5) \qquad \nabla W_R(x) = 2\frac{x}{r} w_R(r) = 2\vec{w}_R(x),$$

$$(2.6) \qquad w_R^2 \leq W_R,$$

$$(2.7) \qquad R^2 \leq W_R(x) \leq 4R^2 \text{ for } r \geq R.$$

One of our key ingredients in the proof is the following generalization of virial identity (1.1):

LEMMA 2.1. *We have: for* $t \in [0, T_m)$,

$$\langle W_R, |\psi(t)|^2 \rangle = \langle W_R, |\psi_0|^2 \rangle + 2\Im t \langle \psi_0, \vec{w}_R \cdot \nabla \psi_0 \rangle - t^2 \mathcal{H}^* \tag{2.8}$$
$$- 2 \int_0^t ds \int_0^s d\tau \mathcal{H}^R(\psi(\tau))$$
$$- \frac{1}{2} \int_0^t ds \int_0^s d\tau \langle \triangle(\nabla \cdot \vec{w}_R), |\psi(\tau)|^2 \rangle$$
$$= \mathrm{I} + \mathrm{II} + \mathrm{III}.$$

Here the functional $\mathcal{H}^R$ *is defined by*:

$$\mathcal{H}^R(f) \equiv \int_{\mathbb{R}^d} m_1(r) |\nabla f(x)|^2 - m_2(r) |v(x)|^{2+4/d} dx, \tag{2.9}$$

where

$$m_1(r) \equiv 1 - w_R'(r), \tag{2.10}$$

$$m_2(r) \equiv \frac{1}{2+d} \left(d - w_R'(r) - \frac{d-1}{r} w_R(r) \right). \tag{2.11}$$

We note that we have $m_2 = 1/3 m_1$ *if* $d = 1$.

For the proof of this proposition, see Ogawa–Y.Tsutsumi [**26**, **27**] (see also [**22**]).

We should note that the following facts:

$$m_j(r) \geqq 0, \quad j = 1, 2, \tag{2.12}$$

$$\Omega_R \equiv \operatorname{supp} m_j(r) = \{ x \in \mathbb{R}^d \mid |x| \geqq R \}, \tag{2.13}$$

$$m_1(r) = 1 \quad \text{and} \quad m_2(r) = \frac{2}{2+4/d}, \quad \text{if } r \geqq 2R, \tag{2.14}$$

$$\|\nabla(m_2)^{1/2}\|_\infty \leqq \frac{K_4}{R} \text{ for some constant } K_4 > 0. \tag{2.15}$$

The third term (III) in (2.8) can be easily handled. There exists a constant $K_5 > 0$ independent of $R > 0$ such that we have

$$|\triangle(\nabla \cdot \vec{w}_R)| \leqq \frac{K_5}{R^2}, \tag{2.16}$$

so that we obtain

$$|\mathrm{III}| \leqq \frac{K_5}{4R^2} \|\psi_0\|^2 t^2. \tag{2.17}$$

So it is absorbed in the term $-\mathcal{H}^* t^2$ of (I), if we choose $R > 0$ sufficiently large. Hence, if we manage to overcome the second term (II) to be absorbed in $-\mathcal{H}^* t^2$ of (I) as well, the right hand side of (2.8) will be dominated by a quadratic form of t whose top term has a negative coefficient, so that we are led to a contradiction.

In [**22**], in order to handle the second term (II) in (2.8), we introduce the following variational value:

$$\text{(2.18)} \qquad \mathcal{N}_R \equiv \inf_{\substack{f \in \mathcal{X} \\ f \neq 0}} \left\{ \int_{|x|>R} |f(x)|^2 dx \,\bigg|\, \mathcal{H}^R(f) \leqq -\frac{1}{4}\mathcal{H}^*, \|f\| \leqq \|\psi_0\| \right\},$$

where: $\mathcal{X} \equiv H_r^1(\mathbb{R}^d)$ the space of all radially symmetric functions in $H^1(\mathbb{R}^d)$ if $d \geqq 2$; $\mathcal{X} \equiv H^1(\mathbb{R})$ if $d = 1$.

LEMMA 2.2. *For sufficiently large $R > 0$, we can find $\mathcal{N}_* > 0$ independent of $R > 0$:*

$$\text{(2.19)} \qquad \mathcal{N}_R \geqq \mathcal{N}_*.$$

PROOF. The key ingredients to prove this fact are:
(i) we have, for $f \in \mathcal{X}$,

$$\text{(2.20)} \quad \int_{\mathbb{R}^d} m_2(x)|f(x)|^{2+4/d} dx[t] \leqq C_d \|f\|_{L^2(\Omega_R)}^{4/d} \int_{\mathbb{R}^d} m_2(x)|\nabla f(x)|^2 dx$$
$$+ C_d \|f\|_{L^2(\Omega_R)}^{2+4/d} \|\nabla(m_2^{1/2})\|_\infty^2,$$

for some universal constant $C_d > 0$ independent of f and m_2; and
(ii) there is a constant $K > 0$ independent of $R > 0$,

$$\text{(2.21)} \qquad \inf_{x \in \Omega_R} \frac{m_1(x)}{C_d m_2(x)} \geqq K.$$

From (2.20), the definition of $\mathcal{H}^R$ and the constraint $\mathcal{H}^R(f) \leqq -1/4\mathcal{H}^*$, we have that, for sufficiently large $R > 0$,

$$\text{(2.22)} \qquad \int_{\mathbb{R}^d} \left\{ m_1(x) - C_d \|v\|_{L^2(\Omega_R)}^{4/d} m_2(x) \right\} |\nabla f(x)|^2 dx < 0.$$

Consequently we get (2.19) by (2.21).

Now, take $R > 0$ large enough at $t = 0$ so that we have

$$\text{(2.23)} \qquad \frac{K_5}{R^2} \|\psi_0\|^2 < \mathcal{H}^*,$$

$$\text{(2.24)} \qquad \int_{\Omega_R} |\psi_0(x)|^2 dx < \mathcal{N}_*,$$

$$\text{(2.25)} \qquad \frac{1}{R^2}\left(\frac{2}{\mathcal{H}^*}\|\nabla \psi_0\|^2 + 1\right)\langle W_R, |\psi_0|^2 \rangle < \mathcal{N}_*.$$

Then we can show, by contradiction, through the generalized virial identity (2.8), that:

LEMMA 2.3. *Under the assumptions (2,23)–(2.25), we have:*

$$\text{(2.26)} \qquad T_m = \sup\left\{ t > 0 \,\bigg|\, \int_{|x|>R} |\psi(x,\tau)|^2 dx < \mathcal{N}_*, \quad 0 \leqq \tau < t \right\}.$$

PROOF. We have,

$$\langle W_R, |\psi(t)|^2 \rangle \leq \langle W_R, |\psi_0|^2 \rangle + 2\Im \langle \psi_0, \vec{w}_R \cdot \nabla \psi_0 \rangle t - \frac{1}{2} t^2 \mathcal{H}^* \tag{2.27}$$

$$\leq -\frac{1}{2} \mathcal{H}^* \left(t - \frac{2}{\mathcal{H}^*} \Im \langle \psi_0, \vec{w}_R \cdot \nabla \psi_0 \rangle \right)^2$$

$$+ \frac{2}{\mathcal{H}^*} |\langle \psi_0, \vec{w}_R \cdot \nabla \psi_0 \rangle|^2 + \langle W_R, |\psi_0|^2 \rangle$$

$$< \left(\frac{2}{\mathcal{H}^*} \|\nabla \psi_0\|^2 + 1 \right) \langle W_R, |\psi_0|^2 \rangle < R^2 \mathcal{N}_*.$$

From (2.26), we thus obtain by the definition of $\mathcal{N}_*$ that

$$-\frac{1}{4} \mathcal{H}^* \leq \mathcal{H}^R(\psi(t)) \text{ for } t \geq 0. \tag{2.28}$$

Consequently, for $t \geq 0$, we have from (2.8) with (2.23) that

$$\langle W_R, |\psi(t)|^2 \rangle \leq \langle W_R, |\psi_0|^2 \rangle + 2\Im \langle \psi_0, \vec{w}_R \nabla \psi_0 \rangle t - \frac{1}{2} t^2 \mathcal{H}^*. \tag{2.29}$$

Using (2.29), we can prove that $T_m < \infty$, i.e., the solution $\psi(t)$ blow-s up in finite time. Noting the fact that Lemma 2.3 with $\mathcal{N}_*$ replaced by arbitrarily small $\varepsilon > 0$ holds valid, we have the tightness of $\{|\psi(x,t)|^2 dx\}_{0 \leq t < T_m}$.

COMMENTS.
1. The variational problem considered in (2.18) is a local analogue of the following:

$$\mathcal{N}_c \equiv \inf_{\substack{f \in H^1(\mathbb{R}^d) \\ f \neq 0}} \{ \|f\|^2 \mid \mathcal{H}(f) \leq 0 \}. \tag{2.30}$$

The infimum of the right hand side is achieved by a nontrivial standing wave solution of (NSC) of the form $\psi(x,t) = Q_g(x) e^{i/t^2}$. Here, $Q_g(x)$ satisfies:

$$\triangle Q_g - Q_g + |Q_g|^{4/d} Q_g = 0, \quad Q_g > 0, \tag{2.31}$$
$$\mathcal{N}_c = \|Q_g\|^2, \quad \mathcal{H}(Q_g) = 0. \tag{2.32}$$

This standing wave is called the *ground state*, since it has the least L^2 norm. For this see Weinstein [**34**] (see also [**20**]).

2. Unfortunately, we have not succeeded in solving the full version of the variational problem proposed here for general case of $d \geq 2$. In general case, $\mathcal{H}^R$ has one more additional term (for details, see [**22**]). However, it partially works so that we can prove:

THEOREM 2.1. *If ψ_0 satisfies*

$$\mathcal{H}(\psi_0) < 0, \tag{2.33}$$

then the corresponding solution $\psi(t)$ of (NSC)–(IV) satisfies

$$\sup_{t \in [0, T_m)} \|\nabla \psi(t)\| = \infty. \tag{2.34}$$

In other words, if ψ_0 satisfies (2.33), then there exists $T_m \in (0, \infty]$ such that the corresponding solution $\psi(t)$ blows-up (in finite time) or grows up (at infinity).

For details, see [**22**] again. One can find another proof of Theorem 2.1 in [**19**] and [**21**].

Anyway, we have from Proposition 1.1 and 1.2 that

PROPOSITION 2.2. *Suppose one of the following three cases*:
(i) $d = 1$ and $\mathcal{H}(\psi_0) < 0$,
(ii) $d \geq 2$, $\mathcal{H}(\psi_0) < 0$ and ψ_0 being radially symmetric,
(iii) $d \geq 1$, $|x|\psi_0 \in L^2(\mathbb{R}^d)$ and $\mathcal{H}(\psi_0) < 0$. Then we have $T_m < \infty$, and there exist a sequence $\{t_n\}$; $t_n \to T_m$ and a positive measure $\mu \in C_b(\mathbb{R}^d)' \equiv$ the dual of $C_b(\mathbb{R}^d) \equiv L^\infty \cap C(\mathbb{R}^d)$ such that

$$\text{(2.35)} \qquad \lim_{n \to \infty} |\psi(x,t)|^2 dx = \nu(dx),$$

in the weak topology of measures (i.e., weakly* in $C_b(\mathbb{R}^d)'$).

In the next section, we shall consider the structure of the limiting measure ν.

3. Structure of the Limiting Measure

We note that the limiting measure ν may depend on the choice of the sequence $\{t_n\}$ such that $t_n \to \infty$ as $n \to \infty$. Nevertheless, for a class of initial data and for some suitably chosen $\{t_n\}$, we can identify the limiting measure ν. In order to show this fact, we need to know the asymptotic profile of the blow-up solution ψ: we investigate the detailed behavior of the scaled down blow-up solutions ψ_n's (as $n \to \infty$) defined by:

$$\text{(3.1)} \qquad \psi_n(x,t) = \lambda_n^{d/2} \overline{\psi(\lambda_n x, t_n - \lambda_n^2 t)},$$

where

$$\text{(3.2)} \qquad t_n \uparrow T_m, \quad \sup_{t \in [0,t_n)} \|\psi(t)\|_{2+4/d} = \|\psi(t_n)\|_{2+4/d},$$

$$\text{(3.3)} \qquad \lambda_n = \frac{1}{\|\psi(t_n)\|_{2+4/d}^{1+2/d}}.$$

It is worth noting here that the scaled down functions ψ_n's also solve (NSC); this is a peculiarity of our equation (NSC) (see e.g., [**Appendix B; 22**]). In [**19, 20, 21**] (see also [**22**]), we showed that

THEOREM 3.1. *For $\{\psi_n\}$ defined by (3.1)–(3.3), we have*:

$$\text{(3.4)} \qquad \psi_n(x,t) \sim \sum_{j=1}^{L} \psi^j(x - \gamma_n^j, t) + \varphi_n(x,t) \quad n \to \infty,$$

in the strong topology of $C\big([0,T]; L^2(\mathbb{R}^d)\big)$ (for any $T > 0$). Here,
(i) $\psi^j(x,t)$'s are solutions of (NSC) in $C_b(\mathbb{R}_+; H^1(\mathbb{R}^d))$ with $\mathcal{H}(\psi^j) = 0$;
(ii) $\varphi_n(x,t)$ solves:

$$\text{(3.5)} \qquad \begin{cases} 2i \dfrac{\partial \varphi_n}{\partial t} + \triangle \varphi_n = 0, & (x,t) \in \mathbb{R}^d \times \mathbb{R}_+, \\ \varphi_n(x,0) = \psi_n(x,0) - \sum_{j=1}^{L} \psi^j(x - \gamma_n^j, 0), & x \in \mathbb{R}^d, \end{cases}$$

that is, $\varphi_n(x,t)$'s are solutions of the free Schrödinger equation; and
(iii) the sequences $\{\gamma_n^1\}, \{\gamma_n^2\}, \ldots, \{\gamma_n^L\}$ are in $\mathbb{R}^d$ such that $\lim_{n\to\infty} |\gamma_n^j - \gamma_n^k| = \infty$ $(j \neq k)$.

In the original world of ψ, we have

$$(3.6) \quad \lim_{n\to\infty} \sup_{t \in [t_n - \lambda_n^2 T, t_n]} \left\| \overline{\psi(\cdot,t)} - \sum_{j=1}^{L} \psi_n^j(\cdot,t) - \widetilde{\varphi}_n(\cdot,t) \right\| = 0,$$

with

$$(3.7) \quad \lim_{n\to\infty} \lambda_n^2 \sup_{t \in [t_n - \lambda_n^2 T, t_n]} \|\widetilde{\varphi}_n(t)\|_{2+4/d}^{2+4/d} = 0,$$

where

$$(3.8) \quad \psi_n^j(x,t) = \frac{1}{\lambda_n^{d/2}} \psi^j\left(\frac{x - \gamma_n^j \lambda_n}{\lambda_n}, \frac{t_n - t}{\lambda_n^2} \right),$$

$$(3.9) \quad \widetilde{\varphi}_n(x,t) = \frac{1}{\lambda_n^{d/2}} \varphi_n\left(\frac{x}{\lambda_n}, \frac{t_n - t}{\lambda_n^2} \right).$$

Furthermore it holds that, for any $T > 0$ and any $f \in C_b(\mathbb{R}^d)$,

$$(3.10) \quad \lim_{n\to\infty} \sup_{t \in [t_n - \lambda_n^2 T, t_n]} \left| \int_{\mathbb{R}^d} \left(|\psi(x,t)|^2 - \sum_{j=1}^{L} |\psi_n^j(x,t)|^2 - |\widetilde{\varphi}_n(x,t)|^2 \right) f(x) \, dx \right| = 0.$$

From (3.10), we have

$$(3.11) \quad \sum_{j=1}^{L} |\psi_n^j(x, t_n - \lambda_n^2 T)|^2 dx \rightharpoonup \sum_{j=1}^{L} \|\psi^j\|^2 \delta_{a^j}(dx),$$

where $a^j = \lim_{n\to\infty} \gamma_n^j \lambda_n$;

$$(3.12) \quad |\widetilde{\varphi}_n(x, t_n - \lambda_n^2 T)|^2 dx \rightharpoonup \mu(dx) \text{ as } n \to \infty,$$

for any $T > 0$, provided that the family of Radon measures $\{|\psi(x,t)|^2 dx\}_{0 \leq t < T_m}$ is tight (see [**22**]). Thus we have:

COROLLARY 3.1. *Assume either* (i), (ii), *or* (iii) *of Proposition* 2.2. *Let ψ be the corresponding blow-up solution of* (NSC)–(IV), *and let $\{t_n\}$ be a time sequence as in* (3.2). *For any $T > 0$, we put*

$$(3.13) \quad s_n \equiv t_n - \lambda_n^2 T.$$

Then there exists a subsequence of $\{s_n\}$ (still denoted by the same letter) that satisfies the following properties: there is a finite number $L \in \mathbb{N}$, a family of finite points $\{a^2, a^2, \ldots, a^L\} \subset \mathbb{R}^d$, and a positive measure $\mu \in C_b(\mathbb{R}^d)'$ such that we have

$$(3.14) \quad |\psi(x, s_n)|^2 dx \rightharpoonup \sum_{j=1}^{L} \|\psi^j(0)\|^2 \delta_{a^j}(dx) + \mu(dx) \text{ as } n \to \infty,$$

in the weak topology of measures. When ψ_0 is radially symmetric, $L=1$ and $a^1=0$ in (3.14). We note that

$$\|\psi_0\|^2 = \sum_{j=1}^{L} \|\psi^j(0)\|^2 + \mu(\mathbb{R}^d). \tag{3.15}$$

We note that the measure μ comes from a part of blow-up solution which has a different nature from the other part (the top term of the asymptotic expansion of the blow-up solution near the blow-up time) producing the Dirac masses, while μ may include Dirac masses.

COMMENT. It is worth while to explain here the role of Theorem 2.1 in the proof of Theorem 3.1. The proof is closely related to the method of concentrated compactness principle (see, e.g., [16]). We can safely say that the analysis investigates how the "dichotomy" (in the terminology of concentrated compactness) occurs in the sequence $\{\psi_n\}$. Theorem 3.1 asserts that ψ_n behaves like a finite superposition of *zero-energy, global-in-time* solutions of (NSC). It is worth while to note again that the scaled function $\tilde{\psi}_n$ also solves (NSC), and satisfies $\|\tilde{\psi}_n(t)\| = \|\psi(t)\|$ and $\mathcal{H}(\tilde{\psi}_n(t)) = \lambda_n^2 \mathcal{H}(\psi(t)) (\to 0$ as $n \to \infty)$. We iteratively construct ψ^j's. Here the important thing is the finiteness of ψ^j's. This follows from $\mathcal{H}(\psi^j)=0$ (for any j), since in this case we have $\|\psi^j\| \geqq \|Q_g\|$ for any j (see (2.30)–(2.32)). If the iteration were not terminated at some finite index, we would have by the construction of ψ^j's that $\limsup_{k\to\infty} \sum_{j=1}^{k} \mathcal{H}(\psi^j(t)) \leqq 0$. Hence Theorem 2.1 in Section 2 ensure that $\mathcal{H}(\psi^j)=0$ *for any j*. For details, see [20] (see also [19, 21, 22]).

Note that the amplitude of Dirac masses in (3.14) is larger than or equal to $\|Q_g\|^2$.

4. Problems and Questions

From (3.6) together with (3.7)–(3.9), we might be able to expect that, in general, the blow-up solution is going to separate into two states as time approaches to the blow-up time; say $\psi_\delta + \psi_\mu$:

$$\psi(x,t) \sim \psi_\delta(x,t) + \psi_\mu(x,t), \text{ as } t \uparrow T_m, \tag{4.1}$$

such that, for some $\{a^1, a^2, \ldots, a^L\} \subset \mathbb{R}^d$ and positive constants $A_j \geqq \|Q_g\|^2$ $(j=1,2,\ldots,L)$,

$$|\psi_\delta(x,t)|^2 dx \rightharpoonup \sum_{j=1}^{L} A_j \delta_{a^j}(dx), \tag{4.2}$$

$$|\psi_\mu(x,t)|^2 dx \rightharpoonup \mu(dx) \text{ as } t \uparrow T_m, \tag{4.3}$$

in the weak topology of measures, and these two states could be distinguished by the rate of blow-up:

$$\frac{\|\nabla \psi_\delta(x,t)\|}{\|\nabla \psi(x,t)\|} \sim 1, \tag{4.4}$$

$$\frac{\|\nabla \psi_\mu(x,t)\|}{\|\nabla \psi(x,t)\|} \to 0 \quad (t \uparrow T_m). \tag{4.5}$$

However, there have remained a possibility that the *number*, L, the *amplitudes*, $\{A_j\}_{j=1}^{L}$, and the *configuration*, $\{a^j\}_{j=1}^{L}(\subset \mathbb{R}^d)$, of singularities depend on the choice of the sequence $\{t_n\}$ ($t_n \uparrow T_m$). Furthermore, the tightness of the family of Radon measures, $\{|\psi(x,t)|^2 dx\}_{0<t<T_m}$, is only known for a class of initial data as we have just seen in the previous sections (see Proposition 2.2).

In the next two sections, we will see that this question is relevant to the problem of determining the rate of blow-up, while it is still an important open problem to determine it in a rigorous mathematical way. Some asymptotic and numerical analyses (see, e.g. [**8, 14, 15, 30, 31**]) suggest that generic blow-up solutions satisfy the so-called log log *law*:

$$\text{(4.6)} \qquad \|\nabla\psi(t)\| \sim \sqrt{\frac{\ln\ln(T_m-t)^{-1}}{T_m-t}}.$$

However, the known explicit blow-up solutions [**1, 17, 35**] violate the condition (4.6); they enjoy $\|\nabla\psi(t)\| \sim (T_m-t)^{-1}$. On the other hand, it is considered that these explicit blow-up solutions are in the extreme case in terms of blow-up rate; it is conjectured that generic blow-up solutions satisfy the log log law of (4.6) [**7**].

In the next section, we introduce a condition on the blow-up rate (see (5.1) in Section 5) which exactly exclude the blow-up solutions which have the same blow-up rate as the explicit blow-up solutions, and under such rate, we shall show that the *number* and the *configuration* of singularities and their *amplitude* are independent of the choice of the sequence $\{t_n\}$. Furthermore, the limit $\lim_{t\uparrow T_m}|\psi(x,t)|^2 dx$ exists along full net under the condition (5.1).

The novelty of our analysis is to consider a evaluation stochastic process, $t \to X_t$, such that

$$\text{(4.7)} \qquad P[X_t \in A] = \int_A |\psi(x,t)|^2 dx, \quad A \in \mathcal{B}(\mathbb{R}^d),$$

where P is a "probability" measure on the trajectory space of $\Gamma \equiv C\big([0,T_m);\mathbb{R}^d\big)$. This type of diffusion was first introduced by Nelson [**24, 25**] for a class of (linear) Schrödinger equation with a potential, and Carlen [**2, 3, 4**] solved Nelson's stochastic differential equation to prove the existence of desired processes. One can easily see that Carlen's argument also works for our problem; we can construct a process satisfying (4.7) for the solution ψ of (NSC)–(IV).

Our analysis employing Nelson's stochastic differential equation also casts another light on the log log law for the blow-up rate: In Section 6. we shall show that, roughly speaking, the condition (5.1) implies (4.6), under some assumptions on the asymptotic shape of the blow-up solution.

COMMENT. Mathematically, we know only the lower bound on the blow-up rate (see [**6, 30**], and see also [**18, 31**]):

$$\text{(4.8)} \qquad \|\nabla\psi(t)\| \geqq \frac{C}{\sqrt{T_m-t}},$$

for some positive constant $C > 0$. Recently, for the case of $d=1$, Perel'man [**28**], in a mathematical way, constructs a blow-up solution of (NSC) near the ground state level which has log log asymptotic behavior under some assumption on the spectrum of her linearized operator.

5. Nelson Diffusions and Blow-Up Phenomena

We shall introduce another condition (see (5.1) below) which yields the tightness of the family of Radon measures $\{|\psi(x,t)|^2 dx\}_{0 \le t < T_m}$, and show that the *number* and the *configuration* of singularities and their *amplitude* are independent of the choice of the sequence $\{t_n\}$ under the condition. Furthermore, we shall see the limit $\lim_{t \uparrow T_m} |\psi(x,t)|^2 dx$ exists along full net. We have:

THEOREM 5.1. *Let ψ be a blow-up solution of* (NSC)–(IV). *Suppose that*

$$(5.1) \qquad \int_0^{T_m} \|\nabla \psi(t)\| \, dt < \infty.$$

Then, there is a finite number $L \in \mathbb{N}$, a family of L-points $\{a^1, a^2, \ldots a^L\} \in \mathbb{R}^d$, L-positive constants $\{A_1, A_2, \ldots A_L\}$ satisfying (1.6), *and a positive measure $\mu \in C_b(\mathbb{R}^d)'$ (the dual of $C_b(\mathbb{R}^d)$) such that we have, along full net $t \uparrow T_m$,*

$$(5.2) \qquad |\psi(x,t)|^2 dx \rightharpoonup \sum_{j=1}^L A_j \delta_{a^j}(dx) + \mu(dx) \quad \text{as } t \uparrow T_m,$$

in the weak topology of measures, where

$$(5.3) \quad A_j \geqq \|Q_g\|^2 = \inf\{\|f\|^2 \mid f \in H^1(\mathbb{R}^d), f \ne 0, \mathcal{H}(f) \leqq 0\} > 0, \quad 1 \leqq j \leqq L.$$

In case of ψ_0 being radially symmetric ($d \geqq 2$), (5.2) *should read with $L = 1$ and $a^1 = 0$.*

We have made an assumption on the rate of blow-up. Of course, as noted in the previous section, it is still an important open problem to determine the rate of blow-up in a rigorous mathematical way. We note that the condition (5.1) exactly exclude the blow-up solutions which have the same blow-up rate as the explicit blow-up solutions, and that if the blow-up solution satisfies log log law (4.6), then we have (5.1).

In [24] (see also [25]), Nelson introduced his stochastic mechanics to give the same prediction as orthodox quantum mechanics does. Viewing the factor $|\psi|^{4/d}$ in the nonlinear term in (NSC) as a time dependent potential, we can consider the so-called Nelson diffusions for our blow-up solutions of (NSC)–(IV). Following Nelson, for our solution ψ, we define the *osmotic velocity* $u(x,t)$ and *current velocity* $v(x,t)$ by:

$$(5.4) \qquad u(x,t) \equiv \begin{cases} \Re \frac{\nabla \psi(x,t)}{\psi(x,t)}, & \text{if } \psi(x,t) \ne 0, \\ 0, & \text{if } \psi(x,t) = 0, \end{cases}$$

$$(5.5) \qquad v(x,t) \equiv \begin{cases} \Im \frac{\nabla \psi(x,t)}{\psi(x,t)}, & \text{if } \psi(x,t) \ne 0, \\ 0, & \text{if } \psi(x,t) = 0. \end{cases}$$

Then the corresponding process is a *weak solution* of the following stochastic differential equation:

$$(5.6) \qquad dX_t = b(X_t, t)dt + dB_t,$$

with the law P:

$$(5.7) \qquad P[X_t \in A] = \int_A |\psi(x,t)|^2 dx, \quad A \in \mathcal{B}(\mathbb{R}^d),$$

where $t \to B_t$ is a standard Brownian motion and

(5.8) $$b(x,t) \equiv u(x,t) + v(x,t).$$

Carlen [**2, 3, 4**] solved stochastic differential equation (5.6)–(5.7) for the linear Schrödinger equation of the form:

(5.9) $$i\frac{\partial}{\partial t}\psi(x,t) + \frac{1}{2}\triangle\psi(x,t) + V(x)\psi(x,t) = 0,$$

for a class of potentials. In his argument, the following two facts are essential: the *finite energy condition*:

(5.10) $$\int_S^T \int_{\mathbb{R}^d} (u^2 + v^2)(x,t)\rho(x,t)\,dx\,dt < \infty,$$

for any finite interval $[S,T] \subset [0, T_m)$, and the *weak continuity equation*:

(5.11) $$\frac{d}{dt}\int_{\mathbb{R}^d} f(x)\rho(x,t)\,dx = \int_{\mathbb{R}^d} \bigl(v(x,t)\cdot \nabla f(x)\bigr)\rho(x,t)\,dx,$$

where ρ is the density for the process X_t defined by:

(5.12) $$\rho(x,t) \equiv |\psi(x,t)|^2.$$

In order to construct the desired diffusion, he consider the corresponding *backward martingale* equation to (5.6):

(5.13) $$\frac{\partial}{\partial t}f(x,t) - \frac{1}{2}\triangle f(x,t) + b_*(x,t)\cdot \nabla f(x,t) = 0,$$

where

(5.14) $$b_*(x,t) \equiv v(x,t) - u(x,t).$$

Using (5.10) and (5.11), he obtained the transition function for the *backward martingale* equation (5.13):

$$P_{t,s} : L^2\bigl(\rho(x,s)dx\bigr) \to\to L^2\bigl(\rho(x,t)dx\bigr)$$

for $S < s < t < T$. This transition functions form a *Markovian propagator*, which enable us to construct the diffusion having the property (5.7). Our blow-up solution ψ also satisfies these conditions (5.10) and (5.11). Thus, we can prove the following:

THEOREM 5.2. *Let u, v, b and ρ be defined through the blow-up solution ψ of* (NSC)–(IV) *as above. We associate $\Gamma \equiv C([0, T_m); \mathbb{R}^d)$ with its Borel σ-algebra $\mathcal{F}$. Let $(\Gamma, \mathcal{F}, \mathcal{F}_t, X_t)$ be evaluation stochastic process $X_t(\gamma) \equiv \gamma(t)$ for $\gamma \in \Gamma$ with natural filtration $\mathcal{F}_t = \sigma(X_s, s \leq t)$. Then there exists a Borel "probability" measure P on Γ such that:*

(i) $(\Gamma, \mathcal{F}, \mathcal{F}_t, X_t, P)$ *is a Markov process;*

(ii) *the image of P under X_t has density $\rho(x,t)$, that is,*

(5.15) $$P[X_t \in dx] = |\psi(x,t)|^2 dx;$$

(iii) *The following process B_t is a $(\Gamma, \mathcal{F}_t, P)$-Brownian motion:*

(5.16) $$B_t \equiv X_t - X_0 - \int_0^t b(X_\tau, \tau)\,d\tau.$$

We are now in a position to prove Theorem 5.1. First, we know that the convergence of processes implies that of their distributions (see, e.g., [**29**]), i.e.,

(5.17) $$\exists \lim_{t \uparrow T_m} X_t \quad \text{a.s.} \implies \exists \lim_{t \uparrow T_m} P[X_t \in dx] \equiv \lim_{t \uparrow T_m} |\psi(x,t)|^2 dx,$$

so that we have the *tightness* of $\{|\psi(x,t)|^2 dx\}_{0 \leq t < T_m}$. Thus, (5.17) together with (3.11) and (3.12) yields the desired result. Therefore, our task is the following lemma:

LEMMA 5.1. *Suppose that (5.1) holds true, that is,*

(5.18) $$\int_0^{T_m} \|\nabla \psi(t)\| \, dt < \infty.$$

Then there exists the following limit:

(5.19) $$X_{T_m} \equiv \lim_{t \uparrow T_m} X_t \quad \text{a.s.}$$

PROOF. The argument here is similar to that of Carlen in [**5**]. For simplicity, we suppose $\|\psi(t)\| = 1$. From (5.16), we have:

(5.20) $$X_t - X_s = \int_s^t b(X_\tau, \tau) \, d\tau + B_t - B_s,$$

for $t, s \in [0, T_m)$. We need to estimate

(5.21) $$P\left[\sup_{T < s, t < T_m} |X_t - X_s| > \varepsilon\right],$$

for any $\varepsilon > 0$, and want to show that this tends to zero as $T \uparrow T_m$. In order to prove this fact, we shall estimate the first term on the right hand side of (5.20):

(5.22) $$P\left[\int_T^{T_m} |b(X_t, t)| \, dt > \varepsilon\right].$$

By L^1-Chebychev inequality, Fubini's theorem and the Schwartz inequality, we have

(5.23) $$\varepsilon P\left[\int_T^{T_m} |b(X_t, t)| \, dt > \varepsilon\right]$$
$$\leq E\left(\int_T^{T_m} |b(X_t, t)| \, dt\right)$$
$$= \int_T^{T_m} E|b(X_t, t)| \, dt \leq \int_T^{T_m} \left(E|b(X_t, t)|^2\right)^{1/2} dt$$
$$= \int_T^{T_m} \left(\int_{\mathbb{R}^d} |b(x,t)|^2 |\psi(x,t)|^2 dx\right)^{1/2} dt$$
$$= \int_T^{T_m} \|\nabla \psi(t)\| \, dt < \infty.$$

Here E denotes the expectation with respect to P.

From this estimate, we can show that, for any $n \in \mathbb{N}$, there exists a time T_n large enough such that

$$(5.24) \quad P\left(\bigcup_{T_n < s,t < T_m} \left[|X_t - X_s| > \frac{1}{n}\right]\right) < \frac{1}{2^n}.$$

Now we put

$$(5.25) \quad A_n \equiv \bigcup_{T_n < s,t < T_m} \left[|X_t - X_s| > \frac{1}{n}\right].$$

Then, by the Borel–Cantelli lemma, we have

$$(5.26) \quad P\left(\bigcap_{n \in \mathbb{N}} \bigcup_{k > n} A_k\right) = 0,$$

which implies that $\lim_{t \uparrow T_m} X_t$ exists on a set of "probability" one. For more details, see [5] or [27].

6. log log Law

In this section, we propose an another story which gives us the log log law for the blow-up rate:

THEOREM 6.1. *Under some ansatz for the blow-up solution ψ, we have:*

$$(6.1) \quad \limsup_{t \uparrow T_m} \sqrt{\frac{T_m - t}{\ln \ln(T_m - t)^{-1}}} \left(\frac{1}{T_m - t} \int_t^{T_m} \|\nabla \psi(\tau)\| \, d\tau\right) \asymp 1.$$

We shall see that the log log law for the blow-up rate is relevant to the so-called *iterated logarithmic law* for the Brownian motion through the Nelson's stochastic differential equation.

By Theorems B and C, we are very close to the following "picture": if $L = 1$,

$$(6.2) \quad \psi(x,t) \approx \frac{1}{\lambda(t)^{d/2}} Q\left(\frac{x}{\lambda(t)}\right) e^{i\theta(x,t)} + \frac{1}{\lambda(t)^{d/2}} r\left(\frac{x}{\lambda(t)}, \eta(t)\right).$$

Here

$$(6.3) \quad \lambda(t) \equiv \frac{\|\nabla \psi(0)\|}{\|\nabla \psi(t)\|};$$

and $Q(x)$ nearly resemble a bound state of (NSC) which satisfies, for some $\omega \gg 1$,

$$(6.4) \quad \begin{cases} \triangle Q - \omega Q + |Q|^{4/d} Q = 0, \\ \nabla Q(0) = 0, \\ Q(0) \neq 0, \\ \lim_{|x| \to \infty} Q(x) = 0, \\ \exists \lim_{|x| \to \infty} \left|\frac{\nabla Q(x)}{Q(x)}\right|, \end{cases}$$

and the remainder term satisfies

$$(6.5) \quad |r(x,t)| \approx |e^{it\triangle} \phi_0| \text{ as } t \uparrow T_m,$$

for some "nice" $\phi_0 \in H^1(\mathbb{R}^d)$. The scaled-down procedure (3.1) destroy the phase factor in ψ profiling the amplitude of the blow-up solution as suggested in [**Appendix E; 22**]. So, we borrow the phase factor θ in (6.2) from the generalized pseudo-conformal transformations (see [**8, 14, 15, 30, 31**]):

(6.6) $$\theta(x,t) \equiv \zeta(t) - a(t)\frac{|x|^2}{2\lambda(t)^2},$$

where

(6.7) $$a(t) = -\lambda(t)\frac{d\lambda(t)}{dt}.$$

We also assume that

(6.8) $$\lim_{t \uparrow T_m} \zeta(t) = \infty,$$

(6.9) $$\lim_{t \uparrow T_m} a(t) = 0,$$

(6.10) $$\lim_{t \uparrow T_m} \eta(t) = \infty.$$

We shall make the following additional assumption on the blow-up rate:

(6.11) $$\frac{d}{dt}\{\|\nabla\psi(t)\|(T_m - t)\} \leqq 0 \iff a(t)(T_m - t)\frac{1}{\lambda^2(t)} \leqq 1 \text{ (by (6.2))}.$$

We take $\omega \gg 1$ and $T_m > 0$ so that we have, for some $R > 1$, that

(6.12) $$\frac{1}{\sqrt{T_m}} \leqq R \leqq \frac{1}{4}\left|\frac{\nabla Q}{Q}\right|\left(\frac{1}{R}\right) \approx \frac{1}{4}\lim_{|x| \to \infty}\left|\frac{\nabla Q(x)}{Q(x)}\right|,$$

and

(6.13) $$P(\widetilde{\Gamma_0(R)}) > 0,$$

where

(6.14) $$\widetilde{\Gamma_0(R)} \equiv \bigcup_{\eta > 0} \bigcap_{\eta < t < T_m} \left\{\gamma \in \Gamma \ \bigg| \ \frac{\lambda(t)}{R} \leqq |\gamma(t)|\right\} \bigcap \Gamma_0(R),$$

with

(6.15) $$\Gamma_0(R) \equiv \bigcup_{\eta > 0} \bigcap_{\eta < t < T_m} \left\{\gamma \in \Gamma \ \bigg| \ |\gamma(t)| \leqq R\frac{\|\nabla\psi(t)\|}{\|\nabla\psi(0)\|}(T_m - t)\right\}.$$

Furthermore we assume

(6.16) $$\frac{d}{dt}\|\nabla\psi(t)\| \leqq 0,$$

for the presentation simplicity of the log log law in Theorem D.

For $\gamma \in \widetilde{\Gamma_0(R)}$, we suppose that the factor η gives us:

(6.17) $$\frac{r((\gamma(t))/(\lambda(t), \eta(t)))}{Q((\gamma(t))/(\lambda(t)))} \to 0 \text{ as } t \uparrow T_m,$$

which could be guaranteed by (6.5). This condition (6.17) yields:

(6.18) $$b(\gamma(t), t) \approx \frac{1}{\lambda(t)}\left(\frac{\nabla Q}{Q}\right)\left(\frac{\gamma(t)}{\lambda(t)}\right) + \frac{a(t)}{\lambda(t)}\frac{\gamma(t)}{\lambda(t)} \text{ as } t \uparrow T_m.$$

Now we take a sample path γ from $\widetilde{\Gamma_0(R)}$, and we shall fix it. We use the following notation below: $X_t \equiv X_t(\gamma) = \gamma(t)$ and $B_t \equiv B_t(\gamma)$.

Lower estimate.

From (5.20) with (5.19), we have

$$(6.19) \quad \frac{|B_{T_m} - B_t|}{T_m - t} \leqq \frac{|X_{T_m} - X_t|}{T_m - t} + \frac{1}{T_m - t}\int_t^{T_m} |b(X_\tau, \tau)|\, d\tau,$$

$$\lesssim R\|\nabla\psi(t)\| + \frac{1}{T_m - t}\int_t^{T_m} \|\nabla\psi(\tau)\|\, d\tau,$$

$$\lesssim \frac{1}{T_m - t}\int_t^{T_m} \|\nabla\psi(\tau)\|\, d\tau.$$

Here, we have used the property of γ and (6.18) with (6.11). Since we know that $\{B_{T_m} - B_{T_m-s}\}_{0 \leqq s < T_m}$ is a Brownian motion (see, e.g., [**11, 12**]), the so-called *iterated logarithmic law* (see, e.g., [**11, 12**] again) implies

$$(6.20) \quad \limsup_{s\downarrow 0} \frac{1}{\sqrt{s \ln\ln\frac{1}{s}}} |B_{T_m} - B_{T_m-s}| < \infty.$$

In other words, we have

$$(6.21) \quad \limsup_{t\uparrow T_m} \sqrt{\frac{T_m - t}{\ln\ln(T_m - t)^{-1}}} \left|\frac{B_{T_m} - B_t}{T_m - t}\right|$$

$$= \limsup_{t\uparrow T_m} \frac{1}{\sqrt{(T_m - t)\ln\ln(T_m - t)^{-1}}} |B_{T_m} - B_t| < \infty \quad \text{a.s.}$$

Hence,

$$(6.22) \quad 1 \lesssim \limsup_{t\uparrow T_m} \sqrt{\frac{T_m - t}{\ln\ln(T_m - t)^{-1}}} \left(\frac{1}{T_m - t}\int_t^{T_m} \|\nabla\psi(\tau)\|\, d\tau\right).$$

Upper estimate.

We have from (5.20) with (5.19) again that

$$(6.23) \quad \left|\frac{1}{T_m - t}\int_t^{T_m} b(X_\tau, \tau)\, d\tau\right| \leqq \frac{|X_{T_m} - X_t|}{T_m - t} + \frac{|B_{T_m} - B_t|}{T_m - t}.$$

On the other hand, we have from (6.11) and (6.12) that:

$$(6.24) \quad \left|\int_t^{T_m} b(X_\tau, \tau)\, d\tau\right|$$

$$\geqq \left|\int_t^{T_m} \frac{1}{\lambda(\tau)}\left(\frac{\nabla Q}{Q}\right)\left(\frac{X_\tau}{\lambda(\tau)}\right) d\tau\right| - \int_t^{T_m} \frac{1}{\lambda(\tau)}\left|a(\tau)\frac{X_\tau}{\lambda(\tau)}\right| d\tau,$$

$$\geqq \left|\int_t^{T_m} \frac{1}{\lambda(\tau)}\left(\frac{\nabla Q}{Q}\right)\left(\frac{X_\tau}{\lambda(\tau)}\right) d\tau\right| - R\int_t^{T_m} \frac{1}{\lambda(\tau)}\left|a(\tau)(T_m - \tau)\frac{1}{\lambda^2(\tau)}\right| d\tau,$$

$$\geqq 4R\int_t^{T_m} \frac{1}{\lambda(\tau)}\, d\tau - R\int_t^{T_m} \frac{1}{\lambda(\tau)}\, d\tau.$$

Hence, we have

$$
\begin{aligned}
(6.25)\quad \frac{3R}{T_m - t}\int_t^{T_m} \frac{1}{\lambda(\tau)}d\tau &\leqq \frac{|X_{T_m} - X_t|}{T_m - t} + \frac{|B_{T_m} - B_t|}{T_m - t},\\
&\leqq \frac{R}{\lambda(t)} + \frac{|B_{T_m} - B_t|}{T_m - t},\\
&\leqq R\int_t^{T_m}\frac{1}{\lambda(\tau)}d\tau + \frac{|B_{T_m} - B_t|}{T_m - t}.
\end{aligned}
$$

In the last inequality, we have used (6.16). Thus,

$$
(6.26)\quad \frac{2R}{T_m - t}\int_t^{T_m}\frac{1}{\lambda(\tau)}d\tau \leqq \frac{|B_{T_m} - B_t|}{T_m - t},
$$

so that

$$
(6.27)\quad \frac{1}{T_m - t}\int_t^{T_m}\|\nabla\psi(\tau)\|\,d\tau \lesssim \frac{|B_{T_m} - B_t|}{T_m - t}.
$$

Consequently, we have

$$
(6.28)\quad \limsup_{t\uparrow T_m}\sqrt{\frac{T_m - t}{\ln\ln(T_m - t)^{-1}}}\left(\frac{1}{T_m - t}\int_t^{T_m}\|\nabla\psi(\tau)\|\,d\tau\right) \lesssim 1.
$$

References

1. J. Bourgain and W. Wang, *Construction of blow-up solutions for the nonlinear Schrödinger equation with critical nonlinearity*, Ann. Scuola Norm. Sup. Pisa Cl. Sci. (4) **25** (1997), no. 1–2, 197–215.
2. E. Carlen, *Conservative diffusions*, Comm. Math. Phys. **94** (1983), 293–315.
3. _____, *Existence and sample path properties of the diffusions in Nelson's stochastic mechanics*, Springer Lecture Notes in Mathematics (S. Albeverio et al., eds.), Stochastic Processes in Mathematics and Physics (Bielefeld, 1984), vol. 1158, Springer-Verlag, Berlin–Heidelberg–New York, 1985, pp. 25–51.
4. _____, *Progress and problems in stochastic mechanics*, Stochastic Methods in Mathematical Physics (W. Karwowski, ed.), Stochastic Processes in Mathematics and Physics, vol. **345**, World Scientific, Singapore, 1989, pp. 3–31.
5. _____, *Potential scattering in stochastic mechanics*, Ann. Inst. H. Poincaré Phys. Théor. **42** (1985), 407–428.
6. T. Cazenave and F. B. Weissler, *The Cauchy problem for the critical nonlinear Schrödinger equation in H^s*, Nonlinear Anal. **14** (1990), 807–836.
7. G. Fibich, private communication, (1997).
8. G. M. Fraiman, *Asymptotic stability of manifold of self-similar solutions in self-focusing*, Soviet Phys. JETP **61** (1985), no. 2, 228–233.
9. J. Ginibre and G. Velo, *On a class of nonlinear Schrödinger equations. I. The Cauchy problem general case*, J. Funct. Anal. **32** (1979), no. 1, 1–32; II. *Scattering theory, general case*, J. Funct. Anal. **32** (1979), no. 1, 33–71.
10. R. T. Glassey, *On the blowing-up solution to the Cauchy problem for nonlinear Schrödinger equations*, J. Math. Phys. **18** (1979), 1794–1797.
11. K. Itô and H. P. McKean, Jr., *Diffusion processes and their sample paths*, 2nd ed., Springer-Verlag, Berlin-New York, 1974.
12. I. Karatzas and S. E. Shreve, *Brownian motion and stochastic calculus*, 2nd ed., Springer-Verlag, Berlin-New York, 1991.
13. T. Kato, *Nonlinear Schrödinger equations*, Lecture Notes in Phys. (H. Holden and A. Jensen, eds.), Schrödinger Operators (Sonderborg, 1988), vol. **345**, Springer, Berlin, 1989, pp. 218–263.

14. D. W. Landman, C. Papanicolaou, C. Sulem, and P.-L. Sulem, *Rate of blow-up for solutions of the nonlinear Schrödinger equation at critical dimension*, Phys. Rev. A **38** (1988), 3837–3843.
15. B. LeMesurier, C. Papanicolaou, C. Sulem, and P. L. Sulem, *Local structure of the self-focusing singularity of the nonlinear Schrödinger equation*, Physica D **32** (1988), 210–226.
16. P. L. Lions, *The concentration-compactness principle in the calculus of variations. The locally compact case.* I, Ann. Inst. H. Poincaré, Anal. Nonlinéaire **1** (1984), no. 2, 109–145; *The concentration-compactness principle in the calculus of variations. The locally compact case.* II, Ann. Inst. H. Poincaré Anal. Nonlinéaire **1** (1984), no. 4, 223–283.
17. F. Merle, *Construction of solutions with exactly k blow-up points for the Schrödinger equation with the critical power nonlinearity*, Comm. Math. Phys. **129** (1990), 223–240.
18. _____, *Lower bounds for the blow-up rate of solutions of the Zakharov equation in dimension two*, Comm. Pure Appl. Math. **49** (1996), no. 8, 765–794.
19. H. Nawa, *Asymptotic profiles of blow-up solutions of the nonlinear Schrödinger equation*, Singularities in Fluids, Plasmas and Optics (R. E. Caflisch and G. C. Papanicolaou, eds.), NATO Adv. Sci. Inst. Series C Math. Phys. Sci., Kluwer Acad. Publ., Dordrecht, 1993, pp. 221–253; its revised manuscript, T.I.T. preprint series NO. 06-93 (#15), 1993.
20. _____, *Asymptotic profiles of blow-up solutions of the nonlinear Schrödinger equation with critical power nonlinearity*, J. Math. Soc. Japan **46** (1994), 557–586.
21. _____, *Formation of singularities in solutions of the nonlinear Schrödinger equation with critical power nonlinearity*, Miniconference on Analysis and Applications (Brisbane, 1993) (G. Martin and B. Thompson, eds.), Proc. Centre Math. Appl. Austral. Nat. Univ., vol. 33, Austral. Nat. Univ., Canberra, 1994, pp. 167–188.
22. _____, *Asymptotic and limiting profiles of blow-up solutions of the nonlinear Schrödinger equation with critical power*, Comm. Pure and Appl. Math. **52** (1999), 193–270.
23. _____, *Limiting profiles and $\log\log$ law for blow-up solutions of the nonlinear Schrödinger equation with critical power*, preprint (1999).
24. E. Nelson, *Derivation of the Schrödinger equation from Newtonian dynamics*, Phys. Rev. **150** (1966), 1079–1085.
25. _____, *Quantum fluctuations*, Princeton University Press, Princeton, 1984.
26. T. Ogawa and Y. Tsutsumi, *Blow-up of H^1-solution for the nonlinear Schrödinger equation*, J. Differential Equations **92** (1991), 317–330.
27. _____, *Blow-up of H^1-solution for the one dimensional nonlinear Schrödinger equation with critical power nonlinearity*, Proc. Amer. Math. Soc. **111** (1991), no. 2, 487–496.
28. G. S. Perel'man (1999), in preparation.
29. A. N. Shiryaev, *Probability*, Graduate Texts in Mathematics, 2nd ed., vol. 95, Springer-Verlag, New York, 1996.
30. A. I. Smirnov and G. M. Fraiman, *Interaction representation in the self-focusing theory*, Physica D **52** (1991), 2–15.
31. C. Sulem and P.-L. Sulem, *The nonlinear Schrödinger equation. Self-focusing and wave collapse*, Appl. Math. Sci., vol. 139, Springer-Verlag, New York, 1999.
32. M. Tsutsumi, *Nonexistence and instability of solutions of nonlinear Schödinger equations*, unpublished.
33. Y. Tsutsumi, *Lower estimates of blow-up solutions for nonlinear Schrödinger equations*, unpublished.
34. M. I. Weinstein, *Nonlinear Schrödinger equations and sharp interpolation estimates*, Comm. Math. Phys. **87** (1983), 511–517.
35. _____, *On the structure and formation singularities in solutions to nonlinear dispersive evolution equations*, Comm. Partial Differential Equations **11** (1986), 545–565.

Graduate School of Mathematics, Nagoya University, Chikusa-ku Nagoya 464-8602, Japan
E-mail address: nawa@math.nagoya-u.ac.jp

Embedded Solitons of the DSII Equation

Dmitri E. Pelinovsky and Catherine Sulem

ABSTRACT. We study the stability of exact standing wave solutions in the focussing Davey–Stewartson II equation. The standing wave solutions correspond to embedded solitons coupled to a linear wave spectrum. Decay of localized waves and propagation of nonlocalized (one-dimensional) solitons is observed due to instability of the exact solutions with respect to variations of the initial data. An outline of the proof of instability of embedded solitons in integrable PDEs is given for the case study.

1. Introduction

Gravity-capillary surface wave packets are described by the Davey–Stewartson (DS) system [**8, 9**],

$$(1.1) \quad iu_t + \lambda u_{xx} + \mu u_{yy} + (\nu_1 n + \nu_2 |u|^2)u = 0, \quad \alpha n_{xx} + n_{yy} + \delta(|u|^2)_{xx} = 0,$$

where $u = u(x, y, t)$ and $n = n(x, y, t)$ are amplitudes of wave envelope and mean flow, respectively, and parameters λ, μ, ν_1, ν_2, α and δ are expressed through parameters of a fluid [**1**]. A review of the well-posedness of the initial value problem can be found in [**25**] and references therein.

This paper is concerned with the hyperbolic-elliptic case of the DS system when $\lambda \mu < 0$ and $\alpha > 0$. In particular, when $\lambda = -\mu$, $\nu_1 = -\nu_2$, $\alpha = 1$, and $\delta = -2$, the hyperbolic-elliptic DS system (1.1) reduces to the integrable DSII equation that possesses two-dimensional solitons (lumps) [**2**]. We generalize the family of lump solutions of the DSII equation and discuss its instability with respect to variations of the initial data.

General standing wave solutions of Eq. (1.1) are prescribed by the substitution,

$$(1.2) \quad u = U(\xi, \eta) e^{i/2((v_x \xi)/\lambda + (v_y \eta)/\mu) + i/4((v_x^2)/\lambda + (v_y^2)/\mu)t + i\Omega t + i\theta}, \quad n = N(\xi, \eta),$$

where $\xi = x - v_x t + x_0$, $\eta = y - v_y t + y_0$, and $U(\xi, \eta)$ and $N(\xi, \eta)$ satisfy coupled equations,

$$(1.3) \quad \lambda U_{\xi\xi} + \mu U_{\eta\eta} - \Omega U + (\nu_1 N + \nu_2 |U|^2)U = 0, \quad \alpha N_{\xi\xi} + N_{\eta\eta} + \delta(|U|^2)_{\xi\xi} = 0.$$

2000 *Mathematics Subject Classification.* Primary: 35Q55; Secondary: 37K40.

We benefited from stimulating discussions with P. Deift and A. Fokas. D.P. acknowledges support from a NATO fellowship provided by NSERC and C.S. acknowledges support from NSERC Operating grant OGP0046179.

This is the final form of the paper.

©2001 American Mathematical Society

Here v_x, v_y, Ω, θ, x_0, and y_0 are arbitrary real parameters.

Existence of standing wave solutions was studied for the hyperbolic-elliptic case by Ghidaglia and Saut [17]. The authors showed that localized solutions of Eq. (1.3) may exist in the parameter range: $\lambda = -\mu$, $\alpha = 1$, $\delta = -2$, $\nu_1 > 0$, and $\nu_2 \in (-2\nu_1, 0)$. An explicit solution was found in the integrable case $\nu_2 = -\nu_1$. It corresponds to $\Omega = 0$ and has the radially symmetric form,

$$(1.4) \qquad U = \frac{2c\sqrt{2\lambda}}{\sqrt{\nu_1}(c^2 + \xi^2 + \eta^2)}, \qquad N = \frac{8\lambda(c^2 - \xi^2 + \eta^2)}{\nu_1(c^2 + \xi^2 + \eta^2)^2},$$

where c is an arbitrary real parameter. This solution is the lump of the integrable DSII equation [2]. From the standing wave solutions, one can construct a blow up solutions by means of lens transformation [20]. However, it is not known whether such solutions are stable with respect to variation of initial data and whether they attract a blow-up of general localized initial data.

We show in this paper that standing wave solutions of the DS system, if they exist, are embedded solitons coupled to a linear wave spectrum. This fact implies that the standing wave solutions are unstable with respect to variations of the initial data. We also generalize the family of exact lump solutions (1.4) to the case $\Omega \neq 0$ and show that the standing wave solutions for $\Omega \neq 0$ are not localized in $\mathcal{R}^2$ and represent a coupled state of a single lump and a one-dimensional soliton of the NLS equation. The corresponding solution is expressed analytically for $\Omega = a^2 \geq 0$ as

$$(1.5) \qquad \begin{aligned} U &= \frac{2c\sqrt{2\lambda}[1 + a(\xi - i\eta)]e^{-a(\xi - i\eta)}}{\sqrt{\nu_1}(c^2 e^{-2a\xi} + \xi^2 + \eta^2)}, \\ N &= \frac{8\lambda(c^2 e^{-2a\xi}(1 + 4a\xi + 2a^2(\xi^2 + \eta^2)) - \xi^2 + \eta^2)}{\nu_1(c^2 e^{-2a\xi} + \xi^2 + \eta^2)^2} \end{aligned}$$

and for $\Omega = -b^2 \leq 0$ as

$$(1.6) \qquad U = \frac{2c\sqrt{2\lambda}[1 + ib(\xi - i\eta)]e^{ib(\xi - i\eta)}}{\sqrt{\nu_1}(c^2 e^{-2b\eta} + \xi^2 + \eta^2)}, \qquad N = \frac{8\lambda(c^2 e^{-2b\eta} - \xi^2 + \eta^2)}{\nu_1(c^2 e^{-2b\eta} + \xi^2 + \eta^2)^2}.$$

In the limit $\Omega \to 0$, i.e. $a, b \to 0$, the standing waves (1.5) and (1.6) reduce to the lump (1.4). The general standing wave solutions (1.5) and (1.6) are also unstable with respect to variation of the initial data. In particular, translation of the exact solutions along to its parameters destroys the coupling between the localized and nonlocalized components and lead to a decay of the localized field and propagation of the nonlocalized (one-dimensional) soliton.

The paper is organized as follows. In Section 2, we discuss the concept of embedded eigenvalues and bifurcations in linear spectral problems. In Section 3, we analyze the characteristic properties of exact standing wave solutions of the DSII equation. In Section 4, we outline the proof of their instability in the nonlinear initial-value problem. Section 5 concludes the paper. Details of the dressing method for finding new exact solutions of the DSII equation are given in Appendix A.

2. Motivations for Instability of Embedded Solitons

The concept of instability of embedded eigenvalues with respect to variation of the potentials is now well understood for linear spectral problems (see, e.g., Refs. [19, 24]). To give a simple description of this phenomenon, let us consider the linear problem $\mathcal{L}\psi = \lambda\psi$ in the space $L^2(\mathcal{R}^n)$ for a general nonselfadjoint matrix operator $\mathcal{L}$. Suppose that $\mathcal{L} = \mathcal{L}_0 + \varepsilon \Delta \mathcal{L}$ and the operator $\mathcal{L}_0$ does not have eigenvalues in λ. Then, there are **two types** of the bifurcation of a new eigenvalue λ_ε for $\varepsilon \neq 0$ (see our recent paper [**22**]).

The **type–I** bifurcation occurs from a resonance at the bottom of the continuous spectrum of the operator $\mathcal{L}_0$. The eigenvalue λ_ε detaches from the bottom and goes to the gap of the continuous spectrum or to the complex plane of λ. The corresponding new bound state appears then from a resonant (nonlocalized) eigenfunction of the operator $\mathcal{L}_0$.

The **type–II** bifurcation occurs from an embedded eigenvalue of the operator $\mathcal{L}_0$. For instance, if the operator $\mathcal{L}_0$ has a one-dimensional continuous spectrum with an embedded eigenvalue λ_0, then the new eigenvalue λ_ε emerges transversely into a complex plane of λ. The corresponding new bound state appears from a localized bound state of the operator $\mathcal{L}_0$.

For both types, the bifurcation occurs under a certain sign-definite condition on the perturbation $\Delta \mathcal{L}$ and may exhibit various approximations for the dependence $\lambda_\varepsilon = \lambda(\varepsilon)$. A bifurcation of a new eigenvalue may not actually occur if a proper eigenvalue is not observed into the same functional space $\psi \in L^2(\mathcal{R}^n)$. In other words, if the perturbation does not violate the requirements of the functional space, the embedded eigenvalue or the resonance of the operator $\mathcal{L}_0$ just disappears upon the perturbation for $\varepsilon \neq 0$. The latter situation is typical in quantum mechanics because the Schrödinger operator $\mathcal{L}$ is self-adjoint and the eigenvalues embedded into a one-dimensional continuous spectrum of $\mathcal{L}_0$ decay exponentially in time for $\varepsilon \neq 0$ [**19, 24**]. Such disappearing eigenvalues may become "visible" into a weaker functional space, e.g. into a space of weighted eigenfunctions with exponential growth at infinity. But if we are restricted by certain "physical" requirements on the construction of the functional space, then the new eigenvalues are invisible and are not to be considered.

These preliminary facts from linear spectral theory are important for understanding the problem of existence and stability of embedded solitons in nonlinear PDEs. We consider the DS system (1.1) as our case study and turn to the problem arising for the standing wave solutions given by Eq. (1.2). For simplicity, we set here $v_x = v_y = 0$ and consider the spectrum of linear waves of the form,

$$u \sim e^{i(k_x x + k_y y) + i\omega t}, \quad n \sim 0,$$

where $\omega = \omega(k_x, k_y)$ is given by

$$\omega = -\lambda k_x^2 - \mu k_y^2.$$

If $\lambda \mu < 0$ as in the hyperbolic-elliptic DS system, the spectrum covers the whole axis of real ω's. Therefore, even if the standing localized waves given by Eq. (1.2) exist for real Ω, there is always a curve in the (k_x, k_y) plane, where $\omega(k_x, k_y) = \Omega$. It indicates that the standing waves for any Ω are embedded in linear spectrum of the hyperbolic-elliptic DS system (1.1). Therefore, we expect that standing localized waves being coupled with linear waves will lose their energy due to radiation and globally decay as time evolves.

Although a direct analogy with the linear problem is simple, it is not clear how to prove the instability of embedded solitons within the nonlinear PDE (1.1). In this paper we use the integrability properties of the DSII equation to reduce the nonlinear problem to the linear Dirac system with a nontrivial two-dimensional essential spectrum. Then, we construct and study exact standing wave solutions of Eq. (1.3) within the linear problem and give the main links for the proof of instability of standing wave solutions in the DSII equation.

3. Exact Standing-Wave Solutions of the DSII Equation

We rewrite the integrable DSII system in the convenient form,

$$(3.1) \qquad iu_t + \frac{1}{2}(u_{xx} - u_{yy}) + (n - |u|^2)u = 0, \quad n_{xx} + n_{yy} - 2(|u|^2)_{xx} = 0.$$

The DSII equation (3.1) was solved through a $\bar{\partial}$ problem of complex analysis [**5, 12, 14**] in the case where the bound states of discrete spectrum are not present. The bound states correspond to algebraically localized multi-lump potentials. They were included in the inverse scattering scheme later in [**3, 13**]. Recently, a perturbation theory of a single lump was developed for the DSII equation by Gadyl'shin and Kiselev [**15, 16**]. They found that a single lump is unstable and disappears under a small variation of initial data. This conclusion was generalized in [**23**] to the case of potentials of the Dirac system with embedded eigenvalues.

Other special solutions localized algebraically and exponentially were constructed for the DSII equation by methods of Liouville–Laplace reductions [**4**], gauge transformations [**7**], Darboux transformations [**18**] and the $\bar{\partial}$ dressing method [**10**]. However, no standing localized waves were identified among these special solutions, and their spectral properties and dynamical role were not studied.

The DSII equation (3.1) appears as a compatibility condition of two linear systems: the two-dimensional Dirac problem,

$$(3.2) \qquad i\mathbf{J}\boldsymbol{\varphi}_y + \boldsymbol{\varphi}_x + \mathbf{P}\boldsymbol{\varphi} = 0,$$

and the time-evolution problem,

$$(3.3) \qquad i\mathbf{J}\boldsymbol{\varphi}_t + \boldsymbol{\varphi}_{xx} + \mathbf{P}\boldsymbol{\varphi}_x + \mathbf{R}\boldsymbol{\varphi} = 0,$$

where

$$\boldsymbol{\varphi}(x,y,t) = \begin{pmatrix} \varphi_1 \\ \varphi_2 \end{pmatrix}, \quad \mathbf{J} = \begin{pmatrix} 1 & 0 \\ 0 & -1 \end{pmatrix}, \quad \mathbf{P} = \begin{pmatrix} 0 & u \\ -\bar{u} & 0 \end{pmatrix}, \quad \mathbf{R} = \begin{pmatrix} r_1 & r_2 \\ -\bar{r}_2 & \bar{r}_1 \end{pmatrix}$$

and

$$r_1 = \frac{1}{2}(\partial_x + i\partial_y)^{-1}(\partial_x - i\partial_y)|u|^2, \quad r_2 = \frac{1}{2}(\partial_x - i\partial_y)u.$$

The dressing method [**27**], can be used to generate exact solutions to the DSII equation (3.1) via the linear systems (3.2) and (3.3). In Appendix A we apply this technique and find the following family of exact solutions,

$$(3.4) \qquad u = \frac{2\bar{w}_j[1 + \bar{\lambda}_j(\bar{z} + \bar{z}_j)]}{|z + z_j|^2 + |w_j|^2},$$

$$(3.5) \qquad n = 2\frac{|w_j|^2[(\lambda_j + \bar{\lambda}_j)^2|z + z_j|^2 + 2(\lambda_j + \bar{\lambda}_j)(z + \bar{z} + z_j + \bar{z}_j) + 2]}{(|z + z_j|^2 + |w_j|^2)^2}$$

$$+ \frac{-(z + z_j)^2 - (\bar{z} + \bar{z}_j)^2}{(|z + z_j|^2 + |w_j|^2)^2}(|z + z_j|^2 + |w_j|^2)^2,$$

where

$$w_j = c_j e^{-(\lambda_j + ik_j)z - i\bar{k}_j\bar{z} - i(\lambda_j^2 + k_j^2 + \bar{k}_j^2)t}.$$

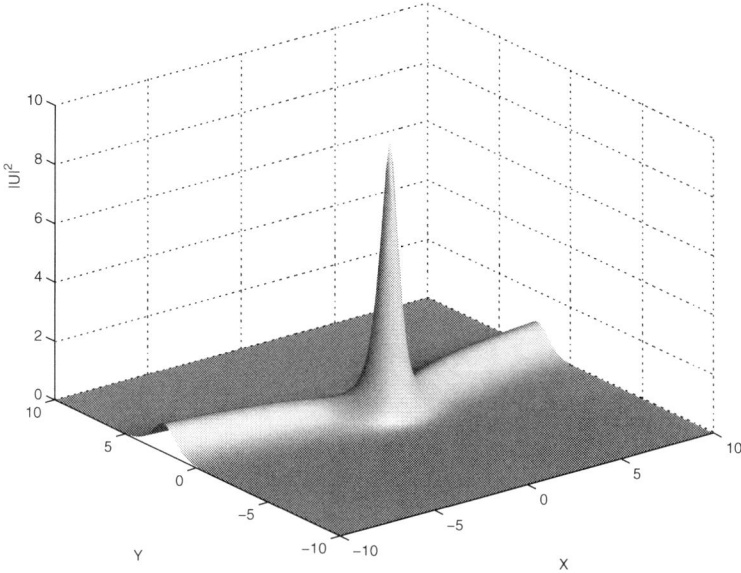

FIGURE 1. The exact standing wave solution (3.4) for $\lambda_j = i$.

The solution (3.4)–(3.5) has four complex parameters λ_j, k_j, z_j, and c_j. Identifying Eqs. (1.2) and (3.4) for parameters $\lambda = -\mu = 1/2$, $\nu_1 = -\nu_2 = 1$, $\alpha = 1$, and $\delta = -2$, we find the connection formulas to the standing wave representation,

$$\bar{\lambda}_j^2 = \Omega, \quad 2k_j = v_x + iv_y, \quad c_j = ce^{-i\theta}, \quad z_j = x_0 + iy_0.$$

There are only two standing wave solutions, when λ_j is real or purely imaginary, respectively. These standing wave solutions are given by Eqs. (1.5) and (1.6).

Generally, the exact solution (3.4)–(3.5) is localized exponentially for $\lambda_j \neq 0$ in all directions except in the direction of the curve,

(3.6) $$\{z = (x, y) \in \mathcal{R}^2 : \operatorname{Re}(\lambda_j z) + \ln|z| = \zeta_r < \infty\},$$

as $|z| \to \infty$. In this direction, the wave profile has a constant value,

(3.7) $$\lim_{|z| \to \infty} |u|^2(z, \bar{z}) = \frac{4|c_j|^2 |\lambda_j|^2 e^{-2\zeta_r}}{(1 + |c_j|^2 e^{-2\zeta_r})^2}.$$

In Fig. 1 we plot the profile of $|u|^2(x, y)$ for $\lambda_j = i$. The picture is stationary in time and can be seen as a nonlinear combination of the DSII lump and a curved one-dimensional soliton.

When the parameter λ_j is complex, the exact solution (3.4)–(3.5) depends on time. The profile $|u|^2(x, y, t)$ for $\lambda_j = 0.25 + i$ is shown in Figs. 2(a-d) at various times: $t = 0$, $t = 2.5$, $t = 5$, and $t = 7.5$. Solutions for arbitrary values of λ display the same type of behaviour. As time evolves, the two-dimensional localized field (the DSII lump) decays, the curvature of the front of the one-dimensional soliton disappears, and the one-dimensional soliton moves with an asymptotically constant propagation speed $\mathbf{v} = (v_x, v_y)$.

It follows from the exact solution (3.4)–(3.5) that the standing waves of the DSII equation display a structural sensitivity to variations of the parameter λ_j. When

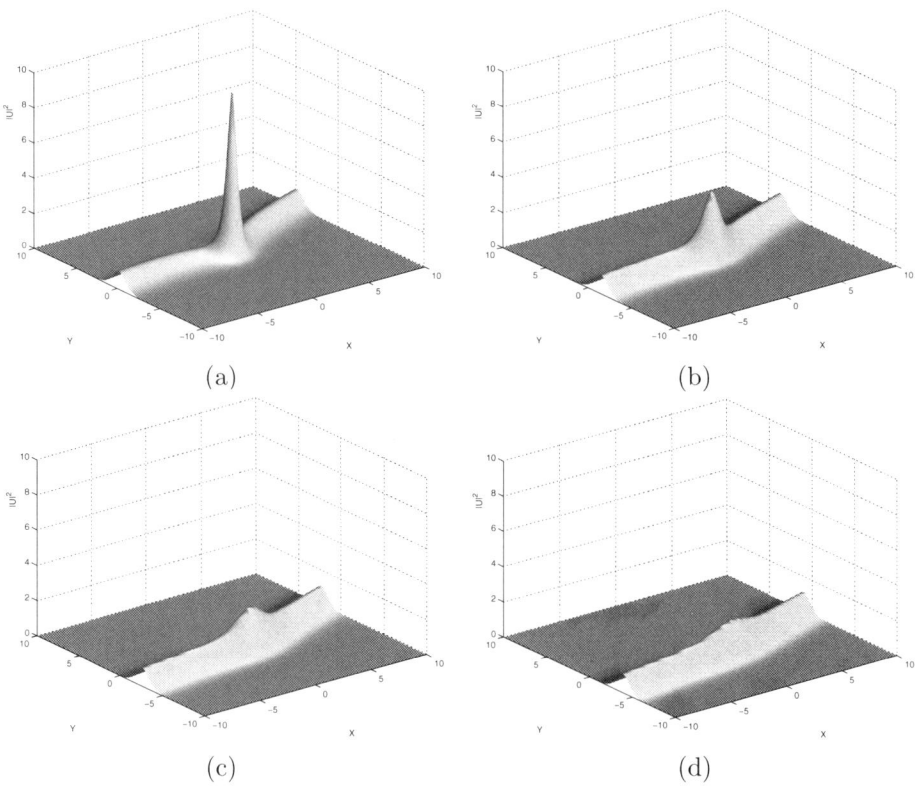

FIGURE 2. The exact solution (3.4) for $\lambda_j = 0.25 + i$, $k_j = 0$, $z_j = 0$, and $c_j = 1$. The shown is the field $|u|^2(x,y,t)$ at different time instances: (a) $t = 0$, (b) $t = 2.5$, (c) $t = 5$, and (d) $t = 7.5$.

this parameter slightly deviates from real or imaginary axes, the standing wave is destroyed and the localized field decays. This observation indicates the instability of standing wave solutions of the DSII equation with respect to the variations of the initial data. As we have discussed in Section 2, this instability occurs because the standing waves of the DSII equations are solitons embedded into the linear spectrum.

4. Proof of Instability of Embedded Solitons

Here we only provide the basic ideas needed to prove the instability of embedded solitons of the DSII equation. We also discuss the connection between the instability of lump potentials and the new standing wave solutions. A detailed proof is described in [**23**] on the basis of a spectral decomposition within the Dirac system (3.2). Another proof was developed by Gadyl'shin and Kiselev [**15, 16**] within a spectral decomposition over squared eigenfunctions.

The proof of instability of embedded solitons is based on the following result.

PROPOSITION. *Suppose $u(x,y)$ supports an embedded double degenerate eigenvalue at $k = k_j$ in the Dirac system (3.2) with two bound states $\Phi_j(x,y)$ and*

$\Phi'_j(x,y)$ and $\Delta u(x,y) \in L^2(\mathcal{R}^2)$ satisfies the constraint,

$$\text{(4.1)} \qquad \iint dx\, dy [\Delta \bar{u} \Phi^{a'}_{1j} \Phi_{1j} + \Delta u \Phi^{a'}_{2j} \Phi_{2j}](x,y) \neq 0,$$

where $\Phi^{a'}_j(x,y)$ is an adjoint bound state. Then, the potential $u^\varepsilon = u(x,y) + \varepsilon \Delta u(x,y)$ does not support embedded eigenvalues for $\varepsilon \neq 0$.

The proof of this proposition consists of three steps.

STEP I. We study the conditions on existence of embedded eigenvalues in the discrete spectrum of the Dirac system. In the case where the potential is a multi-lump solution, the conditions are satisfied by construction but general potentials may also support embedded eigenvalues under a special constraint on the spectral data $b(k, \bar{k})$, where k is complex.

STEP II. We construct a spectral decomposition for the Dirac system (3.2). It consists in constructing an adjoint set of eigenfunctions and proving the orthogonality and completeness relations through the inverse scattering identities. In this way, we found that a scalar spectral decomposition is associated with the matrix Dirac problem (3.2).

STEP III. The spectral decomposition is used for a perturbation theory of the double degenerate embedded eigenvalue $k = k_j$. We reduce the Dirac system asymptotically to a set of algebraic equations for the amplitudes of the bound states $\Phi_j(x,y)$ and $\Phi'_j(x,y)$. The nonexistence of the embedded eigenvalue in the neighbourhood of k_j for $\varepsilon \neq 0$ follows from the fact that the determinant of this system is strictly positive under the constraint (4.1).

A small variation of the initial data for the DSII equation (3.1) thus leads to the decay of localized waves. Local bound estimates for the decay are also reported in [**16**].

Although the embedded eigenvalues disappear in the space of functions $u \in L^2(\mathcal{R}^2)$, a weaker space can support discrete spectrum. Indeed, the standing wave solutions for $\lambda_j \neq 0$ are not in the Sobolev space and they have a nontrivial discrete spectrum construction. Thus, bifurcations from embedded eigenvalues for lump potentials to eigenvalues of the discrete spectrum for nonlocalized wave solutions could be possible in a space of bounded but nonlocalized eigenfunctions. Still these bifurcations are not physically significant if one considers the initial-data problem in the space $L^2(\mathcal{R}^2)$.

5. Conclusion

Embedded solitons of the DSII equation are generally nonlocalized for $\Omega \neq 0$. In the special case, $\Omega = 0$, the embedded solitons become localized and equivalent to lumps of the DSII equation. However, the embedded solitons are unstable with respect to variations of the initial data and we have proved this structural instability using the equivalence with the linear problem.

This result indicates that embedded solitons coupled to linear spectrum should be generally unstable in nonintegrable PDEs. Indeed, there are recently several examples of such solitons arising in nonintegrable generalizations of the NLS equation [**6, 26**]. A further stimulating problem is to develop a proof of instability of embedded solitons within the original nonintegrable PDE model.

Finally, we mention another open problem associated with the lump solitons of the DSII equation. The exact blow-up solutions were found in Ref. [20] by using the lens transformation of the DSII lump. Since we have shown that the embedded solitons are unstable and decay under variations of the initial data, we expect that the exact blow-up solutions are also unstable. Moreover, we expect that the instability of embedded solitons in the initial-value problem is closely related to structural instability of embedded soliton solutions in the general hyperbolic-elliptic DS system (1.1) beyond the integrability case.

Appendix A. The Dressing Method

The dressing method for integrable nonlinear evolution equations was developed by Zakharov and Shabat [27] to generate new hierarchies of integrable equations as well as their exact solutions. Application of the method to the (defocusing) DSII equation was described by Pelinovsky [21] within a canonical formalism and by Dubrovsky [11] for a new $\bar{\partial}$-dressing formalism. Here we modify the scheme of Ref. [21] and construct exact solutions of the (focussing) DSII equation (3.1).

The idea for the dressing method is to relate nontrivial solutions of a linear eigenvalue system associated with a nonlinear PDE and plane-wave eigenfunctions of the same linear problem with no potentials. We develop this idea for the DSII equation (3.1) and suppose that the eigenfunction $\varphi(x,y,t)$ in Eqs. (3.2) and (3.3) can be expressed through the following integral representation with a triangular kernel,

$$(A.1) \qquad \varphi = \varphi_0(x,y,t) + \int^{x} \mathbf{K}(x,x';y,t)\varphi_0(x')\,dx'.$$

Here $\varphi_0(x,y,t)$ is a plane-wave eigenfunction satisfying the same problem with no potentials ($\mathbf{P} = \mathbf{R} = \mathbf{0}$), i.e.

$$(A.2) \qquad i\mathbf{J}\varphi_{0y} + \varphi_{0x} = 0, \quad i\mathbf{J}\varphi_{0t} + \varphi_{0xx} = 0,$$

while the 2×2 matrix function $\mathbf{K}(x,x';y,t)$ can be found from Eqs. (3.2), (3.3), (A.1), and (A.2) to solve the system,

$$(A.3) \qquad \begin{array}{l} i\mathbf{K}_y + \mathbf{J}\mathbf{K}_x + \mathbf{K}_{x'}\mathbf{J} + \mathbf{JPK} = 0, \\ i\mathbf{K}_t + \mathbf{J}\mathbf{K}_{xx} - \mathbf{K}_{x'x'}\mathbf{J} + \mathbf{JPK}_x + \mathbf{JRK} = 0. \end{array}$$

Simultaneously, the matrices of the potentials $\mathbf{P}$ and $\mathbf{R}$ can be related to the matrix function $\mathbf{K}(x,x';y,t)$ at $x' = x$ in the form,

$$(A.4) \qquad \begin{array}{l} u = -2K_{12}(x,x;y,t), \\ r_1 = -2\dfrac{d}{dx}K_{11}(x,x;y,t) - uK_{21}(x,x;y,t), \\ r_2 = -2\dfrac{\partial}{\partial x}K_{12}(x,x';y,t)\Big|_{x'=x} - uK_{22}(x,x;y,t). \end{array}$$

There are also two symmetry constraints following from the symmetry of the potentials $\mathbf{P}$ and $\mathbf{R}$,

$$K_{22} = \overline{K}_{11}, \quad K_{21} = -\overline{K}_{12}.$$

Following Zakharov and Shabat [27], we define the 2×2 matrix function $\mathbf{F}(x,x';y,t)$ by solving the matrix integral equation,

$$(A.5) \qquad \mathbf{F}(x,x';y,t) + \mathbf{K}(x,x';y,t) + \int^{x} \mathbf{K}(x,s;y,t)\mathbf{F}(s,x';y,t)\,ds = 0.$$

The matrix function $\mathbf{F}(x, x'; y, t)$ can be found from Eqs. (A.3) and (A.5) to solve the same linear system with no potentials ($\mathbf{P} = \mathbf{R} = 0$), i.e.

(A.6) $\quad i\mathbf{F}_y + \mathbf{J}\mathbf{F}_x + \mathbf{F}_{x'}\mathbf{J} = 0, \quad i\mathbf{F}_t + \mathbf{J}\mathbf{F}_{xx} - \mathbf{F}_{x'x'}\mathbf{J} = 0.$

The eigenfunctions $\varphi_0(x, y, t)$ and $\mathbf{F}(x, x'; y, t)$ are usually referred to as an undressed solution, while the eigenfunctions $\varphi(x, y, t)$ and $\mathbf{K}(x, x'; y, t)$ represent the dressing transformation. A special class of exact solutions of the linear system (A.6) is obtained by separating the variables,

(A.7) $\quad \mathbf{F}(x, x'; y, t) = \sum_{k=1}^{N} \mathbf{f}_k^+(x, y, t) \mathbf{f}_k^{t-}(x', y, t),$

where N is the order of the exact solution, $\mathbf{f}_k^\pm(x, y, t) \in R^2$ are vector-columns satisfying the system,

(A.8) $\quad \begin{aligned} i\mathbf{J}\mathbf{f}_{ky}^+ + \mathbf{f}_{kx}^+ = 0, &\quad i\mathbf{J}\mathbf{f}_{kt}^+ + \mathbf{f}_{kxx}^+ = 0, \\ i\mathbf{J}\mathbf{f}_{ky}^- + \mathbf{f}_{kx'}^- = 0, &\quad i\mathbf{J}\mathbf{f}_{kt}^- - \mathbf{f}_{kx'x'}^- = 0, \end{aligned}$

and $\mathbf{f}^t$ stands for the transpose vector of $\mathbf{f}$. Solving Eqs. (A.6) for $\mathbf{K}(x, x'; y, t)$,

(A.9) $\quad K(x, x'; y, t) = \sum_{k=1}^{N} \mathbf{g}_k(x, y, t) \mathbf{f}_k^{t-}(x', y, t),$

we find a linear algebraic system for $\mathbf{g}_k(x, y, t)$,

$$\sum_{m=1}^{n} h_{km}(x, y, t) \mathbf{g}_m(x, y, t) = -\mathbf{f}_k^+(x, y, t),$$

where the matrix element $h_{km}(x, y, t)$ is given by

$$h_{km} = \delta_{km} + \int^x \mathbf{f}_k^+ \cdot \mathbf{f}_m^- \, dx,$$

with δ_{km} the Kroneker's symbol. Solving the linear system by the Kramer's rule, we find an explicit solution for $\mathbf{K}(x, x'; y, t)$,

(A.10) $\quad \mathbf{K}(x, x'; y, t) = -\sum_{k=1}^{N}\sum_{m=1}^{N} \frac{\Delta_{mk}(x, y, t)}{\Delta(x, y, t)} \mathbf{f}_m^+(x, y, t) \mathbf{f}_k^{t-}(x', y, t).$

Here $\Delta(x, y, t)$ is the determinant of the matrix consisting of elements $h_{mk}(x, y, t)$ and $\Delta_{mk}(x, y, t)$ is the cofactor of $h_{mk}(x, y, t)$. The determinant representation (A.10) generates a general class of exact solutions with arbitrary functional dependence via Eq. (A.5).

In order to construct the exact standing wave solutions of the DSII equation, we consider only a special case $N = 2$ for the determinant $\Delta(x, y, t)$ and take the particular solution of the system (A.9) in the form,

$$\mathbf{f}_1^+ = \begin{pmatrix} \phi \\ 0 \end{pmatrix}, \quad \mathbf{f}_2^+ = \begin{pmatrix} 0 \\ \bar{\phi} \end{pmatrix}, \quad \mathbf{f}_1^- = \begin{pmatrix} 0 \\ \chi \end{pmatrix}, \quad \mathbf{f}_2^- = \begin{pmatrix} -\bar{\chi} \\ 0 \end{pmatrix}.$$

The functions $\phi = \phi(x, y, t)$ and $\chi = \chi(x, y, t)$ can be found from Eqs. (A.9) to solve the equations,

(A.11) $\quad \begin{aligned} i\phi_y + \phi_x = 0, &\quad i\phi_t + \phi_{xx} = 0, \\ -i\chi_y + \chi_x = 0, &\quad i\chi_t + \chi_{xx} = 0. \end{aligned}$

Then, using Eqs. (A.5) and (A.10) we derive an exact solution of the DSII equation (3.1) in the form,

$$(A.12) \qquad u = 2\frac{\phi\chi}{\Delta}, \quad n = 2\frac{\partial^2}{\partial x^2}\log\Delta, \quad |u|^2 = \left(\frac{\partial^2}{\partial x^2} + \frac{\partial^2}{\partial y^2}\right)\log\Delta,$$

where

$$\Delta = 1 + \left(\int^x \bar{\phi}\chi dx\right)\left(\int^x \phi\bar{\chi} dx\right).$$

A large class of special solutions can be generated from this representation by specifying the functions $\phi(x,y,t)$ and $\chi(x,y,t)$. The following choice leads to a family of exact standing localized wave solutions of the DSII equation (3.1),

$$\phi = e^{p(x+iy)+ip^2t+\nu}, \quad \chi = (x - iy + 2iqt + \theta)e^{q(x-iy)+iq^2t+\sigma},$$

where p, q, ν, σ, and θ are parameters. Using simple algebra, we reduce the resulting solution to the form (3.4) and (3.5), where $\lambda_j = p + \bar{q}$, $k_j = i\bar{q}$, $z = x + iy - 2k_jt$, $z_j = \bar{\theta} - \lambda_j^{-1}$, and

$$c_j = \bar{\lambda}_j e^{-\nu-\sigma}.$$

An equivalent choice for functions $\phi(x,y,t)$ and $\chi(x,y,t)$ is

$$\phi = (x + iy + 2ipt + \theta)e^{p(x+iy)+ip^2t+\varphi}, \quad \chi = e^{q(x-iy)+iq^2t+\psi}.$$

The resulting solution for Eq. (A.12) reproduces the complex conjugate function for $u(x,y,t)$ to the given by Eq. (3.4) and the same form (3.5) for $n(x,y,t)$.

References

1. M. J. Ablowitz and H. Segur, *On the evolution of packets of water waves*, J. Fluid Mech. **92** (1979), 691–715.
2. M. J. Ablowitz and P. A. Clarkson, *Solitons, nonlinear evolution equations and inverse scattering*, Cambridge University Press, Cambridge, 1991.
3. V. A. Arkadiev, A. K. Pogrebkov, and M. C. Polivanov, *Inverse scattering transform method and soliton solutions for Davey–Stewartson II equation*, Physica D **36** (1989), 189–197.
4. _____, *Closed string-like solutions of the Davey–Stewartson equation*, Inverse Problems **5** (1989), L1–L6.
5. R. Beals and R. R. Coifman, *The D-bar approach to inverse scattering and nonlinear evolutions*, Physica D **18** (1986), 242–249.
6. A. R. Champneys, B. A. Malomed, and M. J. Friedman, *Thirring solitons in the presence of dispersion*, Phys. Rev. Lett. **80** (1998), 4169–4172.
7. Y. Cheng, Y. S. Li, and G. X. Tang, *The gauge equivalence of the Davey–Stewartson equation and (2 + 1)-dimensional continuous Heisenberg ferromagnetic model*, J. Phys. A **23** (1990), L473–L477.
8. A. Davey and K. Stewartson, *On three-dimensional packets of surface waves*, Proc. Roy. Soc. London Ser. A **388** (1974), 101–110.
9. V. D. Djordjevic and L. G. Redekopp, *On two-dimensional packets of capillary-gravity waves*, J. Fluid Mech. **79** (1977), 703–714.
10. V. G. Dubrovsky, *The construction of exact multiple pole solutions of some (2 + 1)-dimensional integrable nonlinear evolution equations via the $\bar{\partial}$-dressing method*, J. Phys. A **32** (1999), 369–390.
11. _____, *The application of the $\bar{\partial}$-dressing method to some integrable (2 + 1)-dimensional nonlinear equations*, J. Phys. A **29** (1996), 3617–3630.
12. A. S. Fokas, *Inverse scattering of first-order systems in the plane related to nonlinear multidimensional equations*, Phys. Rev. Lett. **51** (1983), 3–6.
13. A. S. Fokas and M. J. Ablowitz, *On the inverse scattering transform of multidimensional nonlinear equations related to first-order systems in the plane*, J. Math. Phys. **25** (1984), 2494–2505.

14. A. S. Fokas and L. Y. Sung, *On the solvability of the N-wave, Davey–Stewartson and Kadomtsev–Petviashvili equations*, Inverse Problems **8** (1992), 673–708.
15. R. R. Gadyl'shin and O. M. Kiselev, *On solitonless structure of the scattering data for a perturbed two-dimensional soliton of the Davey–Stewartson equation*, Teoret. Mat. Fiz. **161** (1996), 200–208.
16. _____, *Asymptotics of a perturbed soliton for the Davey–Stewartson II equation*, solv-int/9801014.
17. J. M. Ghidaglia and J. C. Saut, *Nonexistence of travelling wave solutions to nonelliptic nonlinear Schrödinger equations*, J. Nonlinear Sci. **6** (1996), 139–145.
18. S. B. Leble, M. A. Salle, and A. V. Yurov, *Darboux transforms for Davey–Stewartson equations and solitons in multidimensions*, Inverse Problems **8** (1992), 207–218.
19. M. Merkli and I. M. Sigal, *A time-dependent theory of quantum resonances*, Comm. Math. Phys. **201** (1999), 549–576.
20. T. Ozawa, *Exact blow-up solutions to the Cauchy problem for the Davey–Stewartson systems*, Proc. Roy. Soc. London Ser. A **436** (1992), 345–349.
21. D. E. Pelinovsky, *On a structure of the explicit solutions to the Davey–Stewartson equations*, Physica D **87** (1995), 115–122.
22. D. E. Pelinovsky and C. Sulem, *Eigenfunction and eigenvalues for the Riemann–Hilbert problem associated to inverse scattering*, Comm. Math. Phys. **208** (2000), 713–760.
23. _____, *Spectral decomposition for a two-dimensional Dirac system associated with the DSII equation*, Inverse Problems **16** (2000), 59–74.
24. A. Soffer and M. I. Weinstein, *Time dependent resonance theory*, Geom. Funct. Anal. **8** (1998), 1086–1128.
25. C. Sulem and P. L. Sulem, *The nonlinear Schrödinger equation: Self-focusing and wave collapse*, Applied Mathematical Sciences, vol. 139, Springer-Verlag, New York, 1999.
26. J. Yang, D. J. Kaup, and B. A. Malomed, *Embedded solitons in the second-harmonic-generating systems*, Phys. Rev. Lett. (1999) (to appear).
27. V. E. Zakharov and A. B. Shabat, *A scheme for integrating the nonlinear equations of mathematical physics by the method of the inverse scattering problem* I, Funct. Anal. Appl. **8** (1974), 226–235.

DEPARTMENT OF MATHEMATICS, UNIVERSITY OF TORONTO, TORONTO, ONTARIO, M5S 3G3, CANADA.

Current address: Department of Mathematics, McMaster University, Hamilton, Ontario, L8S 4K1, Canada.

E-mail address: `dmpeli@math.mcmaster.ca`

DEPARTMENT OF MATHEMATICS, UNIVERSITY OF TORONTO, TORONTO, ONTARIO, M5S 3G3, CANADA.

E-mail address: `sulem@math.toronto.edu`

On the Blow up Phenomenon for the Critical Nonlinear Schrödinger Equation in 1D

Galina Perelman

1. Introduction

Consider the nonlinear Schrödinger equation

(1.1) $$i\psi_t = -\psi_{xx} - |\psi|^{2p}\psi, \quad x \in \mathbb{R},$$

with initial data

$$\psi|_{t=0} = \psi_0 \in H^1.$$

It is well known that for $p \geq 2$ the problem has solutions that blow up in finite time. The case $p = 2$ marks the transition between the global existence and the blow up phenomenon. In this paper we study the participation of nonlinear bound states in singularity formation in the critical case $p = 2$.

The NLS (1.1) has an important solution of special form-soliton: $e^{it}\varphi_0(x)$, where φ_0 is the "ground state solitary wave". We consider the Cauchy problem for (1.1) with initial data close to a soliton:

$$\psi|_{t=0} = \varphi_0 + \chi_0,$$

where χ_0 is small in suitable sense. We show that for a certain class of initial perturbations the solution ψ blows up in finite time T^*, admitting the following asymptotic representation

(1.2) $$\psi(t,x) \sim e^{i\mu(t)}\lambda^{1/2}(t)\varphi_0\big(\lambda(t)x\big), \quad t \to T^*,$$

$$\lambda(t) \sim (T^* - t)^{-1/2}\big(\ln\big|\ln(T^* - t)\big|\big)^{1/2}, \quad \mu(t) \sim \ln(T^* - t)\ln\big|\ln(T^* - t)\big|.$$

Thus, up to a phase factor the formation of the singularity is self-similar with a profile given by the ground state. The behavior (1.2) was predicted in [**FS, KSZ, LPSS1, LPSS2, Ma, SS**].

2000 *Mathematics Subject Classification.* Primary: 35Q55; Secondary: 34L10.
This is the final form of the paper.

2. Preliminary Facts and Formulation of the Result

2.1. The nonlinear equation. We formulate here the necessary facts about Cauchy problem for the equation

$$i\psi_t = -\psi_{xx} - |\psi|^4 \psi, \tag{2.1}$$

with initial data in H^1.

PROPOSITION 2.1. *The Cauchy problem for equation* (2.1) *with initial data* $\psi(0,x) = \psi_0(x)$, $\psi_0 \in H^1$ *has a unique solution ψ in the space* $C\big([0,T^*) \to H^1\big)$ *with some $T^* > 0$ and*

(i) ψ *satisfies the conservation laws*

$$\int dx |\psi|^2 = \text{const}, \quad H(\psi) = \int dx \left[|\psi_x|^2 - \frac{1}{3}|\psi|^6 \right] = \text{const};$$

(ii) *if $T^* < \infty$, then $\|\psi_x\|_2 \to \infty$ as $t \to T^*$ and*

$$\|\psi_x\|_2 \geq c(T^* - t)^{-1/2};$$

(iii) *if $H(\psi_0) < 0$ then $T^* < \infty$.*

Suppose in addition that $x\psi_0 \in L_2$. Then $x\psi \in C\big([0,T^*) \to L_2\big)$ and ψ satisfies the pseudo-conformal conservation law

$$\int dx \big|(x + 2it\partial_x)\psi\big|^2 - \frac{4}{3}t^2 \int dx |\psi|^6 = \text{const}.$$

Equation (2.1) is invariant with respect to transformations:

$$\psi(x,t) \to (a+bt)^{-1/2} e^{i\omega + ibx^2/(4(a+bt))} \psi\left(\frac{x}{a+bt}, \frac{c+dt}{a+bt}\right), \tag{2.2}$$

where $\omega \in \mathbb{R}$, $\begin{pmatrix} a & b \\ c & d \end{pmatrix} \in \text{SL}(2,\mathbb{R})$.

2.2. Exact blow up solutions. The equation (2.1) has a family of soliton solutions

$$e^{i\alpha^2 t/4} \varphi_0(x,\alpha), \quad \alpha > 0,$$

where φ_0 is a positive even smooth decreasing function satisfying the equation

$$-\varphi_{0xx} + \frac{\alpha^2}{4}\varphi_0 - \varphi_0^5 = 0.$$

As $|x| \to \infty$, $\varphi_0 \sim \varphi_\infty e^{-\alpha/2|x|}$.

One has a relation

$$\varphi_0(x,\alpha) = \left(\frac{\alpha}{2}\right)^{1/2} \varphi_0\left(\frac{\alpha}{2}x\right),$$

where $\varphi_0(x)$ stands for $\varphi_0(x,2)$. Applying transformations (2.2) to (2.1) one gets a 3-parameter family of solutions

$$e^{i\mu(t) - i\beta(t)z^2/4} \lambda^{1/2}(t) \varphi_0(z), \quad z = \lambda(t)x, \tag{2.3}$$

where μ, b, λ are given by

$$\lambda(t) = (a+bt)^{-1}, \quad \beta(t) = -b(a+bt), \quad \mu(t) = \frac{c+dt}{a+bt}.$$

Remark that $\lambda(t)$, $\beta(t)$, $\mu(t)$ satisfy the system
$$\lambda^{-3}\lambda_t = \beta, \quad \lambda^{-2}\beta_t + \beta^2 = 0, \quad \lambda^{-2}\mu_t = 1.$$

If $b \neq 0$ solution (2.3) blows up in finite time. It is known that equation (2.1) has no blow-up solutions in the class
$$\{\psi \in H^1(\mathbb{R}), \|\psi\|_2 < \|\varphi_0\|_2\},$$
solutions (2.3) being the only blow-up solutions (up to Galilei invariance) with minimal mass, see [**W1, Me**].

2.3. Extended manifold of blow-up solutions. The 3-parameter family (2.3) can be considered as the boundary $a = 0$ of the 4-parameter family of formal solutions $w(x, \sigma(t))$,
$$w(x,\sigma) = e^{i\mu - i\beta z^2/4} \lambda^{1/2} \varphi(z,a), \quad z = \lambda x,$$
$\sigma = (\mu/2, \lambda, \beta, a)$, $\lambda \in \mathbb{R}_+$, $\beta, \mu, a \in \mathbb{R}$. Here

(2.4) $$\varphi(z,a) = \sum_{k=0}^{\infty} a^k \varphi_k(z),$$

is a formal solution of the equation
$$-\varphi_{zz} + \varphi - \frac{az^2}{4}\varphi - \varphi^5 = 0,$$
all φ_k being even smooth exponentially decreasing (as $|z| \to \infty$) functions.

The $w(x, \sigma(t))$ is a formal solution of (2.1) if $\sigma(t)$ satisfies the system

(2.5) $$\lambda^{-3}\lambda_t = \beta, \quad \lambda^{-2}\beta_t + \beta^2 = a, \quad \lambda^{-2}\mu_t = 1, \quad a_t = 0,$$

which gives, in particular, $\lambda = (d_2 t^2 + d_1 t + d_0)^{-1/2}$, $a = d_1^2/4 - d_2 d_1$. Here d_j are constant.

We shall use the notation $\varphi^N(z,a) = \sum_{k=0}^{N} a^k \varphi_k(z)$,
$$\varphi^N(z,\alpha,a) = \left(\frac{\alpha}{2}\right)^{1/2} \varphi^N\left(\frac{\alpha}{2}x, \frac{16a}{\alpha^4}\right).$$

2.4. Linearization of (2.1) on a soliton. Consider the linearization of (2.1) on the soliton $e^{it}\varphi_0(x)$:
$$i\chi_t = -\chi_{xx} - \varphi_0^4 \chi - 2\varphi_0^4(\chi + e^{2it}\overline{\chi}).$$
Introduce function f: $\chi = e^{it} f$. Then f satisfies the equation
$$i\vec{f}_t = H_0 \vec{f}, \quad \vec{f} = \begin{pmatrix} f \\ \overline{f} \end{pmatrix},$$
$$H_0 = (-\partial_x^2 + 1)\sigma_3 + V(\varphi_0), \quad V(\xi) = -3\xi^4 \sigma_3 - 2i\xi^4 \sigma_2,$$
σ_2, σ_3 being the standard Pauli matrices.

H_0 is considered as a linear operator in $L_2(\mathbb{R} \to \mathbb{C}^2)$ defined on the natural domain. Here and later L_2 stands for the subspace of the standard L_2 consisting of even functions. The operator H_0 satisfies the relations

(2.6) $$\sigma_3 H_0 \sigma_3 = H_0^*, \quad \sigma_1 H_0 \sigma_1 = -H_0.$$

The continuous spectrum of H_0 consists of two semi-axes $(-\infty, -1]$, $[1, \infty)$ and is simple.

The point $E = 0$ is an eigenvalue of the multiplicity 4. By differentiating the solution w with respect to the parameters it is easy to distinguish an eigenfunction $\vec{\xi}_0$

$$\vec{\xi}_0 = i\varphi_0 \begin{pmatrix} 1 \\ -1 \end{pmatrix}, \quad H_0 \vec{\xi}_0 = 0,$$

and three associated functions $\vec{\xi}_j$, $j = 1, 2, 3$,

$$\vec{\xi}_1 = \frac{1}{4}(1 + 2x\partial_x)\varphi_0 \begin{pmatrix} 1 \\ 1 \end{pmatrix}, \quad \vec{\xi}_2 = -i\frac{x^2}{8}\varphi_0 \begin{pmatrix} 1 \\ -1 \end{pmatrix}, \quad \vec{\xi}_3 = \frac{1}{2}\varphi_1 \begin{pmatrix} 1 \\ 1 \end{pmatrix}, \quad H_0\vec{\xi}_j = i\vec{\xi}_{j-1},$$

φ_1 being the second coefficient in the expansion (2.4). φ_1 can be characterized by the equation

$$L_+ \varphi_1 = \frac{x^2}{4}\varphi_0, \quad L_+ = -\partial_x^2 + 1 - 5\varphi_0^4,$$

the operator L_+ is invertible being restricted on the subspace of even function.

One can show that $E = 0$ is the only eigenvalue of H_0 (see [**W2, G, P**], for example).

2.5. Main theorem.

Consider the Cauchy problem for equation (2.1) with initial data

(2.7) $\quad \psi|_{t=0} = \psi_0, \quad \psi_0(x) = e^{-i\beta_0 x^2/4}\big(\varphi^N(x, \beta_0^2) + f_0(x)\big), \quad \beta_0 > 0,$

where $f_0(x) = f_0(-x)$ and f_0 satisfies the estimate

(2.8) $\quad \|f_0\|_X \leq \beta_0^{2N}.$

Here $\|f\|_X = \|f\|_{H^1} + \|xf\|_{L_2}$. Assume that
  (i) β_0 is sufficiently small;
  (ii) N is sufficiently large.
These conditions give, in particular,

$$H\big(\varphi^N(\beta_0^2) + f_0\big) = -\frac{\beta_0^2}{4}e + O(\beta_0^4) < 0, \quad e = \int dx x^2 \varphi_0^2,$$

which together with conformal invariance implies that the solution ψ of Cauchy problem (2.1, 2.7) blows up in finite time $T^* < \infty$. Our main result is the following.

THEOREM 2.1. *The solution ψ of the Cauchy problem (2.1, 2.7) blows up in finite time $T^* = (2\beta_0)^{-1}(1 + o(1))$, as $\beta_0 \to 0$, and there exist $\lambda(t)$, $\mu(t) \in C^1([0, T^*))$,*

(2.9) $\quad \begin{aligned} \lambda(t) &= \mathrm{const}(T^* - t)^{-1/2}\big(\ln\big|\ln(T^* - t)\big|\big)^{1/2}(1 + o(1)), \\ \mu(t) &= \mathrm{const} \ln(T^* - t)\ln\big|\ln(T^* - t)\big|(1 + o(1)), \quad t \to T^*, \end{aligned}$

such that ψ admits the representation

$$\psi(x, t) = e^{i\mu(t)}\lambda^{1/2}(t)\big(\varphi_0(z) + f(z, t)\big), \quad z = \lambda(t)x,$$

where f is small in $L_2 \cap L_\infty$ uniformly with respect to $t \in [0, T^*)$. Moreover, $\|f\|_\infty = o(1)$, as $t \to T^*$. The constants in (2.9) are independent of initial data.

It is worth mentioning that due to the conformal invariance the same result remains valid for initial data of the form

$$\widetilde{\psi}_0(x) = e^{i\omega - ibz^2/4} \lambda^{1/2} \psi_0(z), \quad z = \lambda x,$$

where $\omega \in \mathbb{R}$, $\lambda \in \mathbb{R}_+$, $b > -1/T^*$.

3. Some Words About the Proof

This section contains the outline of the proof. The details can be found in [**P**].

3.1. Splitting of motions. The main idea repeats the main idea of the works [**SW1, SW2, BP**] where the asymptotic stability of solitary waves were considered. We start by introducing some new coordinates for the description of the solution with initial data (2.7). The new coordinates posses an important property: they allow us to split the motion into two parts, the first part being a finite-dimensional dynamics on the manifold of formal solutions $\{w(\cdot, \sigma)\}$ and the second part remains small in some sense for all $t \in [0, T^*)$. To describe these coordinates we need to introduce a modified ground state $\widetilde{\varphi}(z, \alpha, a)$ which is characterized by the equation

$$(3.1) \quad -\widetilde{\varphi}_{zz} + \frac{\alpha^2}{4}\widetilde{\varphi} - \frac{az^2}{4}\theta(hz)\widetilde{\varphi} - \widetilde{\varphi}^5 = 0, \quad h = \sqrt{|a|} > 0,$$

$\alpha, a \in \mathbb{R}$. Here $\theta \in C_0^\infty(\mathbb{R})$, $\theta(\xi) = \theta(-\xi)$,

$$\theta(\xi) = \begin{cases} 1, & |\xi| \leq 2 - \delta \\ 0, & |\xi| > 2 - \delta/2 \end{cases},$$

δ being a sufficiently small fixed number. One has the following proposition.

PROPOSITION 3.1. *For α in some finite vicinity of 2 and for a sufficiently small, equation (3.1) has a unique positive even smooth decreasing solution $\widetilde{\varphi}(z, \alpha, a)$ which is close to $\varphi_0(z, \alpha)$. Moreover,*
  (i) *as $a \to 0$, $\widetilde{\varphi}(z, \alpha, a)$ admits the asymptotic expansion (2.4) in the sense*

$$\left\|\rho_\gamma(\alpha, a)\left(\widetilde{\varphi}(\alpha, a) - \varphi^N(\alpha, a)\right)\right\|_\infty \leq c|a|^{N+1}, \quad \text{for any } \gamma > 0.$$

Here $\rho_\gamma(z, \alpha, a) = \exp\left((1-\gamma)1/h \int_0^{h|z|} ds \sqrt{\alpha^2/4 - s^2\theta(s)/4}\right)$;
  (ii) $\left\|\rho_0(\alpha, a)\widetilde{\varphi}(\alpha, a)\right\|_\infty \leq c.$
The similar formulas are valid for the derivatives of $\widetilde{\varphi}$ with respect to z, α, a.

Introduce a linearized operator $\widehat{H}(a)$ associated to the modified ground state $\widetilde{\varphi}(z, a)$ (as before, $\widehat{\varphi}(z, a) = \widehat{\varphi}(z, 2, a)$)

$$\widehat{H}(a) = \left(-\partial_x^2 + 1 - \frac{az^2}{4}\theta\right)\sigma_3 + V(\widetilde{\varphi}(a)).$$

The continuous spectrum of $\widehat{H}(a)$ is the same as in the case of the operator H_0. The point $E = 0$ is an eigenvalue of $\widehat{H}(a)$ of the multiplicity 2. There are an eigenfunction $\vec{\zeta}_0(a)$

$$\vec{\zeta}_0(a) = i\widehat{\varphi}(a)\begin{pmatrix} 1 \\ -1 \end{pmatrix}, \quad \widehat{H}\vec{\zeta}_0 = 0,$$

and an associated function $\vec{\zeta}_1(a)$

$$\vec{\zeta}_1(a) = \partial_\alpha \widehat{\varphi}(\alpha, a)|_{\alpha=2} \begin{pmatrix} 1 \\ 1 \end{pmatrix}, \quad \widehat{H}\vec{\zeta}_1 = \vec{\zeta}_0.$$

A more detailed description of the discrete spectrum can be obtained by means of the standard WKB method.

In particular, the following proposition can be proved.

PROPOSITION 3.2. *For a sufficiently small, the discrete spectrum of the operator $\widehat{H}(a)$ in some finite vicinity of the point $E = 0$ consists of 0 and two simple eigenvalues $\pm \lambda(a)$, $\lambda(a) = i\sqrt{a}\lambda'(a)$, where $\lambda'(a)$ is a single valued function of a, real for real a. As $a \to 0$, $\lambda'(a) = \lambda_0 + O(a)$, with some $\lambda_0 > 0$. Let $\vec{\zeta}_2(a)$ be an eigenfunction corresponding to $\lambda(a)$ normalized by the condition*

$$\langle \vec{\zeta}_2(a), \vec{\xi}_0 \rangle = -i\langle \vec{\zeta}_0(a), \vec{\xi}_0 \rangle + i\lambda^2(a)\langle \vec{\xi}_2, \vec{\xi}_0 \rangle.$$

Then $\vec{\zeta}_2(a)$ admits the following asymptotic expansion as $a \to 0$

$$\vec{\zeta}_2 = -i\vec{\zeta}_0 - \lambda \vec{\zeta}_1 + i\lambda^2 \vec{\xi}_2 + \lambda^3 \vec{\xi}_3 + a\lambda^2 g_0 \begin{pmatrix} 1 \\ -1 \end{pmatrix} + a\lambda^3 g_1 \begin{pmatrix} 1 \\ 1 \end{pmatrix} + O(a^3),$$

where g_i, $i = 0, 1$ are real smooth exponentially decreasing functions, $(g_0, \varphi_0) = 0$. $O(a^3)$ corresponds to the L_∞-norm with the weight $\rho_\gamma(z, 2, a)$ for some $\gamma > 0$.

In the subspace generated by $\vec{\zeta}_j(a)$, $j = 0, \ldots 3$, $\vec{\zeta}_3 = -\sigma_1 \vec{\zeta}_2$ being an eigenfunction corresponding to the eigenvalue $-\lambda$, we introduce a new basis $\{\vec{e}_j(a)\}_{j=0}^3$:

$$\vec{e}_0 = \vec{\zeta}_0, \quad \vec{e}_1 = \vec{\zeta}_1,$$
$$\vec{e}_2 = -\frac{i}{2\lambda^2}(\vec{\zeta}_2 + \vec{\zeta}_3 + 2i\vec{\zeta}_0), \quad \vec{e}_3 = \frac{1}{\lambda^3}(\vec{\zeta}_2 - \vec{\zeta}_3 + 2\lambda\vec{\zeta}_1),$$

$\vec{e}_2 = e_2\binom{1}{-1}$, $\vec{e}_3 = e_3\binom{1}{1}$, $\bar{e}_j = (-1)^{j-1} e_j$. It follows from Proposition 3.2 that as $a \to 0$,

$$\vec{e}_2 = \vec{\xi}_2 - iag_0 \begin{pmatrix} 1 \\ -1 \end{pmatrix} + O(a^2), \quad \vec{e}_3 = \vec{\xi}_3 + ag_1 \begin{pmatrix} 1 \\ 1 \end{pmatrix} + O(a^2).$$

Return to the Cauchy problem (2.1, 2.7). Using the profile $\widehat{\varphi}$ one can rewrite the initial data ψ_0 in the form: $\psi_0 = e^{-i\beta_0 x^2/4}(\widehat{\varphi}(\beta_0^2) + f_0')$, $\|f_0'\|_X = O(\beta_0^{2N})$. Below we shall omit "′" in the notation of f_0'. Write the solution ψ as the sum

$$(3.2) \quad \psi(x,t) = e^{i\Phi} \lambda^{1/2}(t) \big(\widehat{\varphi}(z, a(t)) + f(z,t)\big), \quad \Phi = \mu(t) - \frac{\beta}{4}z^2, \quad z = \lambda(t)x,$$

where $\sigma(t) = \big(\mu(t)/2, \lambda(t), \beta(t), a(t)\big)$ is an arbitrary curve in $\mathbb{R}_+ \times \mathbb{R}^3$, it is not a solution of (2.5) in general.

The decomposition can be fixed by the orthogonality conditions

$$(3.3) \quad \langle \vec{f}(t), \sigma_3 \vec{e}_j(a(t)) \rangle = 0, \quad j = 0, \ldots, 3.$$

This means that σ has to satisfy the system

(3.4) $$F_j(\psi, \sigma) = 0, \quad j = 0, \ldots 3,$$

$$F_j(\psi, \sigma) = \lambda^{1/2} \langle \vec{\psi}, \sigma_3 e^{i\Phi\sigma_3} \vec{e}_j(\lambda \cdot, a) \rangle - \langle \vec{e}_0(a), \vec{e}_j(a) \rangle = 0, \quad \vec{\psi} = \begin{pmatrix} \psi \\ \bar{\psi} \end{pmatrix}.$$

The solvability of (3.4) for ψ in some small L_2-vicinity of φ_0 is guaranteed by the smoothness of the basis $\vec{e}_j(a)$, $j = 0, \ldots, 3$ and the nondegeneration of the corresponding Jacobi matrix

$$B_0 = \left\{ \frac{\partial F_j}{\partial \sigma_k} \right\} \bigg|_{\substack{\psi=\varphi \\ \sigma=(1,0,0,0)}}.$$

It is not difficult to check that

$$B_0 = -2 \{ \langle \vec{\xi}_k, \sigma_3 \vec{\xi}_j \rangle \}_{k,j=0}^3, \quad \det B_0 = \left| 2 \langle \vec{\xi}_1, \sigma_3 \vec{\xi}_2 \rangle \right|^4 = e^4 \neq 0.$$

So, one can assume that the initial decomposition (2.7) obeys conditions (3.3). To prove the existence of a trajectory $\sigma(t)$ we need the following orbital stability result:

PROPOSITION 3.3.. *For any $\varepsilon > 0$ there exist $\delta > 0$ such that for any ψ_0, $\|\psi_0 - \varphi_0\|_{H^1} \leq \delta$, $E(\psi_0) < 0$, there exists $\mu(t) \in C([0, T^*))$ such that the solution ψ corresponding to the initial data ψ_0 satisfies the inequality*

$$\left\| \psi(t) - \lambda^{1/2}(t) e^{i\mu(t)} \varphi_0(\lambda(t) \cdot) \right\|_2 \leq \varepsilon, \quad 0 \leq t < T^*,$$

where $\lambda(t)$ is given by

$$\lambda(t) = \frac{\|\psi_x(t)\|_2}{\|\varphi_{0x}\|_2}.$$

See [**W2, W3, LBSK**] for the proof.

By (2.8), $\widehat{\psi}_0$, $\widehat{\psi}_0 = \widetilde{\varphi}(\beta_0^2) + f_0$ satisfies the conditions of the above proposition. Thus, the corresponding solution $\widehat{\psi}(t)$ admits the representation

$$\widehat{\psi}(x, t) = e^{i\widetilde{\Phi}} \widetilde{\lambda}^{1/2}(t) \left(\widetilde{\varphi}(z, \tilde{a}(t)) + \tilde{f}(z, t) \right), \quad \widetilde{\Phi} = \widetilde{\mu}(t) - \frac{\widetilde{\beta}}{4} z^2, \quad z = \widetilde{\lambda}(t) x,$$

where $\widetilde{\sigma}(t) = (\widetilde{\mu}(t)/2, \widetilde{\lambda}(t), \widetilde{\beta}, \tilde{a}(t))$, $\widetilde{\sigma}(0) = (0, 1, 0, \beta_0^2)$ is a continuous trajectory satisfying (3.3), $\|\tilde{f}\|_2$, $\widetilde{\lambda}\|\varphi_{0x}\|_2/\|\psi_x(t)\|_2 - 1$, $\widetilde{\beta}$, $\tilde{a}$ being small uniformly with respect to t.

By conformal invariance we can write now the solution $\psi(t)$ of the Cauchy problem (2.1, 2.7) in the form (3.2) where

$$\mu(t) = \widetilde{\mu}(\rho), \quad \lambda(t) = (1 - \beta_0 t)^{-1} \widetilde{\lambda}(\rho),$$

$$\beta(t) = \beta_0 (1 - \beta_0 t) \widetilde{\lambda}^{-2} + \widetilde{\beta}(\rho), \quad a(t) = \tilde{a}(\rho), \quad \rho = \frac{t}{1 - \beta_0 t},$$

$f(z, t) = \tilde{f}(z, \rho)$ satisfying the orthogonality conditions (3.3).

By (i) of Proposition 2.1, λ admits the estimate

(3.5) $$\lambda(t) \geq c(T^* - t)^{-1/2}.$$

Remark that since $\psi(t) \in C^1([0, T^*) \to H^{-1})$ the trajectory $\sigma(t)$ belongs in fact to C^1.

3.2. Differential equations.

We write a system of equations for σ and f in explicit form. Introduce a new time variable τ:

$$\tau = \int_0^t ds \lambda^2(s).$$

By (3.5), $\tau \to \infty$ as $t \to T^*$.

In terms of f (2.1) takes the form

(3.6) $$i\vec{f}_\tau = \widehat{H}(a)\vec{f} + N,$$

where

(3.7)
$$N = N_0(a, f) + N_1(\widehat{\varphi}, f) + l(\sigma)\left(\widehat{\varphi}\begin{pmatrix}1\\1\end{pmatrix} + \vec{f}\right) - ia_\tau \widehat{\varphi}_a \begin{pmatrix}1\\1\end{pmatrix},$$

$$N_0(a, f) = \frac{az^2}{4}(\theta(hz) - 1)\sigma_3\left(\widehat{\varphi}\begin{pmatrix}1\\1\end{pmatrix} + \vec{f}\right),$$

$$N_1(\widehat{\varphi}, f) = -|\widehat{\varphi} + f|^4 \sigma_3\left(\widehat{\varphi}\begin{pmatrix}1\\1\end{pmatrix} + \vec{f}\right) + \widehat{\varphi}^5 \begin{pmatrix}1\\-1\end{pmatrix} - V(\widehat{\varphi})\vec{f},$$

$$l(\sigma) = (\mu_\tau - 1)\sigma_3 + i(\beta - \lambda_\tau \lambda)\left(z\partial_z + \frac{1}{2}\right)$$
$$+ \left(a - \beta_\tau + \beta^2 - 2\beta\frac{\lambda_\tau}{\lambda}\right)\frac{z^2}{4}\sigma_3.$$

Substitute the expression for $\vec{f}_\tau$ from (3.6, 3.7) into the derivative of the orthogonal conditions. The result can be written down as follows:

(3.8) $$\big(A_0(a) + A_1(a,f)\big)\vec{\eta} = \vec{g}(a,f).$$

Here

$$\vec{\eta} = \left(\frac{\mu_\tau - 1}{2}, \frac{\lambda_\tau}{\lambda} - \beta, \beta_\tau - \beta^2 + 2\beta\frac{\lambda_\tau}{\lambda} - a, a_\tau\right),$$

$$A_0 = 2\begin{pmatrix} 0 & 0 & 0 & -(\widehat{\varphi}_a, \widehat{\varphi}) \\ 2(\widehat{\varphi}, \widehat{\varphi}_E) & 0 & -\left(\frac{z^2}{4}\widehat{\varphi}, \widehat{\varphi}_E\right) & 0 \\ 0 & -i\left((z\partial_z + \frac{1}{2})\widehat{\varphi}, e_2\right) & 0 & -i(\widehat{\varphi}_a, e_2) \\ 2(\widehat{\varphi}, e_3) & 0 & -\left(\frac{z^2}{4}\widehat{\varphi}, e_3\right) & 0 \end{pmatrix},$$

$$(A_1\vec{\eta})_j = \langle l(\sigma)\vec{f}, \sigma_3 \vec{e}_j\rangle + ia_\tau \langle \vec{f}, \sigma_3 \vec{e}_{ja}\rangle,$$
$$g_j(a,f) = -\langle N_0 + N_1, \sigma_3 \vec{e}_j\rangle.$$

By Propositions 3.1, 3.2, as $a \to 0$,

(3.9) $$A_0(a) = iB_0 + O(a).$$

In principle the system (3.8) can be solved with respect to the derivatives η and together with equation (3.6) constitutes a complete system for σ, $\vec{f}$:

(3.10) $$i\vec{f}_t = H(a)\vec{f} + N'(a,f),$$
(3.11) $$\vec{\eta} = G(a,f),$$
$$f|_{t=0} = f_0, \quad \sigma|_{t=0} = (0, 1, \beta_0, \beta_0^2).$$

Here $H(a) = \big(-\partial_z^2 + 1 - az^2/4\big)\sigma_3 + V(\widetilde{\varphi}(a))$, $N' = N - az^2/4(\theta - 1)\sigma_3 \vec{f}$.

3.3. Spectral properties of the operator $H(a)$.

To study the behavior of solutions to (3.10, 3.11) we need some information about spectral properties of $H(a)$, $a > 0$, in the limit $a \to 0$. The necessary facts are collected in this subsection. Consider a little bit more general operator $H(\alpha, a)$

$$H(\alpha, a) = \left(-\partial_z^2 + \frac{\alpha^2}{4} - \frac{az^2}{4}\right)\sigma_3 + V(\widetilde{\varphi}(\alpha, a)).$$

We renormalize $H(\alpha, a)$ to made the principal part independent of parameters:

$$H(\alpha, a) = a^{1/2} T(a) \widehat{H}(\alpha, a) T^{-1}(a), \quad (T(a)\vec{f})(z) = f(a^{1/4} z), \quad a > 0.$$

The operator $\widehat{H}(\alpha, a)$ has the form

$$\widehat{H}(\alpha, a) = \left(-\partial_z^2 + E_0 - \frac{z^2}{4}\right)\sigma_3 + \widehat{W}(\alpha, a), \quad E_0 = \frac{\alpha^2}{4a^{1/2}},$$

where $\widehat{W}(\alpha, a) = a^{-1/2} T^{-1}(a) V(\widetilde{\varphi}(\alpha, a)) T(a)$. The continuous spectrum of $\widehat{H}$ coincides with $\mathbb{R}$. In addition the operator $\widehat{H}(\alpha, a)$ can have a finite and finite dimensional discrete spectrum. $\widehat{H}(\alpha, a)$ satisfies the same relations (2.6) as H_0. As a consequence the spectrum is symmetric with respect to transformations $E \to -E$ and $E \to \overline{E}$.

Consider the equation

(3.12) $$(\widehat{H} - E)\psi = 0.$$

We are interested in the solutions with standard behavior as $z \to +\infty$. One can find a basis of solutions $f_j(z, E)$, $j = 1, \ldots, 4$, with the following properties. The solutions f_j are holomorphic functions of E, $E \in \mathbb{C}$, admitting the following asymptotic representations as $z \to +\infty$

$$f_1(z, E) = e^{iz^2/4} z^{\nu(E)} \left[\begin{pmatrix} 1 \\ 0 \end{pmatrix} + o(1)\right],$$

$$f_2(z, E) = e^{-iz^2/4} z^{\overline{\nu(\overline{E})}} \left[\begin{pmatrix} 1 \\ 0 \end{pmatrix} + o(1)\right],$$

$$f_3(z, E) = e^{iz^2/4} z^{\nu(-E)} \left[\begin{pmatrix} 0 \\ 1 \end{pmatrix} + o(1)\right],$$

$$f_4(z, E) = e^{-iz^2/4} z^{\overline{\nu(-\overline{E})}} \left[\begin{pmatrix} 0 \\ 1 \end{pmatrix} + o(1)\right],$$

where $\nu(E) = -1/2 + i(E - E_0)$. These asymptotic formulas can be differentiated any number of times with respect to z, E, α, a.

The solutions f_j satisfy the relations

$$f_{2,4}(z, E) = \overline{f_{1,3}(z, \overline{E})}, \quad f_{3,4}(z, E) = \sigma_1 f_{1,2}(z, -E),$$
$$w(f_1, f_2) = i, \quad w(f_3, f_4) = i, \quad w(f_k, f_j) = 0, \quad k = 1, 2, \ j = 3, 4.$$

Here Wronskian $w(f, g)$ is defined by $w(f, g) = \langle f', g \rangle_{\mathbb{R}^2} - \langle f, g' \rangle_{\mathbb{R}^2}$. It does not depend on z if f and g are solutions of (3.12).

Introduce solutions $g_j(z, E)$, $j = 1, \ldots, 4$, with standard behavior at $-\infty$ by

$$g_j(z, E) = f_j(-z, E).$$

Consider matrix solutions
$$F_1 = (f_1, f_4), \quad F_2 = (f_2, f_3), \quad G_1 = (g_1, g_4), \quad G_2 = (g_2, g_3).$$
One can express F_1 in terms of G_j, $j = 1, 2$:
$$F_1 = G_2 A + G_1 B,$$
$A = A(E)$, $B = B(E)$ are holomorphic functions of E, $E \in \mathbb{C}$. One can get the Wronskian representations for A and B:
$$A = i\sigma_3 W(G_1, F_1), \quad B = -i\sigma_3 W(G_2, F_1).$$
Here $W(F, G) = F^{t'} G - F^t G'$.

The eigenvalues of the operator $\widehat{H}$ lying in the upper half plane $\{\text{im } E > 0\}$ are characterized by the equation
$$\det A(E) = 0.$$
The solutions of this equation in lower half plane $\{\text{im } E \leq 0\}$ are called resonances. One can prove the following result.

PROPOSITION 3.4. *For α sufficiently close to 2 and for $a > 0$ sufficiently small,*
(i) *the point spectrum of $\widehat{H}$ restricted to the subspace of even functions consists of four simple purely imaginary eigenvalues $\pm i E_{1,2}(a)$, $E_j > 0$,*

$$\left| E_1(a) - \frac{\alpha^2}{4} a^{-1/2} \lambda\left(\frac{16a}{\alpha^4}\right) \right| = O(e^{-\gamma/h}), \quad \gamma > 0,$$

$$E_2 = O\left(e^{-(1-\varepsilon)E_0 S_0}\right), \quad S_0 = \int_0^2 ds \sqrt{1 - s^2/4},$$

(ii) *there exists $C_0 > 0$ such that in the strip $\{E: -C_0 < \text{im } E \leq 0\}$ the operator $\widehat{H}(a, \alpha)$ has only one simple resonance $-iE_R(a)$, $E_R > 0$. Moreover, E_R admit asymptotic estimates*

$$E_R - E_2 = O(a^{-2} e^{-2E_0 S_0}).$$

Here and in what follows the letter ε is used as a general notation for small positive constants that depends on the choice of the cut off function θ and tend to zero as $\delta \to 0$. They may change from line to line.

We introduce the operators $\mathbb{F}$, $\mathbb{G}$: $L_2(\mathbb{R} \to \mathbb{C}^2) \to L_2(\mathbb{R} \to \mathbb{C}^2)$:
$$(\mathbb{F}\vec{\Phi})(z) = \frac{1}{\sqrt{2\pi}} \int_\mathbb{R} dE \mathcal{F}(z, E) \Phi(E), \quad (\mathbb{G}\vec{\Phi})(z) = \frac{1}{\sqrt{2\pi}} \int_\mathbb{R} dE \mathcal{G}(z, E) \Phi(E).$$
Here $\mathcal{F}$, $\mathcal{G}$ are solutions of the scattering problem:
$$\mathcal{F} = F_1 A^{-1}, \quad \mathcal{G} = G_1 A^{-1},$$
$$\mathcal{F}(z, E) \sim e^{iz^2/4\sigma_3} z^{-1/2 + i(E - E_0 \sigma_3)} A^{-1}, \quad z \to +\infty,$$
$$\mathcal{F}(z, E) \sim e^{-iz^2/4\sigma_3} |z|^{-1/2 - i(E - E_0 \sigma_3)} + e^{iz^2/4\sigma_3} |z|^{-1/2 + i(E - E_0 \sigma_3)} BA^{-1}, \quad z \to -\infty.$$

The action of the adjoint operators $\mathbb{F}^*$, $\mathbb{G}^*$ is given by
$$(\mathbb{F}^* \psi)(E) = \frac{1}{\sqrt{2\pi}} \int_\mathbb{R} dz \mathcal{F}^*(z, E) \psi(z), \quad (\mathbb{G}^* \psi)(E) = \frac{1}{\sqrt{2\pi}} \int_\mathbb{R} dz \mathcal{G}^*(z, E) \psi(z)).$$

It is not difficult to show that $\mathbb{F}$, $\mathbb{G}$ are bounded in L_2 and satisfy the relations

$$\mathbb{E}\hat{\sigma}_3\mathbb{E}^*\sigma_3 = P^c, \quad \mathbb{E}^*\sigma_3\mathbb{E}\hat{\sigma}_3 = I,$$

where $\mathbb{E}\colon L_2(\mathbb{R}\to\mathbb{C}^2) \times L_2(\mathbb{R}\to\mathbb{C}^2) \to L_2(\mathbb{R}\to\mathbb{C}^2)$,

$$\mathbb{E}\vec{\Phi} = \mathbb{F}\Phi_1 + \mathbb{G}\Phi_2, \quad \vec{\Phi} = (\Phi_1, \Phi_2),$$

$\hat{\sigma}_3 = \begin{pmatrix} \sigma_3 & 0 \\ 0 & \sigma_3 \end{pmatrix}$, P^c being the spectral projection onto the subspace of the continuous spectrum. Moreover, one can prove the following proposition.

PROPOSITION 3.5. *For α sufficiently close to 2 and for $a > 0$ sufficiently small, there exist d_0, $1/2 > d_0 > 0$, independent of α, a, such that*
  (i) *for $e^{-iz^2/4\sigma_3}f \in H^1$, $(\mathbb{E}^*f)(E)$ is a meromorphic function of E in a strip $-d_0 \leq \operatorname{im} E \leq 0$ with the only pole in $-iE_2$ and satisfies the estimate*

$$\|\mathbb{E}^*f\|_{L_2(\mathbb{R}-ib)} \leq ch^{-K_1}\|e^{-iz^2/4\sigma_3}f\|_{H^1}, \quad h^L \leq b \leq d_0.$$

  (ii) *Introduce operators $\mathbb{F}_b$:*

$$(\mathbb{F}_b\Phi)(z) = \frac{1}{\sqrt{2\pi}}\int_{\mathbb{R}} dE\,\mathcal{F}(z, E-ib)\Phi(E).$$

For $h^L \leq b \leq d_0$, they satisfy the inequality

$$\left\|(1+|z|)^{-\nu}\mathbb{F}_b\vec{\Phi}\right\|_2 \leq ch^{-K_2}\|\vec{\Phi}\|_2, \quad \nu > 1/2,$$

the same being true for $\mathbb{F}$ replaced by $\mathbb{G}$. Here K_j, $j = 1, 2$, depend on L but do not depend on α, a.

3.4. Effective equations. In order to derive a system of effective equations consider the main nonlinear terms of (3.9, 3.10). Below it will become clear that the function a depends slowly on τ. More precisely,

(3.13) $$a \sim \ln^{-2}(\tau + \tau^*),$$

with some $\tau^* = O(e^{2S_0/\beta_0}\beta_0^3)$. We shall also see that the contribution f of the continuous spectrum asymptotically is of the order $\Gamma^{1/2}$, $\Gamma = e^{-2S_0/h}$, $h = \sqrt{a}$, (in the uniform norm) and of the order Γ for z not too large. In it is turn the vector η also has order Γ. We shall use these facts while deriving the equations. At this stage we are not worrying about formal justification.

The main terms of N are generated by the expression

(3.14) $$N \sim F_0(a)\begin{pmatrix} 1 \\ -1 \end{pmatrix}, \quad F_0(a) = a\frac{z^2}{4}(\theta - 1)\widehat{\varphi}.$$

Thus, it is clear that in the region $|z| > \operatorname{const} h^{-1}$ the main order term of f is given by the expression

(3.15) $$f \sim -\bigl(L(a) - i0\bigr)^{-1}F_0(a),$$

where $L(a) = -\partial_z^2 + 1 - az^2/4$. The sign "$-$" (in $-i0$) is essential: it means that $e^{-ihz^2/4}(L(a) - i0)^{-1}F_0(a)$ has finite energy.

For the following it is convenient to write $f = f^0 + f^1$, $f^0 = -(L(a) - i0)^{-1}F_0(a)$. It will become clear later that in the region $|z| \geq \operatorname{const} h^{-1}$ f^0 and f^1 are of the order $\Gamma^{1/2}$ and Γ respectively while for $|z| \sim 1$ both f^0 and f^1 have order Γ.

Consider (3.11). The main term of G is given by the expression

$$G \sim A_0^{-1}(a)\vec{g}^0(a),$$

where $g_j^0 = -\langle N_0(a, f_0), \sigma_3 \vec{e}_j \rangle$. So we rewrite (3.11) in the form

(3.16) $$\vec{\eta} = G_0(a) + G_R(a, f).$$

Here $G_0(a) = -A_0^{-1}(a)\vec{g}^0(a)$, G_R being the remainder.

The behavior of $f^0(a)$, $G_0(a)$ in the limit $a \to 0$ is described by the following proposition.

PROPOSITION 3.6. *For $a > 0$ sufficiently small, $f^0(a)$, $G_0(a)$ satisfy the estimates*

$$\|f^0(a)\|_\infty \leq c\Gamma^{1/2-\varepsilon}, \quad \|G_0(a)\| \leq c\Gamma^{1-\varepsilon}.$$

Moreover, G_0^3 admits the following representation

$$G_0^3(a) = -2\nu_0 \Gamma\big(1 + O(a)\big), \quad \nu_0 = \frac{8\varphi_\infty^2}{e}.$$

This asymptotic estimate can be differentiated any number of times with respect to a.

In order to estimate qualitatively the behavior of a, consider the last equation of (3.16) neglecting the remainder G_R:

$$a_\tau = G_0^3(a).$$

We denote by $a_0(\tau)$ solution of this equation with initial data $a_0(0) = \beta_0^2$. It is easy to check that $h_0 = \sqrt{a_0}$ admits the representation

(3.17) $$h_0^{-1}(\tau) = \frac{1}{2S_0}\big(\ln\nu_1(\tau + \tau^*) + 3\ln\ln\nu_1(\tau + \tau^*)\big) + O\left(\frac{\ln\ln(\tau + \tau^*)}{\ln(\tau + \tau^*)}\right),$$

as $\tau + \tau^* \to +\infty$, $\nu_1 = \nu_0/(4S_0^2)$, $\tau^* = (2S_0\nu_0)^{-1}\beta_0^3 e^{2S_0/\beta_0}\big(1 + O(\beta_0)\big)$.

3.5. Equations on the finite interval. Following [**BP**] we consider system (3.10, 3.11) on some finite interval $[0, \tau_1]$ and later investigate the limit $\tau_1 \to \infty$. On the interval $[0, t_1]$, $t_1 = t(\tau_1)$ we approximate the trajectory $\sigma(t)$ by $\sigma_1(t)$ where $\sigma_1(t) = \big(\mu(t)/2, \lambda_1(t), \beta_1(t), a_1(t)\big)$ is the solution of the following Cauchy problem

$$\lambda_1^{-3}\lambda_1' = \beta_1, \quad \lambda_1^{-2}\beta_1' + \beta_1^2 = a_1, \quad a_1' = 0,$$
$$\lambda_1(t_1) = \lambda(t_1), \quad \beta_1(t_1) = a^{1/2}(t_1), \quad a_1(t_1) = a(t_1).$$

We associate to the trajectory σ_1 a new function g

$$g(y, \rho) = e^{iy^2 \triangle} r^{1/2} f(ry, \tau),$$

where $\triangle = (1 - \beta r^2)/4$, $r = \lambda/(\sqrt{\beta_1 \lambda_1})$, $\rho = \int_0^\tau ds\, r^{-2}$. The equation (3.10) in terms of g takes the form

(3.18) $$i\vec{g}_\rho = \widehat{H}(\alpha, a)\vec{g} + \mathcal{N}_0 + \mathcal{N}_1 + \mathcal{N}_2 + \mathcal{N}_3,$$

where $\alpha^2/4 = a^{1/2}r^2$,

$$
\begin{aligned}
\mathcal{N}_0 &= e^{iy^2 \triangle \sigma_3} r^{5/2} F_0(a) \begin{pmatrix} 1 \\ -1 \end{pmatrix}, \\
\mathcal{N}_1 &= e^{iy^2 \triangle \sigma_3} r^{5/2} N_1, \\
\mathcal{N}_2 &= e^{iy^2 \triangle \sigma_3} r^{5/2} \left(l(\sigma)\widetilde{\varphi}\begin{pmatrix} 1 \\ 1 \end{pmatrix} - ia_\tau \widetilde{\varphi}_a \begin{pmatrix} 1 \\ 1 \end{pmatrix} \right), \\
\mathcal{N}_3 &= e^{iy^2 \triangle \sigma_3} r^{5/2} V(\widetilde{\varphi}(a))\vec{f} - \widehat{W}(\alpha,a)\vec{g} + \left(\mu_\rho - \frac{\alpha^2}{4\sqrt{a}} \right) \sigma_3 \vec{g}.
\end{aligned}
$$
(3.19)

Since α, a depend slowly on τ it is natural to rewrite the above equation in terms of spectral representation of $\widehat{H}(\alpha, a)$. Write $\vec{g}$ as the sum $\vec{g} = \vec{h} + \vec{k}$ of the projections on subspaces corresponding to the discrete and continuous spectra of $\widehat{H}(\alpha, a)$:

$$\vec{h} = (\mathbb{F} + \mathbb{G})\sigma_3 \Phi, \quad \Phi = \mathbb{F}^* \sigma_3 \vec{g}.$$

Let us remark that due to the orthogonality conditions (3.3) the four dimensional component k is controlled by h (or equivalently by Φ).

Projecting (3.18) on the subspace of the continuous spectrum of $\widehat{H}(\alpha,a)$ one gets an equation for Φ:

(3.20) $$i\Phi_\rho = E\Phi + D,$$

where $D = D_0 + D_1 + D_2$,

(3.21) $$D_0 = \mathbb{F}^* \sigma_3 \mathcal{N}_0, \quad D_1 = i\mathbb{F}_\rho^* \sigma_3 \vec{g}, \quad D_2 = \sum_{j=1}^{3} \mathbb{F}^* \sigma_3 \mathcal{N}_j.$$

Consider (3.20) on the line $\operatorname{im} E = -b$ with some b, $0 < b \leq d_0$, that will be fixed later, rewriting (3.20) as an integral equation:

(3.22) $$\Phi(\rho) = e^{-iE\rho} \mathbb{F}^*(0)\sigma_3 \vec{g}_0 - i \int_0^\rho ds\, e^{-iE(\rho-s)} D(s), \quad \operatorname{im} E = -b.$$

Here $\mathbb{F}(0) = \mathbb{F}(\alpha(0), a(0))$, $g_0(y) = e^{iy^2 \triangle_0} r_0^{1/2} f_0(r_0 y)$, $\triangle_0 = (1 - \beta_0 r_0^2)/4$, $r_0 = (\beta_1 \lambda_1^2(0))^{-1/2}$.

Remark that if we know Φ restricted to the line $\operatorname{im} E = -b$ we can return to $\vec{h}$ by means of the representation

(3.23) $$\vec{h} = (\mathbb{F}_b + \mathbb{G}_b)\sigma_3 \Phi_b - \sqrt{2\pi}i \Big(\operatorname*{Res}_{E=-iE_R} + \operatorname*{Res}_{E=-iE_2} \Big) (\mathcal{F}(y,E) + \mathcal{G}(y,E))\sigma_3 \Phi(E),$$

where $\Phi_b(E) = \Phi(E - ib)$, the second term being controlled by the first one again due to the orthogonality conditions (3.3).

The relations (3.3, 3.16, 3.22, 3.23) make up the final form of the equations which is used to investigate the dynamical system on the interval $[0, \tau_1]$.

3.6. Main terms of Φ. It follows from (3.13), (3.14) that the main part of D is given by D_0. The contribution of D_0 in (3.22) allows some asymptotic simplifications. After a natural integration by parts one gets

(3.24) $$\Phi = \Phi_0 + \Phi_1, \quad \Phi_0 = -\frac{1}{E} D_0, \quad \Phi_1 = e^{-iE\rho} \Phi_{10} - i \int_0^\rho ds\, e^{-iE(\rho-s)} D'(s).$$

Here $\Phi_{10} = \mathbb{F}^*(0)\sigma_3\vec{g}_0 + E^{-1}D_0(0)$, $D' = D_1 + D_2 + iD_{0\rho}/E$.

In accordance with (3.13) the main order term of Φ is given by Φ_0.

3.7. Estimates of soliton parameters. Introduce a natural system of norms for the components of the solution ψ:

$$s_0(\tau) = \sup_{s \leq \tau} |h(s) - h_0(s)|h_0^{-2}(s),$$

$$s_1(\tau) = \sup_{s \leq \tau} |\beta(s) - h(s)|h_0^{-2}(s)p^{-1}(s;\kappa_1,r_1),$$

$$s_2(\tau) = \sup_{\tau \leq s \leq \tau_1} |\beta(s) - r^{-2}|h_0^{-1}(s)p^{-1}(s;\kappa_2,r_2),$$

$$M_0(\tau) = \sup_{s \leq \tau} \|f(s)\|_\infty p^{-1}(s;\kappa_0,r_0),$$

$$M_1(\tau) = \sup_{s \leq \tau} \|(1+|z|)^{-\nu} f^1(s)\|_\infty p^{-1}(s;\kappa_3,r_3), \quad \nu \geq 2,$$

$$M_2(\tau) = \sup_{s \leq \tau} \|\rho f(s)\|_2 p^{-1}(s;\kappa_4,r_4),$$

where

$$p(\tau;\kappa,r) = e^{-\kappa \int_0^\tau ds h_0(s)} + e^{-rS_0/h_0(\tau)}, \quad \rho = e^{-(1-\delta_1)/h_0} \int_0^{h_0|z|} ds\sqrt{1 - \frac{s^2}{4}\theta(s)},$$

$\kappa_4 = d_0/4$, $\kappa_0 = \kappa_3 = 7\kappa_4/8$, $\kappa_1 = 3\kappa_4/2$, $\kappa_2 = 5\kappa_4/4$, $r_0 = 3/4$, $r_1 = 15/8$, $r_2 = 7/4$, $r_3 = 4/3$, $r_4 = 3/2$, $\delta_1 > 0$ is supposed to be a sufficiently small fixed number.

At last, set

$$\hat{s}_j = s_j(\tau_1), \quad j = 0,1, \quad \hat{s}_2 = s_2(0), \quad \widehat{M}_j = M_j(\tau_1).$$

Consider equation (3.10). It follows immediately from (3.7, 3.8, 3.9) and from Proposition 3.6 that

(3.25) $\quad |\eta| \leq W(M,s)\big[\Psi_0(M)e^{-2\kappa_3 \int_0^\tau ds h_0(s)} + e^{-(2-\varepsilon)S_0/h_0(\tau)}\big],$

(3.26) $\quad |G_R| \leq W(M,s)\Psi_1(M)\big[e^{-3\kappa_3/2 \int_0^\tau ds h_0(s)} + e^{-3r_4 2S_0/h_0(\tau)}\big],$

$$\Psi_0(M) = M_2 M_0^4 + \beta_0^2 M_1^2 + M_2^2, \quad \Psi_1(M) = e^{-\gamma/\beta_0} + M_2 M_0^4 + M_2^2,$$

with some $\gamma > 0$. We use $W(M,s)$ as a general notation for functions of M_j, $j = 0,1,2$, s_k, $k = 0,1,2$, defined on $\mathbb{R}^6$, which are bounded in some finite neighborhood of 0 and may acquire the infinite value $+\infty$ outside some larger neighborhood. It will be assumed that W does not depend on β_0. In all the formulas where W appear it would be possible to replace them by some explicit expressions but such expressions are useless for our aims.

Using (3.25, 3.26) and Proposition 3.6 it is not difficult to prove the following inequalities

(3.27)
$$s_0 \leq W(M,s)\beta_0^{-4}\Psi_1(M),$$
$$s_1 \leq W(M,s)\big(e^{-\gamma/\beta_0} + \beta_0^{-3}\Psi_0(M)\big),$$
$$s_2 \leq W(\widehat{M},\hat{s})\big(e^{-\gamma/\beta_0} + \beta_0^{-3}\Psi_0(\widehat{M})\big), \quad \gamma > 0.$$

3.8. Estimates of D_j.
Consider the nonlinear terms of (3.20). Using Propositions 3.1, 3.5 one gets for D_0:

$$\|D_0\|_{L_2(\mathbb{R}-ib)} \leq W(M,s) e^{-(1-\varepsilon)S_0/h_0(\tau)}. \tag{3.28}$$

The direct calculations allow to obtain the following estimate for D':

$$\|D'\|_{L_2(\mathbb{R}-ib)} \leq W(\widehat{M},\hat{s}) \Psi_2(\widehat{M}) h_0^{-K} p(\tau;\kappa_2,r_2), \tag{3.29}$$

where $\Psi_2(M) = M_0^2 + M_1^2 + M_2^2$.

In this subsection and the next one we use letter K as a general notation for nonnegative numbers independent of parameters that may change from line to line.

3.9. Estimates of f in L_2.
As it have been already mentioned, due to the orthogonality conditions (3.3) the function f can be controlled by the restriction of Φ on the line $\operatorname{im} E = -b$. More precisely, one has the following estimate

$$\|\rho f\|_2 \leq W(M,s) h_0^{-K} \big[\|\Phi_1\|_{L_2(\mathbb{R}-ib)} + |\beta - r^{-2}| + |\beta - h| + e^{-(2-\varepsilon)S_0/h_0(\tau)} \big].$$

It follows from (3.24, 3.27–3.29) that

$$\|\Phi_1\|_{L_2(\mathbb{R}-ib)} \leq W(\widehat{M},\hat{s}) h_0^{-K} \big[\|e^{-i\beta_0 z^2/4} f_0\|_{H^1} + \Psi_2(\widehat{M}) \big] p(\tau,\kappa_2,r_2),$$

provided $b > \kappa_2$. Combining these two estimates and taking into account (3.27), (2.8) one gets

$$M_2 \leq W(\widehat{M},\hat{s}) \beta_0^{-K_0} \big[\beta_0^{2N} + \Psi_2(\widehat{M}) \big], \tag{3.30}$$

with some $K_0 \geq 0$.

3.10. Estimates of f in L_∞.
We represent f by the sum $\vec{f} = e^{i\beta z^2 \sigma_3/4}(\tilde{f}^0 + \tilde{f}^1)$, where $\tilde{f}^0 = e^{-ihz^2\sigma_3/4} \vec{f}^0(a)$. Then $\tilde{f}^1$ satisfies the equation

$$i\tilde{f}_\tau^1 = (-\partial_z^2 + \mu_\tau)\sigma_3 \tilde{f}^1 - i\frac{\lambda_\tau}{\lambda}\left(\frac{1}{2} + z\partial_z\right) \tilde{f}^1 + \mathcal{H}_0 + \mathcal{H}_1, \tag{3.31}$$

where $\mathcal{H}_0 = \mathcal{H}_{00} + \mathcal{H}_{01} + \mathcal{H}_{02}$,

$$\mathcal{H}_{00} = -i\tilde{f}_\tau^0 + (\mu_\tau - 1)\sigma_3 \tilde{f}^0 + i\left(h - \frac{\lambda_\tau}{\lambda}\right)\left(\frac{1}{2} + z\partial_z\right)\tilde{f}^0,$$

$$\mathcal{H}_{01} = e^{-i\beta z^2 \sigma_3/4} N_1,$$

$$\mathcal{H}_{02} = e^{-i\beta z^2 \sigma_3/4}\left(l(\sigma)\widehat{\varphi}\begin{pmatrix}1\\1\end{pmatrix} - ia_\tau \widehat{\varphi}_a \begin{pmatrix}1\\1\end{pmatrix}\right)$$
$$+ (e^{-i\beta z^2 \sigma_3/4} - e^{-ihz^2 \sigma_3/4}) F_0(a) \begin{pmatrix}1\\-1\end{pmatrix}.$$

At last, $\mathcal{H}_1 = e^{-i\beta z^2 \sigma_3/4} V(\widetilde{\varphi}(a)) \vec{f}$.

We rewrite (3.31) as an integral equation

$$\tilde{f}^1 = U(\tau,0)\vec{\chi}_0 - i\int_0^\tau ds\, U(\tau,s)\big(\mathcal{H}_0(s) + \mathcal{H}_1(s)\big), \tag{3.32}$$

where $\chi_0 = e^{-i\beta_0 z^2/4}(f_0 - f^0(\beta_0^2))$, $U(\tau,s)$ being the propagator corresponding to the equation $if_\tau = (-\partial_z^2 + \mu_\tau)\sigma_3 f - i\lambda_\tau \lambda^{-1}(1/2 + z\partial_z)f$.

It follows from (3.32) that

$$
\|\tilde{f}^1\|_\infty \le c\left[\lambda^{-1/2}(\tau)\|\hat{\chi}_0\|_1 + \int_0^\tau ds \left(\frac{\lambda(s)}{\lambda(\tau)}\right)^{1/2}\|\widehat{\mathcal{H}}_0\|_1 \right. \tag{3.33}
$$
$$
\left. + \int_0^\tau ds \frac{\lambda^{-1/2}(\tau)\lambda^{-1/2}(s)}{\sqrt{t(\tau)-t(s)}}\|\mathcal{H}_1\|_1\right],
$$

where $\hat{\chi}_0$ ($\widehat{\mathcal{H}}_0$) stands for the Fourier transform of χ_0 ($\mathcal{H}_0$).

The first term in the right hand side of (3.33) can be estimated as follows

$$
\lambda^{-1/2}(\tau)\|\hat{\chi}_0\|_1 \le W(\beta_0^{-1}M, s)\big[\|f_0\|_X + e^{-\gamma/\beta_0}\big]p(\tau; \kappa_3, r_3). \tag{3.34}
$$

The direct calculations show

$$
\|\widehat{\mathcal{H}}_0\|_1 \le W(M, s)\big[e^{-\gamma/\beta_0} + \Psi_3(M)\big]p(\tau; \kappa, r), \quad \kappa = \frac{15}{16}\kappa_4, \quad r = \frac{17}{12}.
$$

Here $\Psi_3 = M_0|H(\psi_0)|^{1/2} + M_2 + (M_0 + M_1)^2$. Thus, the contribution of $\mathcal{H}_0$ in the right hand side of (3.33) admits the estimate

$$
\int_0^\tau ds \left(\frac{\lambda(s)}{\lambda(\tau)}\right)^{1/2}\|\widehat{\mathcal{H}}_0\|_1 \le W(\beta_0^{-1}M, s)\big[e^{-\gamma/\beta_0} + \beta_0^{-1}\Psi_3(M)\big]p(\tau; \kappa_3, r_3). \tag{3.35}
$$

The third term of (3.33) can be majorated by the expression

$$
W(\beta_0^{-1}M, s)\beta_0^{-1}M_2,
$$

which together with (3.27, 3.34, 3.35) gives

$$
M_0, M_1 \le W(\beta_0^{-1}M, s)\big[\beta_0^{2N} + \beta_0^{N-1}M_0 + \beta_0^{-1}\big(M_2 + (M_0 + M_1)^2\big)\big]. \tag{3.36}
$$

Here we have made use of (2.8) and of the obvious estimate $|H(\psi_0)| \le c\beta_0^{2N}$.

3.11. Estimates of majorants. Consider the inequalities (3.27, 3.30, 3.36). Introduce new scales:

$$
\widehat{M}_j = \beta_0 \widehat{\mathbb{M}}_j, \; j = 0, 1, \quad \widehat{M}_2 = \beta_0^{K_0+2}\widehat{\mathbb{M}}_2.
$$

Remark that one can choose the function W to be spherically symmetric and monotone. Then in terms of $\widehat{\mathbb{M}}_j$ the inequalities (3.27, 3.30, 3.36) can be written in the form

$$
\beta_0 \hat{s}_0, \; \hat{s}_1, \; \hat{s}_2 \le W(\widehat{\mathbb{M}}, \hat{s})\big[e^{-\gamma/\beta_0} + \beta_0(\widehat{\mathbb{M}}_0 + \widehat{\mathbb{M}}_1)^2 + \beta_0^{2K_0+1}\widehat{\mathbb{M}}_2^2\big],
$$
$$
\widehat{\mathbb{M}}_2 \le W(\widehat{\mathbb{M}}, \hat{s})\big[\beta_0^{2N-2K_0-2} + \beta_0^2\widehat{\mathbb{M}}_2^2 + \beta_0^{-2K_0}(\widehat{\mathbb{M}}_0 + \widehat{\mathbb{M}}_1)^2\big],
$$
$$
\widehat{\mathbb{M}}_0 + \widehat{\mathbb{M}}_1 \le W(\widehat{\mathbb{M}}, \hat{s})\big[\beta_0^{2N-1} + \beta_0^{N-1}(\widehat{\mathbb{M}}_0 + \widehat{\mathbb{M}}_1) + \beta_0^{K_0}\widehat{\mathbb{M}}_2\big].
$$

Fix the ball $\|\widehat{\mathbb{M}}\|^2 + \|\hat{s}\|^2 \le R$ where $W(\widehat{\mathbb{M}}, \hat{s})$ is a bounded by a constant. Then the above inequalities can be simplified

$$
\beta_0 \hat{s}_0, \; \hat{s}_1, \; \hat{s}_2 \le W_1\big[\beta_0^{4N-1} + \beta_0^{2K_0+1}\widehat{\mathbb{M}}_2^2\big],
$$
$$
\widehat{\mathbb{M}}_2 \le W_2\big[\beta_0^{2N-2K_0-2} + \widehat{\mathbb{M}}_2^2\big], \tag{3.37}
$$
$$
\widehat{\mathbb{M}}_0 + \widehat{\mathbb{M}}_1 \le W_3\big[\beta_0^{2N-1} + \beta_0^{K_0}\widehat{\mathbb{M}}_2\big],
$$

where W_j, $j = 1, 2, 3$, some constants that do not depend on β_0 provided $N > 1$, β_0 is sufficiently small.

Choosing $N > 1 + K_0$ one gets that for β_0 sufficiently small the solution of (3.37) can belong either to a small neighborhood of 0 or to some domain whose distance from 0 is bounded uniformly with respect to β_0. Since all $\widehat{\mathbb{M}}_j$, s_j are continuous functions of τ_1 and for $\tau_1 = 0$ are small only the first possibility can be realized. As a consequence, one finally obtains

$$M_0, M_1 \leq c\beta_0^{2N-K_0-1}, \quad M_2 \leq c\beta_0^{2N-K_0},$$
$$\beta_0 \hat{s}_0, \hat{s}_1 \leq c\beta_0^{4N-2K_0-3}, \quad \tau \leq \tau_1.$$

The constant c here does not depend either on β_0 or on τ_1. Since τ_1 is arbitrary these estimates are valid, in fact, for $\tau \in \mathbb{R}$. The statement of the Theorem 2.1 is a simple consequence of the above inequalities and (3.25).

References

[BW] J. Bourgain and W. Wang, *Construction of blow-up solutions for the nonlinear Schrödinger equation with critical nonlinearity*, Ann. Scuola Norm. Sup. Pisa Cl. Sci. **25** (1997), no. 4, 197–215.

[BP] V. S. Buslaev and G. S. Perelman, *Scattering for the nonlinear Schrödinger equation: States close to a soliton*, English transl., St. Petersburg Math. J. **4** (1993), 1111–1143. (Russian)

[CW] T. Cazenave and F. B. Weissler, *The structure of solutions to the pseudo-conformal invariant nonlinear Schrödinger equation*, Proc. Roy. Soc. Edinburgh Sect. A **117** (1991), 251–273.

[GV1] J. Ginibre and G. Velo, *On a class of nonlinear Schrödinger equations* I, II, J. Funct. Anal. **32** (1979), 1–71.

[GV2] _____, *On a class of nonlinear Schrödinger equations* III, Ann. Inst. H. Poincaré Phys. Théor. **28** (1978), 287–316.

[Gl] R. Glassey, *On the blowing up of solutions to the Cauchy problem for nonlinear Schrödinger operators*, J. Math. Phys. **8** (1977), 1794–1797.

[Gr] M. Grillakis, *Linearized instability for nonlinear Schrödinger and Klein—Gordon equations*, Comm. Pure Appl. Math. **41** (1988), no. 6, 747–774.

[KSZ] N. Kosmatov, V. Schvets, and V. Zakharov, *Computer simulation of wave collapse in nonlinear Schrödinger equation*, Phys. D **52** (1991), 16–35.

[BSK] E. W. Laedke, R. Blaha, K. H. Spatschek, and E. A. Kuznetsov, *On the stability of collapse in the critical case*, J. Math. Phys. **33** (1992), no. 3, 967–973.

[LPSS1] B. J. LeMesurier, G. Papanicolau, C .Sulem, and P. Sulem, *Focusing and multi-focusing solutions of the nonlinear Schrödinger equation*, Phys. D **31** (1988), 78–102.

[LPSS2] _____, *Local structure of the self-focusing singularity of the nonlinear Schrödinger equation*, Phys. D **32** (1988), 210–226.

[Ma] V. M. Malkin, *Dynamics of wave collapse in the critical case*, Phys. Lett. A **151** (1990), 285–288.

[Me] F. Merle, *Determination of blow-up solutions with minimal mass for nonlinear Schrödinger equation with critical power*, Duke Math. J. **69** (1993), 427–453.

[P] G. Perelman, *On the formation of singularities in solutions of nonlinear Schrödinger equation with critical power nonlinearity*, in preparation.

[SF] A. I. Smirnov and G. M. Fraiman, *The interaction representation in the self-focusing theory*, Phys. D **51** (1991), 2–15.

[SW1] A. Soffer and M. I. Weinstein, *Multichannel nonlinear scattering theory for nonintegrable equations* I, Comm. Math. Phys. **133** (1990), 119–146.

[SW2] _____, *Multichannel nonlinear scattering theory for nonintegrable equations* II, J. Differential Equations **98** (1992), 376–390.

[SS] C. Sulem and P. Sulem, *Focusing nonlinear Schrödinger equation and wave-packet collapse*, Nonlinear Anal. **30** (1997), no. 2, 833–844.

[W1] M. I. Weinstein, *Modulation stability of ground states of nonlinear Schrödinger equation*, SIAM J. Math. Anal. **16** (1985), 472–491.

[W2] _____, *Lyapunov stability of ground states of nonlinear dispersive evolution equations*, Comm. Pure Appl. Math. **39** (1986), 51–68.

[W3] _____, *Nonlinear Schrödinger equations and sharp interpolation estimates*, Comm. Math. Phys. **87** (1983), 567–576.

[W4] _____, *On the structure and formation of singularities in solutions to nonlinear dispersive evolution equations*, Comm. Partial Differential Equations **11** (1986), no. 5, 545–565.

[W5] _____, *Solitary waves of nonlinear dispersive evolution equations with critical power nonlinearities*, J. Differential Equations **69** (1987), 192–203.

CENTRE DE MATHÉMATIQUES, ÉCOLE POLYTECHNIQUE, F–91128 PALAISEAU CEDEX, FRANCE

E-mail address: perelman@math.polytechnique.fr

Vorticity for the Ginzburg–Landau Model of Superconductors in a Magnetic Field

Sylvia Serfaty

1. Introduction to the Model

According to the model introduced by Ginzburg and Landau in the 50's, the energy of a cylindrical superconductor put in a prescribed external magnetic field h_{ex} is given, in a suitable normalization, by the functional

$$J(u, A) = \frac{1}{2} \int_\Omega \left|(\nabla - iA)u\right|^2 + \left|\nabla \times A - h_{\text{ex}}\right|^2 + \frac{\kappa^2}{2}\left(1 - |u|^2\right)^2.$$

A lot of mathematical work arose on this functional in the past few years. Yet, a complete description of its behavior (at least of its minimizers) according to the parameter h_{ex}, is far from being achieved. In particular, there were almost no results concerning vortex-configurations. For a mathematical and physical presentation of the problem, one may also refer to [7, 12].

1.1. Notations. The superconductor is supposed to be a vertical cylinder of section Ω, smooth, bounded and simply connected domain of $\mathbb{R}^2$. The applied magnetic field is also assumed to be vertical, uniform and constant, of intensity h_{ex}. The complex-valued function u, also called "order parameter" in physics, is a pseudo wave-function describing the local state of the superconductor. The superconductivity phenomenon is understood at a microscopic level through the existence of pairs of superconducting electrons, called "Cooper pairs". The modulus of $u(x)$ is then the local density of Cooper pairs, while its phase determines the superconducting current. $|u(x)| \leq 1$ and, at points where $|u(x)| \simeq 1$, the material is in its superconducting phase, whereas, where $|u(x)| \simeq 0$, it is in its normal phase. The "mixed phase" is characterized by the coexistence of the two phases. There are normal regions of characteristic size $1/\kappa$, called vortices, surrounded by superconducting phase. u vanishes at the center of a vortex and the winding-degree of its phase around it is called the degree of the vortex.

The $A = (A_1, A_2)$ is the vector-potential of the magnetic field. The induced magnetic field in the material h, is deduced by $h = \nabla \times A$ or $h = \operatorname{curl} A$. $\nabla_A = \nabla - iA$ will be the covariant derivative $\nabla_A u = (\partial_{x_1} u - iA_1 u, \partial_{x_2} u - iA_2 u)$.

2000 *Mathematics Subject Classification.* Primary: 35Q99; Secondary: 35J50.
This is the final form of the paper.

The κ is the Ginzburg–Landau parameter, a dimension-less constant depending on the material only. If κ is small (roughly $\kappa < 1/\sqrt{2}$), the superconductor is of type–I, and if κ is high (roughly $\kappa > 1/\sqrt{2}$), it is of type–II. Type–I and type–II superconductors have very different behaviors; here, we reduce to type–II superconductors with high κ also called "extreme type–II superconductors". κ is often high in reality, for example in the case of high critical-temperature superconductors. We will set
$$\varepsilon = \frac{1}{\kappa},$$
and let $\varepsilon \to 0$ to get asymptotic expansions. This reinforces the discrimination between $|u| \simeq 0$ and $|u| \simeq 1$, and leads to small-size vortices.

The stationary configurations are the critical points of J, satisfying the Ginzburg–Landau equations

(G.L.) $$\begin{cases} -\nabla_A^2 u = \kappa^2 u\big(1 - |u|^2\big), \\ -\nabla^\perp h := -(-\partial_{x_2} h, \partial_{x_1} h) = (iu, \nabla_A u) := j. \end{cases}$$

They also satisfy the following boundary conditions:
$$\begin{cases} (\nabla u - iAu).n = 0 & \text{on } \partial\Omega, \\ h = h_{\text{ex}} & \text{on } \partial\Omega. \end{cases}$$

This theory is called in theoretical physics a $\mathbb{U}(1)$-gauge theory (Abelian gauge theory). This implies that the energy, and the physically relevant quantities, such as h, the superconducting current j and the zeros of u, are invariant under the $\mathbb{U}(1)$-gauge transformations

(1.1) $$\begin{cases} u \mapsto e^{i\phi} u & \text{for } \phi \in H^2(\Omega, \mathbb{R}), \\ A \mapsto A + \nabla\phi. \end{cases}$$

1.2. Physical behavior of type–II superconductors. We only sketch the main aspects that are important in our study. For more details, one can refer to the many physics references, such as [**9, 22, 23**], and [**13**].

Type–II superconductors are materials that exhibit two main properties, below a (low) critical temperature T_c. The first one is that they then have zero-resistivity and thus permanent nondecaying currents can be observed. The second one, on which we specially focus, is their particular behavior when they are submitted to an external magnetic field.

If $h_{\text{ex}} = 0$, it is easy to see that $(u \equiv 1, A \equiv 0)$ minimizes the energy. For small h_{ex} ($h_{\text{ex}} < H_{c_1}$), this situation is preserved in the sense that u is approximately equal to a constant of modulus 1, the superconductor is in the superconducting phase everywhere, and the magnetic field h does not penetrate it. More precisely, it satisfies approximately the following equation
$$\begin{cases} -\Delta h + h = 0 & \text{in } \Omega, \\ h = h_{\text{ex}} & \text{on } \partial\Omega. \end{cases}$$

This state is called the Meissner phase, and the corresponding solution "Meissner solution".

For a critical value $h_{\text{ex}} = H_{c_1}$, a phase transition occurs and vortices appear. They are energetically favorable, but repeal one another. The competition between these two effects rules their number and positions. Indeed, they tend to form a triangular lattice, called the Abrikosov lattice. Naming a_i the centers of these vortices, the magnetic field approximately satisfies

$$(1.2) \quad \begin{cases} -\Delta h + h = 2\pi \sum_i \delta_{a_i} & \text{in } \Omega, \\ h = h_{\text{ex}} & \text{on } \partial\Omega, \end{cases}$$

an equation introduced by London on a phenomenological basis, and that can be derived formally from the Ginzburg–Landau equations.

The superconductor is then said to be in the mixed phase.

Raising h_{ex} again, vortices become more and more numerous, until for $h_{\text{ex}} = H_{c_2}$, the second critical field, a second phase transition occurs, after which the superconductor is everywhere in the normal phase, and is thoroughly penetrated by the magnetic field. ($u \equiv 0, h \equiv h_{\text{ex}}$).

The value of H_{c_1} is defined with an energy argument: it is the field for which the energy of the Meissner solution becomes equal to the energy of a single-vortex solution, i.e. the value of h_{ex} for which vortices become energetically favorable.

In the physics literature, it is known that $H_{c_1} = O(\log \kappa)$ (DeGennes formula) and $H_{c_2} = O(\kappa^2)$.

2. Mathematical Approaches to Vortices

In [**5**], F. Bethuel, H. Brezis, and F. Hélein studied a simplified energy

$$(2.1) \quad F(u) = \frac{1}{2} \int_\Omega |\nabla u|^2 + \frac{1}{2\varepsilon^2} (1 - |u|^2)^2,$$

corresponding to setting $A = 0$, and suppressing the magnetic fields. Then, one needs a mechanism to create vortices. For this purpose, they imposed a Dirichlet boundary condition $u = g$ on $\partial\Omega$, where g is a S^1-valued map of positive degree d.

Because of the penalization term $1/\varepsilon^2 \int_\Omega (1 - |u|^2)^2$, and $\varepsilon \to 0$, $|u|$ must tend to 1 almost everywhere. But, because of its degree $d > 0$ on $\partial\Omega$, it has to vanish d times (with multiplicity). Thus, they forced vortices to appear because of the boundary condition (whereas, in our problem, they arise because of the external field). They obtained a very precise description of these vortices, their location, and the limits of the solutions as $\varepsilon \to 0$. What is most interesting for our study is to know that the minimal energy $F(u_\varepsilon)$ diverges in $\pi d |\log \varepsilon|$, that there are d vortices of degree 1 in the minimizing configurations, and that each of them "costs" approximately $\pi |\log \varepsilon|$. They proved an expansion of the form

$$\min F \sim \pi d |\log \varepsilon| + W(a_i),$$

where the "renormalized energy" W governs the positions of the vortices, and contains a repulsion term $-\pi \sum_{i \neq j} \log |a_i - a_j|$.

In their proof, they defined what we call a "vortex-structure" of a solution u, by covering $\{x / |u(x)| < 1/2\}$ by a bounded number of balls of radius ε.

In our work, we use other constructions of vortex-structures, borrowed from Almeida and Bethuel [**3**], Jerrard [**11**], and Sandier [**14**]. They allow to define vortices for almost arbitrary maps, and not only for solutions of the Ginzburg–Landau equations.

3. Branches of Solutions Around H_{c_1}

3.1. The local minimization. The main technical difficulty, compared to [**5**], is that we do not have an a-priori determination of the number (and degrees) of vortices of the minimizing solutions. This is why we restrict to configurations for which there exists a bound on the number of vortices. Setting $\mathcal{M} > 0$, we study and minimize J over the open domain

$$D_\mathcal{M} = \left\{ (u, A) \in H^1(\Omega, \mathbb{C}) \times H^1(\Omega, \mathbb{R}^2) \Big/ \right.$$
$$\left. F(u) = \frac{1}{2} \int_\Omega |\nabla u|^2 + \frac{1}{2\varepsilon^2}\left(1 - |u|^2\right)^2 < \mathcal{M} |\log \varepsilon| \right\}.$$

Configurations in $D_\mathcal{M}$ satisfy the suitable condition to use the construction of [**3**]. Thus, we can define vortices for them; and through an analysis of the type of [**5**], we know that this domain roughly corresponds to configurations having less than $\mathcal{M}/\pi$ vortices.

3.2. Computing H_{c_1}. We recall that H_{c_1} is defined as the first value of h_{ex} for which the minimal energy over 1-vortex configurations becomes equal to the minimal energy over vortex-free configurations.

We define ξ_0 to be the unique solution of

(3.1) $\qquad \begin{cases} -\Delta \xi_0 + \xi_0 = -1 & \text{in } \Omega, \\ \xi_0 = 0 & \text{on } \partial\Omega. \end{cases}$

ξ_0 is a smooth negative function depending only on Ω. It will account for boundary effects. We also define

$$\Lambda = \left\{ x \in \Omega / |\xi_0(x)| = \max |\xi_0| \right\}, \quad C_1 = \left(2 \max |\xi_0|\right)^{-1}.$$

THEOREM 3.1 (see [**19, 20**]).

$$H_{c_1} = \frac{1}{2 \max |\xi_0|} \log \kappa + O(1) \quad \text{as } \kappa \to \infty.$$

For $\varepsilon < \varepsilon_0(\mathcal{M})$,
- if $h_{\text{ex}} < H_{c_1}$, any minimizer in $D_\mathcal{M}$ is solution of (G.L.) and is vortex-free $\left(|u| \geq 1/2\right)$. (Meissner solution),
- if $H_{c_1} \leq h_{\text{ex}} \leq H_{c_1} + O(1)$, any minimizer in $D_\mathcal{M}$ is solution of (G.L.) and has vortices a_i of degree 1, tending to distinct points in Λ as $\varepsilon \to 0$.

This theorem asserts first that $H_{c_1} \simeq C_1 \log \kappa$, with $C_1 > 1/2$, which agrees with the physical theoretical predictions; C_1 is given implicitly and can be estimated as a function of the size of the domain, for example.

Furthermore, it shows that, while passing H_{c_1}, vortices become energetically favorable, and appear near Λ. In the case of a disc or convex domain, one can show that Λ is reduced to a single point (the center for a disc); then, a single vortex appears at H_{c_1}.

We also obtained the following

THEOREM 3.2 (see [**21**]). *The vortex-free Meissner solution is unique among those which are energy-minimizing over vortex-free configurations.*

Then, in a subsequent work with E. Sandier, we were able to prove, by a different method, that, up to H_{c_1}, the global minimizer of the energy (and not only local minimizer over $D_\mathcal{M}$ as in Theorem 3.1), is the vortex-free solution found in Theorem 3.1 (and unique, according to Theorem 3.2).

THEOREM 3.3 (Sandier–Serfaty, see [**15**]). *There exists $H'_{c_1} \sim H_{c_1}$ such that, if $h_{\mathrm{ex}} < H'_{c_1}$, for small ε, the global minimizer of the energy is the vortex-less solution of Theorem 3.3.*

3.3. Above H_{c_1}.

We use again our idea of local minimization, and minimize over domains

$$U_n = \left\{ (u,A) / n\pi |\log \varepsilon| < F(u) < \pi\left(n + \frac{1}{2}\right) |\log \varepsilon| \right\}.$$

The idea is that U_n roughly corresponds to n-vortices configurations, so that, when minimizing over it, we can hope to catch an n-vortices solution.

THEOREM 3.4 ([**21**]). *Suppose that $\Omega = B(0, R)$, and that $h_{\mathrm{ex}}(\kappa)$ is such that*

$$h_{\mathrm{ex}} \xrightarrow[\kappa \to \infty]{} +\infty, \quad h_{\mathrm{ex}} \leq \kappa^\alpha,$$

*for some small $\alpha > 0$. Then, for all $n \in \mathbb{N}$, there exists $\varepsilon_0(n)$, such that, $\forall \varepsilon < \varepsilon_0$, there exists a **stable solution** (u, A) of (G.L.), such that:*

1. *u has **exactly n** vortices of degree 1, centered at a_i.*
2. *$|a_i| \to 0$ as $\varepsilon \to 0$, and if we let $\tilde{a}_i = a_i \sqrt{h_{\mathrm{ex}}}$, the $\tilde{a}_i$ tend to minimize*

$$w(x_1, \ldots, x_n) = -\pi \sum_{i \neq j} \log|x_i - x_j| + \pi \xi_0''(0) \sum_i |x_i|^2.$$

3. *The energy of the solution is*

$$J(u, A) = \frac{h_{\mathrm{ex}}^2}{2} \int_\Omega |\xi_0| + \pi n \left(|\log \kappa| - \frac{h_{\mathrm{ex}}}{C_1} \right) + \frac{\pi}{2}(n^2 - n) \log h_{\mathrm{ex}} + \min w + Q_n + o(1),$$

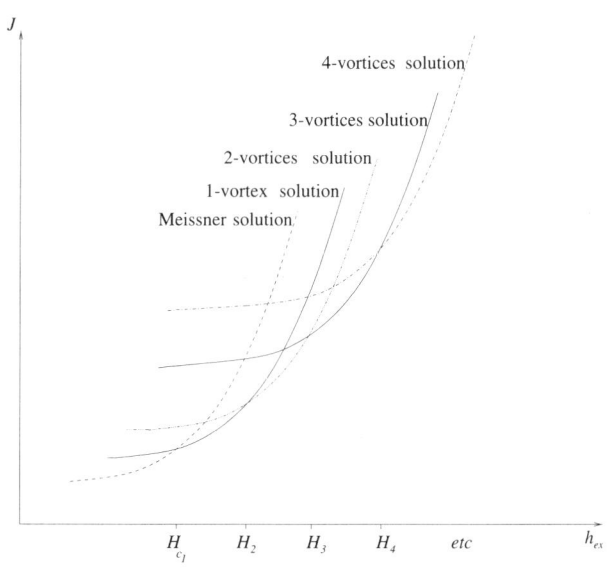

FIGURE 1. The branches of solutions.

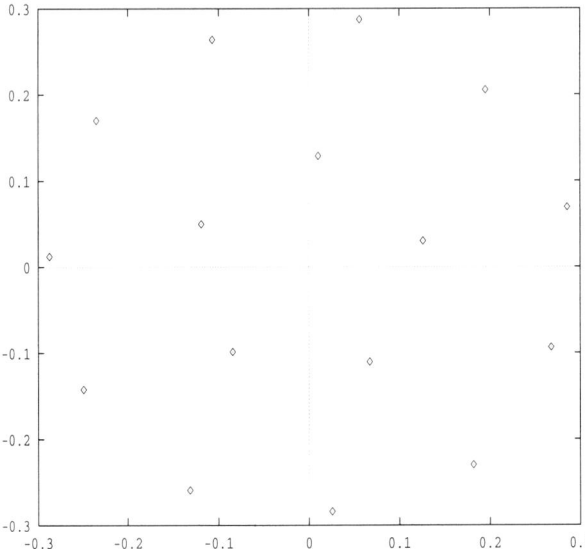

FIGURE 2. Results of the numerical optimization of [10] for w, cases $n = 16$ and 21. Reprinted with the permission from SIAM.

where Q_n is a constant depending only on n.

4. The energies of the $(n-1)$-vortices and n-vortices solutions are equal for $h_{\mathrm{ex}} = H_n$, where

$$H_n \sim \frac{1}{2 \max |\xi_0|} \bigl(\log \kappa + (n-1) \log \log \kappa \bigr).$$

We thus have the existence of branches of stable (in the sense that they locally minimize the energy) n-vortices solutions, for arbitrary n, much below H_{c_1}, and much above H_{c_1}. We thus immediately deduce the multiplicity of solutions for certain fixed h_{ex}.

The pictures of these branches is approximately given by Figure 1, thanks to the explicit formula of 3.

In addition, we know that the vortices of these solutions are concentrated around the center, at the scale $1/\sqrt{h_{\mathrm{ex}}}$. Their positions are governed by the renormalized energy w. This is due to a competition between the coulombian repulsion of the vortices showed in [5] and the confinement effect due to the external field and ξ_0. As in [5], we have thus reduced the infinite-dimensional minimization problem to a finite-dimensional one. The minimization of w has been studied, for example in [10]. It seems that the minimizing configurations are regular polygons centered at the origin for small n, and form some kinds of lattices concentrated around the center for higher n. In Figure 2 and 3, some of the numerically computed configurations are shown.

This is interesting since such configurations seem to correspond to physical observations, in superconductors and in superfluids (ruled by a very similar model), cf. [22, 24] for example. To our opinion, it constitutes the birth of the Abrikosov

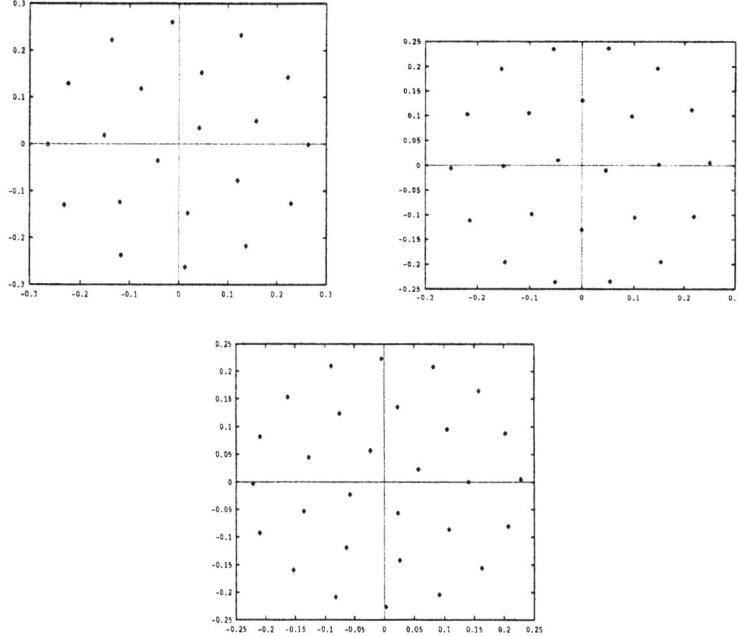

FIGURE 3. Results of the numerical optimization of [**10**] for w, cases $n = 24$ and 29. Reprinted with the permission from SIAM.

lattice. Although they concern high-κ superconductors, these results are also in good agreement with studies on nonhigh κ superconductors, performed in [**18**] and [**2**].

3.4. Sketch of the proofs. The method consists in splitting the energy as follows:

$$J(u, A) = \frac{h_{\text{ex}}^2}{2} \int_\Omega |\xi_0| + F(u) + 2\pi h_{\text{ex}} \sum_i d_i \xi_0(a_i) + O(1),$$

where the (a_i, d_i)'s are the vortices of u (defined through the method of [**3**]) with their degrees, and $F(u)$ is still the functional (2.1). $h_{\text{ex}}^2/2 \int_\Omega |\xi_0|$ is the energy of the Meissner (vortex-free) solution. $F(u)$ counts the intrinsic cost of vortices (about $\pi|d_i| \, |\log \varepsilon|$ per vortex), and their repulsion. $\xi_0 \leq 0$ is the solution of (3.1), the term $2\pi h_{\text{ex}} \sum_i d_i \xi_0(a_i)$ counts the energetic gain of each vortex, and tells that the most favorable is for vortices to be near the point(s) of minimum of ξ_0.

4. Studying Global Minimizers

From now on, we focus only on global minimizers of the energy over the whole functional-space, and we wish to determine their vorticity.

We assume that the applied magnetic field $h_{\text{ex}}(\kappa)$ is such that $h_{\text{ex}} \to \infty$ and $h_{\text{ex}} \ll \kappa^2$ (or $h_{\text{ex}} \ll H_{c_2}$). In addition we assume that

$$\lambda := \lim_{\varepsilon \to 0} \frac{|\log \varepsilon|}{h_{\text{ex}}(\varepsilon)},$$

exists and is finite.

We now consider a global-minimizer of the energy (u, A), with $h = \operatorname{curl} A$ its induced magnetic field. (u, A) satisfies (G.L.), and, thanks to a method inspired from [14], we can, at least for $\lambda > 0$, define vortices of u, a_i, with their degrees d_i, (depending on ε). We can thus define a vorticity-measure

$$\mu_\varepsilon := \frac{2\pi}{h_{\text{ex}}} \sum_i d_i \delta_{a_i},$$

and we wish to determine the behavior of μ_ε in the limit $\varepsilon \to 0$.

THEOREM 4.1 (Sandier–Serfaty, [17]). *As $\varepsilon \to 0$,*

(4.1) $$\frac{h}{h_{\text{ex}}} \rightharpoonup h_* \quad \big(\text{weakly in } H^1(\Omega)\big), \text{ minimizer of}$$

(4.2) $$E(f) = \frac{\lambda}{2} \int_\Omega |-\Delta f + f| + \frac{1}{2} \int_\Omega |\nabla f|^2 + |f - 1|^2,$$

over $V = \{f \in H^1_1(\Omega), -\Delta f + f \text{ is a Radon measure}\}$, and solution of the free-boundary problem

(P) $$\begin{cases} -\Delta h_* + h_* = 0 & \text{in } \Omega \setminus \omega, \\ h_* = 1 - \frac{\lambda}{2} & \text{in } \omega, \\ \frac{\partial h_*}{\partial n} = 0 & \text{on } \partial \omega, \\ h_* = 1 & \text{on } \partial \Omega. \end{cases}$$

Moreover, if $\lambda > 0$, we have

$$\mu_\varepsilon = \frac{2\pi}{h_{\text{ex}}} \sum_i d_i \delta_{a_i} \to \mu := -\Delta h_* + h_* = \left(1 - \frac{\lambda}{2}\right) \mathbf{1}_\omega, \quad \frac{2\pi}{h_{\text{ex}}} \sum_i |d_i| \delta_{a_i} \to \mu.$$

We get an asymptotic expansion of the energy

$$\min J \sim \min E.$$

The weak convergence (4.1) is strong only if $\lambda = 0$, otherwise the lack of convergence is described through a defect measure:

$$\left|\nabla\left(\frac{h}{h_{\text{ex}}}\right)\right|^2 \to |\nabla h_*|^2 + \lambda\mu.$$

Through this result, we extract a limiting functional E, which is simpler than J and accounts for the behavior of the magnetic field and of the vortex-density in the limit $\varepsilon = 0$, both being automatically related. Notice that, contrarily to J, this limiting functional is convex and thus has a unique critical point, given by the free-boundary problem mentioned above. In this problem, ω is a sub-domain of Ω, determined uniquely by the system (P), because this system is over-determined in terms of h_*. The practical interpretation, is that, roughly speaking, when ε is small, there are two regions in the sample: a central region ω in which there is a uniform vortex-density equal to $1/2\pi(h_{\text{ex}} - 1/2|\log \varepsilon|)$, surrounded by a vortex-free region. This free-boundary problem (P) is reminiscent of free-boundary problems formally derived from Ginzburg–Landau, such as that of [8].

Trying to solve problem (P), we get that $\omega = \varnothing$ if and only if

$$\frac{1}{\lambda} = \lim_{\varepsilon \to 0} \frac{h_{\text{ex}}}{|\log \varepsilon|} < \frac{1}{2 \max |\xi_0|},$$

which is totally coherent with the estimate of H_{c_1} given in Theorem 3.1: there are no vortices if and only if $h_{\text{ex}} < H_{c_1}$.

In the case of $\lambda = 0$, i.e. when the applied magnetic field is much bigger than the first critical field H_{c_1}, then $\omega = \Omega$ and $h_* = 1$, hence the theorem above asserts that h/h_{ex} converges to 1, strongly in H^1, and that the limiting vortex-density is uniform, equal to $1/2\pi h_{\text{ex}}$. We get more precise results in the following theorems:

THEOREM 4.2 (Sandier–Serfaty, [16]). *Assume*
$$H_{c_1} \ll h_{\text{ex}} \ll \kappa^2.$$

If (u, A) minimizes J, then, $|.|$ denoting the area,
$$J(u, A) \sim \frac{1}{2}|\Omega| h_{\text{ex}} \log \frac{1}{\varepsilon \sqrt{h_{\text{ex}}}},$$

and for any V open in Ω, denoting J_V the functional restricted to V,
$$J_V(u, A) \sim \frac{1}{2}|V| h_{\text{ex}} \log \frac{1}{\varepsilon \sqrt{h_{\text{ex}}}}.$$

COROLLARY.
$$\frac{h}{h_{\text{ex}}} \xrightarrow[\varepsilon \to 0]{} 1 \quad \text{in } H^1(\Omega).$$

THEOREM 4.3 (Sandier–Serfaty, [16]). *Under the same hypotheses, for $\varepsilon < \varepsilon_0$, there exists a family of disjoint closed disks B_i centered at a_i, of* radii $< 1/\sqrt{h_{\text{ex}}}$ *such that*
- $|u| > 1/2$ *on* ∂B_i,
- *if* $d_i = \deg(u, \partial B_i)$,

$$\mu_\varepsilon := \frac{1}{h_{\text{ex}}} 2\pi \sum_i d_i \delta_{a_i} \xrightarrow[\varepsilon \to 0]{} dx \,|_\Omega,$$

in the sense of measures, where dx denotes the Lebesgue measure. Moreover,
$$2\pi \sum_i |d_i| \sim 2\pi \sum_i d_i \sim h_{\text{ex}} |\Omega|,$$

and most of the energy is contained in these disks:
$$J_{\Omega \setminus \cup B_i}(u, A) \ll J(u, A).$$

Thus, although the analysis in the case $\lambda = 0$ is more delicate, (because h_{ex} can become very high, and thus the number of vortices too), we are still able to extract a vortex-density convergence, and to prove that there is a uniform vortex-repartition and a uniform energy-density repartition in this case.

The formula giving the expansion of the energy in Theorem 4.3 can be understood as follows: there are about $|\Omega| h_{\text{ex}}/2\pi$ vortices; $1/\varepsilon \sqrt{h_{\text{ex}}}$ is the ratio between the characteristic distance between vortices $1/\sqrt{h_{\text{ex}}}$ and the characteristic size of a vortex ε; $\pi \log 1/\varepsilon \sqrt{h_{\text{ex}}}$ is the cost of a vortex seen as isolated in a cell of size $\sqrt{2\pi/h_{\text{ex}}}$ (this can be seen from an analysis of the type of [5]); multiplying the cost by the number of cells $(|\Omega| h_{\text{ex}})/2\pi$ yields the formula.

References

1. A. Abrikosov, *On the magnetic properties of superconductors of the second type*, Soviet Phys. JETP **5** (1957), 1174–1182.
2. E. Akkermans and K. Mallick, *Vortices in Ginzburg–Landau billiards*, J. Phys. A **32** (1999), no. 41, 7133–7143.
3. L. Almeida and F. Bethuel, *Topological methods for the Ginzburg–Landau equations*, J. Math. Pures Appl. (9) **77** (1998), 1–49.
4. H. Berestycki, A. Bonnet, and J. Chapman, *A semi-elliptic system arising in the theory of type-*II *superconductivity*, Comm. Appl. Nonlinear Anal. **1** (1994), no. 3, 1–21.
5. F. Bethuel, H. Brezis, and F. Hélein, *Ginzburg–Landau vortices*, Birkhäuser, Basel, 1994.
6. F. Bethuel and T. Rivière, *Vortices for a variational problem related to superconductivity*, Ann. Inst. H. Poincaré, Anal. Non Linéaire **12** (1995), 243–303.
7. _____, *Vorticité dans les modèles de Ginzburg–Landau pour la supraconductivité*, Séminaire sur les Équations aux Dérivées Partielles (1993–1994), Exp. no. XVI, École Polytechnique, Palaiseau, 1994.
8. S. J. Chapman, J. Rubinstein, and M. Schatzman, *A mean-field model of superconducting vortices*, European J. Appl. Math. **7** (1996), no. 2, 97–111.
9. P.-G. DeGennes, *Superconductivity of metal and alloys*, Benjamin, New York–Amsterdam, 1966.
10. S. Gueron and I. Shafrir, *On a discrete variational problem involving interacting particles*, SIAM J. Appl. Math. **60** (1999), no. 1, 1–17.
11. R. Jerrard, *Lower bounds for generalized Ginzburg–Landau functionals*, SIAM J. Math. Anal. **30** (1999), no. 4, 721–746.
12. J. Rubinstein, *Six lectures on superconductivity*, Boundaries, Interfaces, and Transitions (Banff, 1995), CRM Proc. Lecture Notes, vol. 13, Amer. Math. Soc., Providence, RI, 1998, pp. 163–184.
13. D. Saint-James, G. Sarma, and E. J. Thomas, *Type-*II *superconductivity*, Pergamon Press, 1969.
14. E. Sandier, *Lower bounds for the energy of unit vector fields and applications*, J. Funct. Anal. **152** (1998), no. 2, 379–403.
15. E. Sandier and S. Serfaty, *Global minimizers for the Ginzburg–Landau functional below the first critical magnetic field*, Ann. Inst. H. Poincaré Anal. Non Linéaire **17** (2000), no. 1, 119–145.
16. _____, *On the energy of type-*II. *Superconductors in the mixed phase*, Preprint, 1999.
17. _____, *A rigorous derivation of a free-boundary problem arising in superconductivity*, Ann. Sci. École Norm. Sup. (4) **33** (2000), 561–592.
18. V. A. Schweigert, F. M. Peeters, and P. S. Deo, *Vortex phase diagram for mesoscopic superconducting disks*, Phys. Rev. Lett. **81** (1998), no. 13, 2783–2786.
19. S. Serfaty, *Local minimizers for the Ginzburg–Landau energy near critical magnetic field.* I, Commun. Contemp. Math. **1** (1999), no. 2, 213–254.
20. _____, *Local minimizers for the Ginzburg–Landau energy near critical magnetic field.* II, Commun. Contemp. Math. **1** (1999), no. 3, 295–333.
21. _____, *Stable configurations in superconductivity: Uniqueness, multiplicity and Vortex-nucleation*, Arch. Ration. Mech. Anal. **149** (1999), no. 4, 329–365.
22. D. Tilley and J. Tilley, *Superfluidity and superconductivity*, 2nd edition, Adam Hilger Ltd., Bristol, 1986.
23. M. Tinkham, *Introduction to superconductivity*, 2nd ed., McGraw-Hill, McGraw-Hill Book Co., New York–Toronto, Ont.–London, 1964, 1996.
24. E. J. Yarmchuk, M. J. V. Gordon, and R. E. Packard, *Observation of stationary Vortex arrays in rotating superfluid helium*, Phys. Rev. Lett. **43** (1979), no. 3, 214–217.

CMLA, ENS CACHAN, 61 AVE DU PRÉSIDENT WILSON, 94235 CACHAN CEDEX, FRANCE
E-mail address: serfaty@cmla.ens-cachan.fr

Dissipation Through Dispersion

A. Soffer

ABSTRACT. In this note I describe a time dependent approach to analyze dispersion mediated energy transfer for both linear and nonlinear systems.

Examples, applications and a discussion of some open problems are included.

1. Introduction

Dissipative processes are fundamental to our understanding of many physical properties of large systems. They are important in approach to equilibrium, measurement processes, heat transfer etc.

On the other hand, physical theories are expected to conserve energy, so the resulting dynamics is Hamiltonian.

To reconcile the two, seemingly contradictory processes, we invoke radiation damping, or dispersion of the energy from one ("small") system to a large, infinite dimensional system.

I'll describe a new, time dependent approach to analyze this dispersion mediated energy transfer, and some examples and applications. This approach turns out to be surprisingly general and applied to diverse problems including resonances in quantum mechanics, asymptotic stability of Nonlinear Hamiltonian PDE's, Atoms-laser systems, nonlinear optics and more. Besides describing some of these results, I'll mention a few open problems in different fields, of related nature.

The class of models we are interested here are of the type "small" + infinite systems [**W, Da, Se, AE**].

Typically the small system will have finite dimensional phase space, and the infinite part will be dispersive field, or medium. A classical example, due to Lamb [**La**] is the spring + string system, in which a particle of mass m on a spring, is also coupled to an infinite string.

The system is Hamiltonian, yet, if one looks at (finite energy) states, the asymptotic behavior of the mass on the spring is dissipative, due to energy losses to infinity via the string. In some sense all the models considered here are generalizations of this system. This particular model can be solved exactly. Here I'll describe

---

2000 *Mathematics Subject Classification*. Primary: 81Q99; Secondary: 81U99.
Supported in part by the National Science Foundation.
This is the final form of the paper.

©2001 American Mathematical Society

a method to deal with the general case of finite dimensional system coupled to a dispersive (not necessarily linear) medium [**SW1–6, MSW, KW**].

All the models I'll describe are at zero temperature. See the discussion in Sections 4, 5 concerning the finite temperature case.

2. Models, Examples

A beautiful consequence of dissipation phenomena is the Einstein relation. Suppose p, the momentum of a particle coupled to a heat bath satisfies

$$\frac{dp}{dt} = -\alpha p + \eta(t), \tag{1}$$

where $\eta(t)$ is a random noise, with zero average $\langle \eta \rangle = 0$ and

$$\langle \eta(t)\eta(s) \rangle = 2Dg(t-s), \tag{2}$$

as correlation.

By solving for $\langle p^2 \rangle$, after approximating g by a δ-function we get

$$\langle p^2 \rangle = \frac{D}{\alpha}. \tag{3}$$

If the white noise term $\eta(t)$ is modeling a bath at temperature T, we have from thermal equilibrium

$$\frac{\langle p^2 \rangle}{2M} = \frac{1}{2}kT, \tag{4}$$

and therefore

$$D = Mk\alpha T, \tag{5}$$

which is the Einstein relation.

Here M is the mass of the particle.

Deriving this result directly from the Hamiltonian of the system is a major problem. See Section 5.

Dispersive systems can be cast in the following Schrödinger form:

$$i\frac{\partial}{\partial t}\Psi = K\Psi + NL(\Psi),$$

where Ψ is in general a vector valued function and K is a linear matrix Hamiltonian, self-adjoint operator.

$NL(\Psi)$ stands for a general perturbation/coupling which can be linear or non-linear. For integrable systems see [**DZ**].

EXAMPLE 1. Consider a quantum particle coupled to an infinite system, described in some "mean field" approximation; the resulting equation is, for the effective wave function Ψ:

$$i\frac{\partial \Psi}{\partial t} = (-\Delta + V)\Psi + f(\Psi).$$

Here $f(\Psi)$ can be of the general form

$$f(\Psi) = \int R(x-y)|\Psi(y)|^p \Psi(x)\, dy,$$

for some real valued function R.

When $-\Delta + V$ has bound states the large time behavior of this system is incredibly complicated.

EXAMPLE 2. Nonlinear wave equations.
The equation
$$(\Box + V(x,t))u = f(u),$$
with $V(x,t)$ real, the nonlinear, nonhomogeneous wave equation, appears naturally in many applications e.g., in nonlinear optics models or when linearizing around solitary type solutions of homogeneous systems, as well as in the study of blackhole radiation. The original model of Lamb is of the above form with $f(u) = 0$ for linear string, and $f(u) \neq 0$ if the string is nonlinear.

EXAMPLE 3. Resonances in QM.
It may sound surprising that the resonance problem in QM is of the type discussed above. In fact, one can reformulate the resonance problem in many cases as follows
$$i\frac{\partial \Psi}{\partial t} = H_0 \Psi + W\Psi,$$
with H_0 has embedded eigenvalues in the continuum, and W is a perturbation that couples the eigenvalues to the continuum see [**HS**] and cited ref. In general the eigenvalues disintegrate, exponentially fast into radiation, up to small corrections. For other time dependent analysis of this problem see e.g. [**Hu, PF, GS, CLR**] and cited ref. Again the Lamb's model is exactly of this type, where now H_0 stands for the Hamiltonian describing the motion of a mass on a spring and a string decoupled from each other. The W is then the coupling term between the mass and the string. The point spectrum of the matrix H_0 comes from the basic frequency of the mass-spring system, and the continuous spectrum from the wave equation of infinite string.

3. Linear Theory

The discussion of the last example of Section 1 is the basis for the approach we utilize here; the dissipative behavior should appear as the exponential decay of a "resonance", and is therefore valid only for a finite time scale. In this way we resolve the apparent impossibility of closed Hamiltonian systems to describe dissipative behavior.

Of course, to fulfill the above promise, one needs a general enough resonance theory, which is time dependent, and gives explicit rate of exponential decay and bounds on the correction terms, **uniformly in time**. We first consider linear systems; examples include atom + radiation systems, finite dimensional linear systems of masses and springs coupled to linear wave equations (e.g. a mechanical clock, coupled to sound or water waves if floating ..., in the linear approximation). When the environment is in a finite temperature (e.g. as in Einstein relation case) such formulation is still possible in terms of Liouville dynamics on the space of density matrices [**JP1–2, Da, Se**].

Consider then the following system:

The Hamiltonian of the system is given by a self-adjoint operator acting on a Hilbert space $\mathcal{H}$, $H = H_0 + W$ and let Ψ_0 be an embedded eigenvector of H_0, with eigenvalue λ_0
$$H_0 \Psi_0 = \lambda_0 \Psi_0, \quad \|\Psi_0\|_{\mathcal{H}} = 1.$$
The main result established in [**SW4**] for this case is

THEOREM 1. *Let $H = H_0 + W$ be a self-adjoint operator acting on a Hilbert space $L^2(R^n)$. $H_0 \Psi_0 = \lambda_0 \Psi_0$, $\lambda_0 \in \Delta \subset I$; $I-$ the continuous spectral interval of H_0. Suppose that H_0 satisfies conditions (H) and W satisfies conditions (W) below. Then*

(a) *$H = H_0 + W$ has no eigenvalues in the interval Δ.*
(b) *The spectrum of H in Δ is purely absolutely continuous; in particular we have local decay for H;*

$$\left\| \langle x \rangle^{-\sigma} e^{-iHt} g_\Delta(H) \phi \right\|_{L^2} \leq c \langle t \rangle^{-r+1} \left\| \langle x \rangle^\sigma \phi \right\|_{L^2},$$

for σ as in (H), (W).

  Here we use the notation $\langle x \rangle = \sqrt{1 + |x|^2}$ $g_\Delta(H)$ is a smooth characteristic function of the interval Δ, with argument the operator H.

(c) *For ϕ_0 in the range of $g_\Delta(H)$, we have for $t \geq 0$*

$$e^{-iHt} \phi_0 = \left(1 + O(W)\right)\left[e^{-\Gamma t} a(0) \Psi_0 e^{-i\theta t} + e^{-iH_0 t} \phi_d(0)\right] + R(t)$$

$a(0), \psi_d(0)$ are determined by the initial data ϕ_0.

$$\theta = \lambda_0 + (\Psi_0, W\Psi_0) - \Lambda + O(W^3) \equiv \omega - \Lambda + O(W^3),$$
$$\Lambda = \left(W\Psi_0, P.V.(H_0 - \omega)^{-1} W\Psi_0\right),$$
$$\Gamma = \pi\left(W\Psi_0, \delta(H_0 - \omega)(I - P_0)W\Psi_0\right)$$

P_0 is the Projection on Ψ_0.

$$\left\| \langle x \rangle^{-\sigma} R(t) \right\|_{L^2} \leq c |||W|||, \quad \left\| \langle x \rangle^{-\sigma} R(t) \right\|_{L^2} \leq c |||W|||^\varepsilon \langle t \rangle^{-r+1},$$
$$t \geq |||W|||^{-2(1+\delta)}.$$

Conditions (H).

(H1) H_0 is a self-adjoint operator with dense domain $\mathcal{D} \subset L^2$.
(H2) λ_0 is a simple eigenvalue embedded in $P_c \mathcal{D}$, and eigenfunction Ψ_0 normalized to 1. P_c is the projection on the continuous spectral part of H_0.
(H3) These exists an open interval Δ around λ_0, with no other eigenvalues.
(H4) For some $\sigma > 0$ we have local decay with some $r \geq 2 + \varepsilon$, $\varepsilon > 0$:

$$\left\| \langle x \rangle^{-\sigma} e^{-iH_0 t} P_c^\# f \right\|_{L^2} \leq C \langle t \rangle^{-r} \left\| \langle x \rangle^\sigma f \right\|_{L^2},$$

where $P_c^\# = g_{\tilde{\Delta}}(H_0)$, with $\tilde{\Delta} \supset \Delta$, $\tilde{\Delta}$ avoiding any other threshold or eigenvalue of H_0 (except of course λ_0).
(H5) For some choice of c, $\langle x \rangle^\sigma (H_0 + c)^{-1} \langle x \rangle^{-\sigma}$ can be made small enough.

Conditions (W).

(W1) W is symmetric, $H = H_0 + W$ is self-adjoint on $\mathcal{D}$.
(W2) for some $\sigma > 0$, as above,

$$|||W||| \equiv \|\langle x \rangle^{2\sigma} g_\Delta(H_0) W\| + \|\langle x \rangle^\sigma W g_\Delta(H_0) \langle x \rangle^\sigma\| + \|\langle x \rangle^\sigma W (H_0)^{-1} \langle x \rangle^{-\sigma}\| < \infty.$$

(W3) for some $\delta_0 > 0$
$$\Gamma \geq \delta_0 |||W|||^2.$$

(W4) $|||W||| < \delta_1 |\Delta|$ for some $\delta_1 > 0$, sufficiently small, depending only on the properties of H_0, in particular the constants in the local decay estimates, but not on $|\Delta|$.

COROLLARY. *Let $H_* = H - (\operatorname{Re}\omega)I_d$. Then, for any $T > 0$ there is a constant c_T*

$$\left|(\Psi_0, e^{-iH_*t}\Psi_0) - e^{-\Gamma t}\right| \leq c_T |||W||| \ as \ |||W||| \to 0.$$

REMARKS.
1. In applications to open systems, H_0 will generally be the direct sum

$$H_0 = H_s \oplus H_r,$$

where H_s has discrete spectrum and H_r with continuous spectrum.

Then, the eigenvalues of H_s generally become embedded and therefore unstable under the perturbation W which is the coupling between the system and reservoir.

This extends to the finite temperature case where we now need the Liouvillian to replace the Hamiltonian [**JP1–2**].
2. The corollary gives the time scales on which we see the exponential behavior. Generally it is of order $|||W|||^{-2}$ times a (large) constant.
3. A useful and important generalization of this theorem was recently proved [M-Sig]: it allows degenerate eigenvalue λ_0 and it only requires that

$$\|W\Psi_0\|,$$

be small; this is very important in applications, especially for N-body systems. (The price to pay is that now local decay is needed for H and not H_0, and this is hard to prove in the presence of infraparticles [**BFSS**]).

The method of proof of this theorem is time-dependent. As such it applies to nonautonomous Hamiltonians and nonlinear dispersive equations. For the linear time dependent potential case see [**SW3, SW6, MSW, KW**]. Nonlinear cases are discussed in the next section.

4. Nonlinear Theory—Asymptotic Stability for Dispersive Systems

Finite dimensional Hamiltonian systems do not have asymptotic stable points (due to Liouville's theorem). The nonlinear systems describing solitary type solutions have such stability: radiation goes away from the solitons to infinity leaving them to move freely. This is a general phenomena: kinks, vortices, black holes, binary stars coupled to gravity waves, all expected (and observed) to behave like that.

This phenomena is therefore intrinsically infinite dimensional; our aim is to understand this type of phenomena as a time dependent nonlinear resonance scattering, in which a small system (solitons, kinks, etc. ...) coupled to a big system (radiation = dispersive equations) is stabilized as time approaches infinity. This problem can be studied by the methods described here:

Consider the following model, the nonhomogeneous NLKG equation:

$$(\Box + m^2 + V(x))u(x,t) = \lambda F(u), \quad x \in \mathbb{R}^3,$$
$$\Box = \partial_t^2 - \Delta,$$

and we assume that $-\Delta + V(x)$ has one bound state (isolated eigenvalue) with

$$B^2 - \Delta + V(x) + m^2 > 0, \quad B^2\varphi = \Omega^2\varphi, \qquad \varphi \in L^2,$$

so that the linear system ($\lambda = 0$) has the following localized time periodic solution
$$u(x,t) = R\cos(\Omega t + \theta)\varphi(x).$$
It is easy to verify that the above system is Hamiltonian (when $F(u)$ is real valued) and that 0 is the state of lowest energy. One would guess that as the nonlinearity is turned on, $\lambda \neq 0$, the lump with energy Ω^2 will disintegrate. In fact this is what is proved in [**SW5**] by time dependent methods.

It turns out that as the nonlinearity is turned on (and small)
$$u(x,t) \sim a(t)\cos\big(\Omega t + \rho(t)\big)\varphi(x) + \eta(x,t),$$
$$a(t) \sim c\left(1 + \frac{3\lambda^2}{4\Omega}\Gamma t\right)^{-1/4},$$
$$\rho(t) \sim k_1 t^{1/2} + k_2 \ln t + O(t^{-1/2}),$$
$$\|\eta(\cdot,t)\|_{L^8(\mathbb{R}^3)} \leq O(t^{-3/4}),$$
$$\Gamma \equiv \frac{\pi}{3\Omega}\big(P_c(B)\varphi^3, \delta(B-3\Omega)P_c(B)\varphi^3\big).$$
Here $f(u) = \lambda u^3 + O(u^4)$.

Some ideas of the proof.

Begin with the ansatz
$$u(x,t) = a(t)\varphi(x) + \eta,$$
and orthogonality $(\varphi, \eta) = 0 \ \forall t$
$$a(0) = \big(\varphi, u(0)\big), \quad a'(0) = \big(\varphi, u'_t(0)\big),$$
$$\eta(0,x) = P_c(B)u(0), \quad \partial_t \eta(0,x) = P_c(B)u'_t(0).$$

Using the equations of motion one can then derive a system of equations for $a(t)$ and η; a simplified version of these equations look like this
$$a'' + \Omega^2 a = \lambda k a^3 + 3\lambda a^2(\varphi^3, \eta) + \ldots,$$
$$\partial_t^2 \eta + (-\Delta + V)\eta + m^2\eta = \lambda a^3 \varphi^3 + \ldots.$$

Ignoring the $\ldots$ terms it has the following Hamiltonian structure:
$$\mathcal{E}_{\text{appx.}} = \frac{1}{2}\int \big[(\partial_t u)^2 + |\nabla u|^2 + Vu^2\big]\, dx + \frac{1}{2}\left(a'^2 + \Omega^2 a^2 - \frac{1}{2}\lambda k a^4\right),$$
$$\lambda a^3 \int \chi(x)\eta(x,t)\, dx.$$

It is therefore an anharmonic oscillator coupled to the linear dispersive equation given by the nonhomogeneous KG equation, and coupled linearly in η and nonlinearly in a.

Then, we solve the η equation in terms of the propagator of the linear KG equation (with potential) and treat the $\lambda a^3 \varphi^3$ (and $\cdots$) as source term. Then plug it into the a equation.

Then, by a series of decompositions and integration by parts, we isolate the leading a^m terms in the resulting a equation. The integration by parts leads to evaluation of Green's functions as energies belonging the the continuum of B, which through ε—prescription determines the sign of the dissipation term, according to the direction of time! In particular, as $t \to +\infty$ we get

$$a'' + \left(\Omega^2 + O(|a|^2)\right)a = -\Gamma|a|^4 a + \sum_{j \geq 3} O(a^i) \quad \Gamma > 0,$$

and $-\Gamma$ is replaced by $+\Gamma$ for $t \to -\infty$.

We then prove by a method based on repeated integration by parts that it is possible to change variables $a \to A$ in such a way that the $\sum_{j \geq 3} O(a^i)$ become higher than the leading $|a|^4 a$ term. We therefore conclude that, the dissipative term $-\Gamma|a|^4 a$ dominates the large time behavior, which then gives the decay of $a(t) \sim t^{-1/4}$, and the rest of the results.

REMARKS.
1. For Γ small and t not too large, the decay will "look" like exponential.
2. Some of the mathematical consequences of this analysis include global existence for small data in H^2 for this equation, and asymptotic stability of 0 (since the above estimates hold for **all** initial conditions around zero).

    The first results based on the time dependent discussed in this article are for the nonlinear Schrödinger equation with one bound state [**SW1–2**]. This was later developed for other nonlinear equations [**PeW, BP, SW5, KeW**]. A different way of estimating the large time behavior and the decoupling between the system and radiation is employed in [**KS**], where the Huygens principle for wave equation allows the decoupling to be achieved in essentially finite time.
3. Time independent resonance theory was used for the study of stability in nonlinear equations in [**CH, Si**].

5. A Comment on Irreversibility, Entropy and More

Strictly speaking, proving exponential decay does not necessarily determines the direction of time, or show irreversibility. Since the system is Hamiltonian, changing the direction of time, will return the system to its original form [**Gu, Le**]. However, we do not see such things in nature: Cuckoo clocks do not absorb sound waves and move exponentially faster, solitons and kinks do not swallow radiation to become more wobbly or energetic, and boats do not rock faster by absorbing water waves.

For systems with finite dimensional phase-space, as particle systems, the way it is understood is that the measure of the set of initial conditions which lead to exponential growth is absurdly small, compared with the typical initial conditions. This can be quantified by entropy considerations. Therefore, changing the direction of time and making some arbitrary small change in the state, will lead to a state very different from the initial one [**Le**]. I expect a similar behavior for the systems considered here. However the above argument, based on entropy etc. ... does not seem to apply, since we now deal with systems in infinite dimensional phase-space (that of radiation). The results discussed in the previous sections show that the degrees of freedom of the small observed system move exponentially fast away from the initial state, and that is believed to be a critical aspect in the proofs of irreversibility through "loss of memory" of the initial state [**Gu**].

Some Open Problems

Linear resonance theory in $\mathcal{QM}$.

(a) Consider $H = -\Delta + V(x)$, with $V(x)$ a function with a maximum at some point x_0. $V(x)$ is smooth and vanishing at infinity. Prove, by time dependent methods the appearance of a resonance corresponding to this critical point. Such results were proved, by complex distortion techniques, but to apply the time dependent approach requires new ideas, since there is no obvious way of rewriting the problem as $H_0 + W$, with H_0 having an embedded eigenvalue.

(b) Prove the appearance of a resonance due to the existence of quasimodes: solutions of the eigenvalue equation $H\Psi = z\Psi$, with z complex, and Ψ not in L^2. [G-S]

REMARK. Problems (a) and (b) are in fact related. In [**GS**] (a) is used to construct quasimodes.

In this respect it may be interesting to improve [**GS**] to show the decay in time of the remainder terms.

Linear resonances in field theory.

(α) Apply the theorem of time dependent resonance theory for the case of nonrelativistic QED Hamiltonians [**BFS**].

Derive the detailed equations of multilevel state molecule, and transition rates. What is the effect of infra-red contributions to these rates?

(β) Apply the theorems of time dependent resonance theory to Liouville operator describing an atom coupled to a field at finite temperature T. See [**JP1–2**].

(γ) Consider the cases (α) or (β), even without massless particles, when the small molecule is replaced by a "chain" of N-molecules coupled to each other, and N large [**HL, AE, Se**].

In this case, for N-sufficiently large the distance between embedded eigenvalues is smaller than $|||W|||$, and hence the standard analysis fails.

(δ) Understand the Einstein relation: in this case the small system is replaced by a free particle. Therefore there is no embedded eigenvalue corresponding to the small system (the spectrum of the small system is now continuous $[0, \infty)$). Nevertheless, the methods may be versatile enough to treat such cases [in the study of time dependent perturbations of Quantum Hamiltonians, embedded eigenvalues play no role [**SW3, SW6, MSW, KW**]].

Nonlinear scattering problems.

1. Extend the results of [**BP, PW, SW1–2**] to multisoliton systems. The only results in this direction is the case of two-solutions in 1-dim NLS in [**Pe**].
2. Asymptotic stability of Kinks. The main difficulty is that now we have to deal with one dimensional NLKG equation, with long range nonlinear scattering and linear part with nongeneric spectrum (zero energy resonance)
3. Schrödinger and KG vortices: in this case the nonlinear long range part is much more serious than in (1).
4. Nonlinear scattering on nonflat domains, e.g. around a black hole. The metric in this case results in bound states and resonances. See [**LS**], and cited ref.

Statistical problems.
1. Introduce, in a useful way, the analog of entropy and entropy production to analyze radiation mediated dissipation. Is there an analog of the H theorem for radiation phenomena?
2. Suppose we let one of the systems considered in this article move in forward time, for an interval T, change the direction of time and make a small perturbation of the state. Will the small system's energy grow exponentially in time? Will a small change result in a big difference for the small system? The expressions we have for the remainder terms to exponential decay are pretty explicit. Can it be used? Can it be demonstrated numerically?

References

[AE]   L. Allen and J. H. Eberly, *Optical resonance and two-level atoms*, Dover Publ., New York, 1987.

[BFS]   V. Bach, J. Fröhlich, and I. M. Sigal, *Mathematical theory of nonrelativistic matter and radiation*, Lett. Math. Phys. **34** (1995), no. 3, 183–201.

[BFSS]   V. Bach, J. Fröhlich, I. M. Sigal, and A. Soffer, *Positive commutators and the spectrum of Pauli–Fierz Hamiltonians of atoms and molecules*, Comm. Math. Phys. **207** (1999), no. 3, 557–587.

[BP]   V. S. Buslaev and G. S. Perel'man, *Scattering for the nonlinear Schrödinger equation: States close to a soliton*, St. Petersburg Math. J. **4** (1993), 1111–1142.

[CLR]   O. Costin, J. L. Lebowitz, and A. Rokhlenko, *On the complete ionization of a periodically perturbed quantum system*, this volume, 2001.

[CH]   J. D. Crawford and P. D. Hislop, *Application of the method of spectral deformation to the Vlasov–Poisson system*, Ann. Physics (1989), 265–317.

[Da]   E. B. Davies, *Quantum theory of open systems*, Academic Press, London–New York, 1976.

[DZ]   P. Deift and X. Zhou, *Long time behavior of the nonfocusing Schrödinger equation—A case study*, Lectures in Math. Sci., vol. 5, Univ. of Tokyo, 1994.

[GS]   C. Gérard and I. M. Sigal, *Space-time picture of semiclassical resonances*, Comm. Math. Phys. **145** (1992), 281–328.

[Gu]   M. Gutzwiller, *Chaos in classical and quantum mechanics*, Interdisciplinary Applied Mathematics, vol. 1, Springer-Verlag, New York, 1990.

[HL]   K. Hepp and E. Lieb, *On the superradiant phase transition for molecules in a quantized radiation field: The Dicke maser model*, Ann. Physics **76** (1973), 360–404.

[HS]   P. Hislop and I. M. Sigal, *Introduction to spectral theory. With applications to Schrödinger operators*, Applied Mathematical Sciences, vol. 113, Springer-Verlag, New York, 1996.

[Hu]   W. Hunziker, *Resonances, metastable states and exponential decay laws in perturbation theory*, Comm. Math Phys. **132** (1990), 177–188.

[JP1]   V. Jakšić and C.-A. Pillet, *On a model for quantum friction. I. Fermi's golden rule and dynamics at zero temperature*, Ann. Inst. H. Poincaré Phys. Théor. **62** (1995), 47–68.

[JP2]   _____, *On a model for quantum friction. II. Fermi's golden rule and dynamics at positive temperature*, Comm. Math. Phys. **176** (1996), 619–644.

[KeW]   P. Kerrekidis and M. I. Weinstein, *Dynamics of discrete kinks*, Physica D (to appear).

[KW]   E. Kirr and M. I. Weinstein, *Parametrically excited Hamiltonian partial differential equations* (1999), submitted.

[KS]   A. Komech and H. Spohn, *Soliton like asymptotics for a classical particle interacting with a scalar wave field*, Nonlinear Anal. **33** (1998), no. 1, 13–24.

[LS]   I. Laba and A. Soffer, *Global existence and scattering theory for the nonlinear Schrödinger equation on Schwarzschild manifold*, Helv. Phys. Acta **72** (1999), no. 4, 274–294.

[La]   H. Lamb, *On a peculiarity of the wave-system due to the free vibrations of a nucleus in an extended medium*, Proc. London Math. Soc. (3) **32** (1900), 208–211.

[Le]   J. Lebowitz, *Boltzmann's entropy and time's arrow*, Phys. Today **46** (1993), 32–38.

[MS]   M. Merkli and I. M. Sigal, *A time dependent theory of quantum resonances*, Comm. Math. Phys. **201** (1999), 549–576.

[MSW] P. Miller, A. Soffer, and M. Weinstein, *Metastability of breather modes of time dependent potentials*, Preprint (1999).

[PeW] R. L. Pego and M. I. Weinstein, *Asymptotic stability of solitary waves*, Comm. Math. Phys. **164** (1994), 305–349.

[Pe] G. Perelman, *Some results on the scattering of weakly interacting solitons for nonlinear Schrödinger equations*, Spectral Theory, Microlocal Analysis, Singular Manifolds, Math. Top., vol. 14, Akademie Verlag, Berlin, 1997, pp. 78–137.

[PF] P. Pfeifer and J. Fröhlich, *Generalized time-energy uncertainty relations and bounds on lifetimes of resonances*, Rev. Modern Phys. **67** (1995), no. 5, 759–779.

[PW] C.-A. Pillet and C. E. Wayne, *Invariant manifolds for a class of dispersive, Hamiltonian partial differential equations*, J. Differential Equations **141** (1997), 310–326.

[Se] G. L. Sewell, *Quantum theory of collective phenomena*, Monographs on the Physics and Chemistry of Materials, Oxford Science Publications, The Clarendon Press, Oxford University Press, New York, 1986.

[Si] I. M. Sigal, *Nonlinear wave and Schrödinger equations. I. Instability of time periodic and quasi periodic solutions*, Comm. Math. Phys. **153** (1993), 297–320.

[SW1] A. Soffer and M. Weinstein, *Multichannel nonlinear scattering theory for nonintegrable equations*, Lecture Notes in Phys. (M. Balabane, P. Lochak, and C. Sulem, eds.), Integrable Systems and Applications (Île d'Oléron, 1988), vol. 342, Springer-Verlag, Berlin–New York, 1988, pp. 312–327.

[SW2] _____, *Multichannel nonlinear scattering for nonintegrable equations* I,, Comm. Math. Phys. **133** (1990), 119–146; II, J. Differential Equations **98** (1992), 376–390.

[SW3] _____, *Nonautonomous Hamiltonians*, J. Statist. Phys. **93** (1998), 359–391.

[SW4] _____, *Time dependent resonance theory*, Geom. Funct. Anal. **8** (1998), 1086–1128.

[SW5] _____, *Resonances, radiation damping and instability in Hamiltonian nonlinear wave equation*, Invent. Math. **136** (1999), 9–74.

[SW6] _____, *Ionization and scattering for short lived potentials*, Lett. Math. Phys. **48** (1999), no. 4, 339–352.

[We] U. Weiss, *Dissipative systems*, Path Integrals from peV to TeV (Florence, 1998), World Sci. Publishing, River Edge, NJ, 1999, pp. 27–29.

MATHEMATICS DEPARTMENT, RUTGERS UNIVERSITY, PISCATAWAY, NJ 08854-8019 (USA)
E-mail address: `soffer@math.rutgers.edu`

Quantum Tunneling at Positive Temperature

B. Vasilijevic

1. Introduction

In this talk I describe, in a somewhat informal manner, results of a joint work with I. M. Sigal [12, in preparation]. Here we address the problem of tunneling at positive temperatures. The mathematical framework we develop here rests on the notion of *free resonance energy*, $F(\beta)$ (where β is inverse temperature). This quantity plays the role of the resonance energy at zero temperature. We prove that the probability of escape of the particle from a trap (or a well) due to tunneling can be expressed as $P(t) = 1 - p(t)$, where:

$$(1.1) \qquad p(t) = e^{-\Gamma t/\hbar}(1 + O(\Gamma)).$$

Here $\hbar$ is the Planck constant divided by 2π and

$$(1.2) \qquad \Gamma = -2\operatorname{Im}(F(\beta)),$$

the "width" of the free energy. The formula above is known in physics literature and is used in condensed matter physics [1,5,6] and cosmology [7,8,13], but it was never analyzed systematically, not to mention rigorously. Moreover, our definition of $F(\beta)$ is given in terms of the Schrödinger operator itself rather than through its spectral characteristics as it is done in the above mentioned works.

Next we give a semiclassical bound on Γ:

$$(1.3) \qquad 0 < \Gamma \leq C_\beta \hbar^{-q} e^{-S_\beta/\hbar},$$

for some $q \geq 0$. Here S_β is the action of the instanton of period $\hbar\beta$.

Since the notion of temperature pertains to equilibrium states of systems with infinite number of degrees of freedom while the tunneling is obviously a nonequilibrium process and a quantum particle has only three degrees of freedom we have to clarify what we mean by tunneling at a positive temperature. The latter term refers to the process of tunneling of a particle which is either in a contact with a reservoir (i.e. a system of infinite degrees of freedom, e.g. photon or phonon gas) which at time $t = 0$ is in a state of equilibrium at temperature T, or which is initially thermalized. In the latter case the temperature is introduced through the initial condition and has no other effect on the dynamics of the system. This corresponds

---

2000 *Mathematics Subject Classification.* Primary: 81; Secondary: 47.
The research is partially supported by NSERC under Grant NA7901.
This is the final form of the paper.

©2001 American Mathematical Society

to the following physical situation: the system is prepared by putting it in contact with a reservoir at temperature T, and at $t = 0$ reservoir is removed and the particle system is left to evolve on its own, or, put differently, the effect of the reservoir on the particle is ignored. This physical situation is of interest in its own right as well as for the reason of giving a good approximation to the process of tunneling with a thermal reservoir included. Indeed, usually the coupling between the particle and the reservoir is, on one hand, sufficiently strong so that the reservoir maintains the particle "inside the well" in a state of (approximate) equilibrium and, on the other hand, is sufficiently weak so that it yields only a small perturbation to the tunneling process. In other words, usually we have:

$$(1.4) \qquad T_{\text{tunneling}} \gg T_{\text{relaxation}} \gg T_{\text{part}},$$

where $T_{\text{tunneling}} = \Gamma^{-1}$ and $T_{\text{relaxation}} = (\text{coupling constant})^{-1}$ are the characteristic times of the tunneling and relaxation, respectively, and T_{part} is the characteristic time of the particle system, say $\hbar/\Delta E$, where ΔE is a mean gap between energy levels. Thus (for small enough temperatures) the coupling between the particle and the reservoir during the process of tunneling can be neglected in the leading approximation [4].

2. Resonance Partition Function and Resonance Free Energy

In this section we introduce our key concept—the resonance free energy. To motivate our approach we review first the standard case of a quantum system in a confining (real) potential $V(x)$, i.e. $V(x) \to \infty$ as $|x| \to \infty$ and is bounded below. The key point here is that the Schrödinger operator $H = -\hbar^2 \Delta + V(x)$ is self-adjoint and the heat semigroup $e^{-\beta H}$ is of the trace-class for $\beta > 0$. As it is usual in dealing with systems at positive temperature, we introduce the (quantum mechanical) partition function as:

$$(2.1) \qquad Z(\beta) = \text{tr}\left(e^{-\beta H}\right).$$

All other thermodynamic quantities can be expressed in terms of $Z(\beta)$. Of particular interest to us is the quantity known as the (Helmholtz) free energy which is defined as:

$$(2.2) \qquad F(\beta) = -\frac{1}{\beta} \ln Z(\beta).$$

It plays the role of the ground state energy for open systems (i.e. characterizes the stable equilibrium). Hence the following theorem (for a proof see [3]) is not surprising:

THEOREM 2.1 (Feynman–Kac). *As $\beta \to \infty$,*

$$(2.3) \qquad F(\beta) \to \inf\left[\sigma(H)\right],$$

i.e. as temperature goes to zero free energy converges to the ground state energy of the system.

At this point, let us make a short digression to the topic of spectral deformations. Start with a strongly continuous, unitary, one-parameter semigroup $\mathcal{U} = \{U_\theta \mid \theta \in \mathbb{R}\}$, obtained by means of a smooth diffeomorphism $\varphi_\theta(x)$ in the following manner:

$$(2.4) \qquad (U_\theta \psi)(x) = \left(\det(D\varphi_\theta)\right)^{1/2} \psi\left(\varphi_\theta(x)\right).$$

Using this deformation family we construct a family of deformed Hamiltonians:

(2.5) $$H_\theta := U_\theta H U_\theta^{-1},$$

having explicit expression:

(2.6) $$H_\theta = -\nabla \cdot A_\theta(x) \cdot \nabla + g_\theta(x) + V_\theta(x).$$

Example:
$$\varphi_\theta(x) = e^\theta x, \quad U_\theta \psi(x) = e^{d\theta/2} \psi(e^\theta x), \quad U_\theta \phi = -e^{-2\theta}\Delta + V(e^\theta x).$$

It is then shown in [11] that (under certain conditions on $V(x)$ and $\varphi_\theta(x)$) the deformed Hamiltonians can be analytically continued (in θ) into a certain complex region D_ρ. For θ complex, U_θ is not unitary any more and therefore the spectrum of H_θ depends on θ, hence the term 'spectral deformations'.

Now we consider an abstract framework for the situation in which a particle in question is not confined and can (and does) escape to infinity. The latter process is characterized by the presence of continuum in the spectrum of the corresponding Schrödinger operator. Consequently, the heat semigroup $e^{-\beta H}$ is not of trace-class for any β and the notions of partition function and free energy do not make sense. However there are situations described in the next definition in which closely related concepts can be introduced.

DEFINITION 2.2 (Resonance partition function). *If for a given Hamiltonian H there is a spectral deformation family $\mathcal{U} = \{U_\theta \mid \theta \in D_\rho\}$ and an open set $\Omega \subset D_\rho \cap \mathbb{C}^+$ such that $e^{-\beta H_\theta}$ is a trace-class operator for $\theta \in \Omega$ then define:*

(2.7) $$Z(\beta) := \mathrm{tr}(e^{-\beta H_\theta}), \quad \theta \in \Omega.$$

$Z(\beta)$ *will be called the resonance partition function of H. By the definition it is a complex function of θ and H. It is shown in the next lemma that it is independent of θ.*

LEMMA 2.3. *Let $\mathcal{U} = \{U_\theta \mid \theta \in D_\rho\}$ be a spectral deformation family for H_θ (spectrally deformed Hamiltonians) and suppose there is an open set $\Omega \subseteq D_\rho$ such that $e^{-\beta H_\theta}$ is a trace-class operator for $\theta \in \Omega$. Then the function $Z_U(\beta, \theta) = \mathrm{tr}(e^{-\beta H_\theta})$ is independent of θ. Furthermore, if $\mathcal{U}'$ is another spectral deformation family such that $e^{-\beta H'_\theta}$ is of trace-class for $\theta \in \Omega' \cap \Omega$ then $Z_U(\beta, \theta) = Z_{U'}(\beta, \theta)$, i.e. the resonance partition function does not depend on the choice of deformation family in Definition 2.2 as long as $\Omega' \cap \Omega \neq \emptyset$.*

With the above definition of partition function it is now straightforward to define the *resonance free energy* for H as a corresponding free energy:

DEFINITION 2.4 (Resonance free energy).

(2.8) $$F(\beta) = -\frac{1}{\beta} \ln Z(\beta),$$

where we take the principal branch of the logarithm.

3. Feynman–Kac Theorem for Resonances

In this way we have extended the definition of the partition function to unstable systems which admit spectral deformations leading to deformed Hamiltonians H_θ such that $e^{-\beta H_\theta}$ is of the trace-class. It is now natural to ask if there is an analog to the Feynman–Kac theorem in this case. The answer is yes under a certain condition.

THEOREM 3.1 (Weak Feynman–Kac for resonances). *Let H_θ be such that $e^{-\beta H_\theta}$ is of the trace-class and let $\sigma(H_\theta) = \{\lambda_n \mid n \in \mathbb{Z}_+\}$ with λ_n's ordered so that they are nondecreasing in the real part. If there is only one (possibly degenerate) eigenvalue λ_0 satisfying*:

(3.1) $$\mathrm{Re}(\lambda_0) = \inf\{\mathrm{Re}(\sigma(H_\theta))\},$$

then as $\beta \to \infty$,

(3.2) $$F(\beta) \to \lambda_0,$$

i.e. resonance free energy goes to the ground state resonance energy of the system on zero temperature (there is no true ground state since system is unstable).

THEOREM 3.2 (Strong Feynman–Kac for resonances). *Let the assumptions on H_θ be the same as in the Theorem 3.1. Then (in uniform topology)*:

(3.3) $$\lim_{\beta \to \infty} e^{-\beta(H_\theta - \lambda_0)} = P_{0\theta}.$$

4. Conditions and Results

In this section we describe the class of Schrödinger operators which we consider and present our main results. The Schrödinger operators are given by the differential expressions:

(4.1) $$H = -\hbar^2 \Delta + V(x),$$

extended to functions on $L^2(\mathbb{R}^d)$. Here we use dimensionless variables, $\hbar$ is a small (dimensionless) parameter descending from either the Planck constant or from the inverse coupling constant and $V(x)$, the potential, is a real C^2 function such that the operator H (or, more precisely, its closure) is self-adjoint on its natural domain.

Now we formulate a set of conditions on the potentials $V(x)$, various subsets of which are required for different results of this paper:

(A) Either:
(A′) $V(x) \geq -Cx^2 - C'$ for some $C, C' < \infty$, or
(A″) There exist constants $C, C_1 < \infty$ such that:
    (a) $V(x) \leq C_1 - 1$
    (b) $V''(x) \geq C(V(x) - C_1)$
    (c) $|\nabla V(x)|^2 \leq C(-V(x) + C_1)$.
(A‴) $\exists R' > 0$ and $r > 5/2$ such that $V(x) = V_1(x) + V_2(x)$ with the potentials $V_i(x)$ having the following properties for $|x| \geq R'$: $V_1(x)$ is exterior analytic in the sense of [11] and is $o(|x|^r)$ and $V_2(x)$ is positive-homogeneous of degree r and $V_2(R\widehat{x}) \leq -\delta'$ for some $\delta' > 0$ and all $\widehat{x} = |x|^{-1}x$.
(B) $V(x)$ is exterior analytic in $|x| \geq R$ for some $R > 0$ in the sense of [11].
(C) Volcano-type trapping conditions:
- $V(0) = 0$ and it is the only local minimum,
- $V^*_{\mathrm{bar}} := \min_{\gamma \in \mathcal{P}}(\max_{x \in \gamma} V(x)) > 0$ where:
$$\mathcal{P} = \{\gamma \colon [0,1] \to \mathbb{R}^d \mid \gamma(0) = 0,\ |\gamma(1)| > R\},$$
i.e. the set of paths from the origin through the "mountains" and into the (far away) "low lands".
- $\limsup_{|x| \to \infty} V(x) < V(0)$.
(D) There are numbers $0 < \delta < V_{\mathrm{bar}} < V^*_{\mathrm{bar}}$ with δ being sufficiently small such that:

(a) Condition (B) is satisfied with $R = R(V_{\text{bar}})$, where $R(\varepsilon)$ is the radius of the largest ball, $B_{R(\varepsilon)}$, inside the domain $\text{int}(S_0(\varepsilon))$ (a definition of the latter is given below) and

(b) for any $\varepsilon \in (\delta, V_{\text{bar}})$, $\text{Re}(V_\theta(x)) \geq \varepsilon$ in $B^c_{R(\varepsilon)}$, where $V_\theta(x)$ is a complex deformation of $V(x)$ exterior to the surface $S_0(\varepsilon)$ [**11**].

Note that condition (A''') with $R' < R$ implies condition (B).

The potentials satisfying conditions (A'') and (A''') will be called *admissible unstable potentials*.

Under condition (A) the operator H is self-adjoint with C_0^∞ its core. Indeed, the self-adjointness under condition (A') is a standard fact (see [**10**, Theorem X.38]), and under condition (A'') is proven in [**12**].

THEOREM 4.1. *Assume conditions* (A'') *and* (A'''). *Then there is a spectral deformation family,* $\mathcal{U}$, *defined explicitly in* [**12**] *and sketched below, for the operator H such that the resonance free energy, $F(\beta)$, exists for the pair $(H, \mathcal{U})$ and for any $\beta > 0$.*

SKETCH OF THE DEFORMATION FAMILY. Take a compact set S (for later analysis we should take $S = \text{int}(S_0(\varepsilon))$, see below) and an open ball (around the origin) B in $\mathbb{R}^d$ such that $S \subset B$ and denote by B^c the complement of B. Then, by the smooth Urysohn Lemma, we construct a smooth function $j \colon \mathbb{R}^d \to [0, 1]$ with the following properties:

(4.2) $$j(S) = 0,$$

(4.3) $$j(B^c) = 1.$$

For a vector field that generates the smooth diffeomorphism we choose:

(4.4) $$v(x) = j(x)x.$$

Note that for, $x \in B^c$, $v(x) = x$. Hence we get for all $x \in B^c$:

(4.5) $$\varphi_\theta(x) = e^\theta x.$$

Now, as we analytically continue from $\theta \in \mathbb{R}$ to a complex sector we look for the values of $\phi = \text{Im}(\theta)$ for which the real part of the deformed potential $V_\theta(x) = V(\varphi_\theta(x))$ is confining i.e. going to infinity as $|x| \to \infty$ (this is needed for the proof of existence of resonance free energy, see [**12**]. Since $|V_1(x)| = 0(|V_2(x)|)$ as $|x| \to \infty$ we are only concerned with the behavior of $V_2(x)$ under deformation. Thus for $x \in B^c$ using (A'''):

(4.6) $$V_{2\theta}(x) = V_2(e^\theta x) = e^{r\theta} V_2(x) = e^{ir\phi} V_2(x),$$

where we have used (4.5) and set $\theta = i\phi$. Therefore, in order to have $\text{Re}(V_{2\theta}(x)) \to \infty$ we need $\text{Re}(e^{ir\phi}) < 0$ i.e.

(4.7) $$\phi > \frac{\pi}{2r}.$$

Next, we consider operators H which satisfy condition (A), i.e. satisfy (A''') and either condition (A') or condition (A''). In the first case we associate with H an auxiliary operator H' satisfying (A'') (and the other conditions as required) defined as $H' = -\hbar^2 \Delta + V'(x)$, where $V'(x)$ is given by:

(4.8) $$V'(x) = V(x) - \chi(\{x \geq R\})|x|^3,$$

where $\chi(S)$ denotes a characteristic function of the set S.

Now we introduce some objects necessary for our analysis. For potentials satisfying (C) we will denote the inner and outer classical turning sets at energy $0 < E < V^*_{\text{bar}}$ by $\text{CTS}_i(E)$ and $\text{CTS}_0(E)$, respectively. Note that in the case of volcano-type potentials they are closed (hyper)surfaces of codimension 1 satisfying:

$$(4.9) \qquad \text{CTS}_i(E) \cup \text{CTS}_0(E) = \{x \mid V(x) = E\}.$$

By the Agmon distance between points a and b at an energy E, denoted $A(a,b,E)$, we understand the geodesic distance between a and b in the Agmon metric $ds^2 = \bigl(V(x)-E\bigr)_+ dx^2$. We can also define, in a standard way, Agmon distance between two subsets of $\mathbb{R}^d$, B and C, at energy E and denote it $A(B,C,E)$.

Next we fix an energy level $0 < \varepsilon < V^*_{\text{bar}}$ and a *deformation border*:

$$(4.10) \qquad S_0(\varepsilon) = \text{CTS}_0(\varepsilon + 2\delta),$$

where $0 < \delta \ll 1$ and is independent of $\hbar$. Clearly, $S_0(\varepsilon)$ is a closed (hyper)surface of codimension 1, and therefore it separates $\mathbb{R}^d$ into two disjoint sets: its interior $\bigl(\text{int}\bigl(S_0(\varepsilon)\bigr) \supset S_0(\varepsilon)\bigr)$ and its exterior $\bigl(\text{ext}\bigl(S_0(\varepsilon)\bigr)\bigr)$.

Furthermore, we introduce a *reference potential*, $V_0(x)$, (and corresponding Hamiltonian H_0) satisfying:

- $V_0(x) = V(x)$ for $x \in \text{int}\bigl(S_0(\varepsilon)\bigr)$.
- $V_0(x)$ is strictly increasing to ∞ with $|x|$ for $x \in \text{ext}\bigl(S_0(\varepsilon)\bigr)$.
- $(\exists p \in \mathbb{Z}_+)|x|^{p-2} \leq V_0(x) \leq |x|^p$ for all $x \in \text{ext}\bigl(S_0(\varepsilon)\bigr)$.

Let eigenvalues, eigenfunctions and eigenprojections of the reference Hamiltonian $H_0 := -\hbar^2 \Delta + V_0(x)$ be denoted by E_{0k}, ψ_{0k} and P_{0k}, respectively.

Finally we specify the energy cut-off $\varepsilon(\beta)$. Define energy E_β as the solution to:

$$(4.11) \qquad \frac{d}{dE}\bigl(2A\bigl(0,CTS_0(E),E\bigr)\bigr) = -\hbar\beta,$$

and then define:

$$(4.12) \qquad \varepsilon(\beta) = E_\beta + C\sqrt{\hbar},$$

where $0 < C = O(1)$ and $e^{-C^2} \ll 1$ (for justification see [**12**]).

Next we formulate precisely the problem we consider. By the *Gibbs state in the well* we mean the standard Gibbs state at (inverse) temperature β for H_0:

$$(4.13) \qquad \rho_G = \frac{e^{-\beta H_0}}{Z_G(\beta)} = \sum_{n=0}^{\infty} = \frac{e^{-\beta E_{0n}} P_{0n}}{Z_G(\beta)},$$

where $Z_G(\beta) = \text{Tr}(e^{-\beta H_0})$.

However, this state is not localized in the well since the eigenstates of H_0 corresponding to energies above V^*_{bar}, "leak" over the barrier. Therefore, we take for our initial state the following localized Gibbs state:

$$(4.14) \qquad \rho_0 = \sum_{E_{0n} < \varepsilon(\beta)} \frac{e^{-\beta E_{0n}} P_{0n}}{Z_0(\beta)},$$

where

$$(4.15) \qquad Z_0(\beta) = \sum_{E_{0n} < \varepsilon(\beta)} e^{-\beta E_{0n}},$$

is the *truncated partition function*, for $\varepsilon(\beta) < V^*_{\text{bar}}$.

Note that $(V^*_{\text{bar}} - \varepsilon(\beta))$ need not be small compared to V^*_{bar}. States at the energy closer to V^*_{bar} undergo thermal escapes and not tunneling [1].

Now we can formulate the problem. The *retaining probability* $p(t)$ (which is actually the probability that the particle is still in the well at time t) is:

$$p(t) = \text{tr}(\chi_w \rho(t)), \tag{4.16}$$

with $\chi_w = \chi(\text{int}(S_0(\varepsilon)))$, characteristic function of $\text{int}(S_0(\varepsilon))$, and where $\rho(t)$ is the density operator at time t, i.e. the solution to the von Neumann equation:

$$\dot{\rho}(t) = \frac{i}{\hbar}[H, \rho(t)], \tag{4.17}$$

with the initial condition

$$\rho(0) = \rho_o. \tag{4.18}$$

Note, finally, that the solution to this initial value problem is

$$\rho(t) = e^{-iHt/\hbar} \rho_0 e^{iHt/\hbar}. \tag{4.19}$$

Now we have:

THEOREM 4.2. *Assume conditions* (A)–(D) *on the Hamiltonian* H. *Let* $t \ll \hbar \Gamma^{-1}$, *where* $\Gamma = -2\,\text{Im}(F(\beta))$, $F(\beta)$ *is the resonance free energy of* H *if* H *satisfies* (A″), (A‴), (B)–(D), *and of* H', *associated to* H *as above, if* H *satisfies* (A′), (B)–(D). *Then*:

$$p(t) = e^{-\Gamma t/\hbar}(1 + O(\Gamma)). \tag{4.20}$$

DISCUSSION. Note that this result implies the following expression for the decay probability:

$$P(t) = 1 - p(t) = \frac{\Gamma t}{\hbar}\left(1 + O\left(\frac{\hbar}{t}\right)\right), \tag{4.21}$$

and so the estimates are useful only for $t \gg \hbar$.

Denote $S_\beta = 2A(E_\beta) + \beta\hbar E_\beta$ where E_β is the solution to equation:

$$\frac{d}{dE}(2A(E)) = -\hbar\beta. \tag{4.22}$$

Thus S_β is the action of potential $-V(x)$ along the trajectory of period $\hbar\beta$. This trajectory corresponds to particle moving in the potential $V(x)$ but in imaginary time $t \to i\tau$. Such a particle is called *an instanton*.

THEOREM 4.3. *Assume conditions* (A″), (A‴), (B)–(D). *Then*:

$$\left|\text{Im}(F(\beta))\right| \leq C_\beta \hbar^{-r} e^{-(S_\beta - \varepsilon)/\hbar}, \tag{4.23}$$

where S_β is the action of the instanton of period $\hbar\beta$ and:

$$\varepsilon \leq 4A(E_\beta)\left(\inf_{S_0(\varepsilon)} \|\nabla V(x)\|\right)^{-1} \delta = O(\delta) \ll 1, \tag{4.24}$$

is a small constant independent of $\hbar$.

References

1. I. Affleck, *Quantum-statistical metastability*, Phys. Rev. Lett. **46** (1981), no. 6, 388–391.
2. S. Coleman, *Aspects of symmetry*, Cambridge University Press, 1985.
3. J. Glimm and A. Jaffe, *Quantum physics*, Springer-Verlag, 1987.
4. H. Grabert, U. Weiss, and P. Hanggi, *Quantum tunneling in dissipative systems at finite temperature*, Phys. Rev. Lett. **52** (1984), no. 25, 2193–2196.
5. A. I. Larkin and Yu. N. Ovchinnikov, *Quantum-mechanical tunneling with dissipation. The pre-exponential factor*, Zh. Èksper. Teoret. Fi **86** (1984), 719–726.
6. A. J. Leggett, *Quantum tunneling in the presence of an arbitrary linear dissipation mechanism*, Phys. Rev. B **30** (1984), no. 3, 1208–1218.
7. A. Linde, *Particle physics and inflationary cosmology*, Harwood, 1990.
8. _____, *Stochastic approach to tunneling and baby universe formation*, Nuclear Phys. B **372** (1992), 421–442.
9. A. Pazy, *Semigroups of linear operators*, Springer-Verlag, 1983.
10. M. Reed and B. Simon, *Methods of modern mathematical physics* II, Academic Press, 1975.
11. I. M. Sigal, *Sharp exponential bounds on resonances states and width of resonances*, Adv. in Appl. Math. **9** (1988), no. 2, 127–166.
12. I. M. Sigal and B. Vasilijevic, in preparation.
13. A. Vilenkin and E. P. S. Shellard, *Cosmic strings and other topological defects*, University Press, Cambridge, 1994.

DEPARTMENT OF MATHEMATICS, UNIVERSITY OF TORONTO, TORONTO, ONTARIO, CANADA, M5S 3G3

E-mail address: vasil@math.toronto.edu